Myxobacteria

Genomics, Cellular and Molecular Biology

Edited by

Zhaomin Yang

Biological Sciences
Virginia Tech
Blacksburg, VA
USA

and

Penelope I. Higgs

Department of Ecophysiology
Max Planck Institute for Terrestrial Microbiology
Marburg
Germany

Caister Academic Press

Caister Academic Press
Norfolk, UK

www.caister.com

British Library Cataloguing-in-Publication Data
A catalogue record for this book is available from the British Library

ISBN: 978-1-908230-34-8 (hardback)
ISBN: 978-1-908230-96-6 (ebook)

Cover design adapted from various images as follows. Fruiting bodies from *Chondromyces crocatus* (front cover, top right), *Stigmatella aurantiaca* (front cover, middle right), *Phaselicystis flava* (front cover, bottom left) and *Pyxidicoccus* sp. (front cover, bottom right) courtesy of Rolf Müller (Department of Pharmaceutical Biotechnology, Saarland University, Saarbrücken, Germany). Scanning confocal laser microscopy light image of *Myxococcus xanthus* fruiting bodies surrounded by peripheral rods (front cover, centre), courtesy of Penelope I. Higgs. Co-crystal MglA-MglB homologs: a G-protein and its associated GAP (front cover, centre forward) and phylogenetic tree depicting the three orders of Myxobacteriacea (front cover, centre rear), courtesy of Lotte Søgaard-Andersen. *Cystobacter ferrugineus* (back cover, top left) and *Chondromyces crocatus* (back cover, bottom right) courtesy of Yue-zhong Li (Shandong University, China).

Printed and bound in Great Britain

Contents

	Contributors	v
	Preface	ix
1	Whence Comes Social Diversity? Ecological and Evolutionary Analysis of the Myxobacteria Gregory J. Velicer, Helena Mendes-Soares and Sébastien Wielgoss	1
2	Genome Evolution and Content in the Myxobacteria Stuart Huntley, Kristin Wuichet and Lotte Søgaard-Andersen	31
3	*Myxococcus xanthus* Vegetative and Developmental Cell Heterogeneity Penelope I. Higgs, Patricia L. Hartzell, Carina Holkenbrink and Egbert Hoiczyk	51
4	Cell Cycle Regulation in *Myxococcus xanthus* During Vegetative Growth and Development: Regulatory Links Between DNA Replication and Cell Division Anke Treuner-Lange, Lotte Søgaard-Andersen and Mitchell Singer	79
5	Social Interactions Mediated by Outer Membrane Exchange Daniel Wall	91
6	Developmental Gene Regulation Ramya Rajagopalan, Zaara Sarwar, Anthony G. Garza and Lee Kroos	105
7	Abundance and Complexity of Signalling Mechanisms in Myxobacteria José Muñoz-Dorado, Penelope I. Higgs and Montserrat Elías-Arnanz	127
8	Computational Biology: From Observation to Statistical Image Analysis to Modelling and Back to Biology Cameron W. Harvey, Oleg A. Igoshin, Roy D. Welch, Mark Alber and Lawrence J. Shimkets	151
9	The Mechanism of A-Motility Jennifer Luciano, Beiyan Nan, David R. Zusman and Tâm Mignot	173

10 Type IV Pili and Exopolysaccharide-dependent Motility in *Myxococcus xanthus* 183
Zhaomin Yang, Chengyun Li, Carmen Friedrich and Lotte Søgaard-Andersen

11 Sensory Regulation of *Myxococcus xanthus* Motility 199
Emilia M.F. Mauriello, Beiyan Nan and David R. Zusman

12 The Biophysics of *Myxococcus xanthus* Motility 211
Fabian Czerwinski and Joshua Shaevitz

Index 227

Colour Plate A1

Contributors

Mark Alber
Department of Applied and Computational Mathematics and Statistics;
Department of Physics
University of Notre Dame
Notre Dame, IN
USA

malber@nd.edu

Fabian Czerwinski
Department of Physics and the Lewis-Sigler Institute for Integrative Genomics
Carl Icahn Laboratories
Princeton, NJ
USA

fczerwin@princeton.edu

Montserrat Elías-Arnanz
Departamento de Genética y Microbiología
Universidad de Murcia
Murcia
Spain

melias@um.es

Carmen Friedrich
Department of Ecophysiology
Max Planck Institute for Terrestrial Microbiology
Marburg
Germany

carmen.friedrich@mpi-marburg.mpg.de

Anthony G. Garza
Department of Biology
Syracuse University
Syracuse, NY
USA

agarza@syr.edu

Patricia L. Hartzell
Department of Biological Sciences
University of Idaho
Moscow, ID
USA

hartzell@uidaho.edu

Cameron W. Harvey
Department of Physics;
Department of Applied and Computational Mathematics and Statistics
University of Notre Dame
Notre Dame, IN
USA

charvey2@nd.edu

Penelope I. Higgs
Department of Ecophysiology
Max Planck Institute for Terrestrial Microbiology
Marburg
Germany;
Department of Biological Sciences
Wayne State University
Detroit, MI
USA

pihiggs@wayne.edu

Egbert Hoiczyk
W. Harry Feinstone Department of Molecular Microbiology and Immunology
Johns Hopkins Bloomberg School of Public Health
Baltimore, MD
USA

ehoiczyk@jhsph.edu

Carina Holkenbrink
Department of Ecophysiology
Max Planck Institute for Terrestrial Microbiology
Marburg
Germany

holkenbr@mpi-marburg.mpg.de

Stuart Huntley
Department of Ecophysiology
Max Planck Institute for Terrestrial Microbiology
Marburg
Germany

huntley@mpi-marburg.mpg.de

Oleg A. Igoshin
Department of Bioengineering
Rice University
Houston, TX
USA

igoshin@rice.edu

Lee Kroos
Department of Biochemistry and Molecular Biology
Michigan State University
East Lansing, MI
USA

kroos@msu.edu

Chengyun Li
Department of Biological Sciences
Virginia Tech
Blacksburg, VA
USA

chengyun@vt.edu

Jennifer Luciano
Laboratoire de Chimie Bactérienne
Institut de Microbiologie de la Méditerranée
Aix-Marseille Université
Marseille
France

jennifer.luciano@gmail.com

Emilia M.F. Mauriello
Laboratoire de Chimie Bactérienne
Institut de Microbiologie de la Méditerranée
Aix-Marseille Université
Marseille
France

emauriello@imm.cnrs.fr

Helena Mendes-Soares
Institute for Bioinformatics and Evolutionary Studies
University of Idaho
Moscow, ID
USA

mmendessoares@idaho.edu

Tâm Mignot
Laboratoire de Chimie Bactérienne
Institut de Microbiologie de la Méditerranée
Aix-Marseille Université
Marseille
France

tmignot@ifr88.cnrs-mrs.fr

José Muñoz-Dorado
Departamento de Microbiología
Facultad de Ciencias
Universidad de Granada
Granada
Spain

jdorado@ugr.es

Beiyan Nan
Department of Molecular and Cell Biology
University of California
Berkeley, CA
USA

nanbeiyan@berkeley.edu

Ramya Rajagopalan
Department of Biochemistry and Molecular Biology
Michigan State University
East Lansing, MI
USA

ramya@msu.edu

Zaara Sarwar
Department of Biology
Syracuse University
Syracuse, NY
USA

Zaara.Sarwar@med.nyu.edu

Joshua Shaevitz
Department of Physics and the Lewis-Sigler Institute for Integrative Genomics
Carl Icahn Laboratories
Princeton, NJ
USA

jshaevitz@gmail.com

Lawrence J. Shimkets
Department of Microbiology
University of Georgia
Athens, GA
USA

shimkets@uga.edu

Mitchell Singer
Department of Microbiology and Molecular Genetics
University of California Davis
Davis, CA
USA

mhsinger@ucdavis.edu

Lotte Søgaard-Andersen
Department of Ecophysiology
Max Planck Institute for Terrestrial Microbiology
Marburg
Germany

sogaard@mpi-marburg.mpg.de

Anke Treuner-Lange
Department of Ecophysiology
Max-Planck Institute for Terrestrial Microbiology
Marburg
Germany

Anke.Treunerlange@mpi-marburg.mpg.de

Gregory J. Velicer
Institute of Integrative Biology
Department of Environmental Systems Science
ETH Zürich
Zürich
Switzerland

gregory.velicer@env.ethz.ch

Daniel Wall
Department of Molecular Biology
University of Wyoming
Laramie, WY
USA

dwall2@uwyo.edu

Roy D. Welch
Department of Biology
Syracuse University
New York, NY
USA

rowelch@syr.edu

Sébastien Wielgoss
Institute of Integrative Biology
Department of Environmental Systems Science
ETH Zürich
Zürich
Switzerland

sebastien.wielgoss@env.ethz.ch

Kristin Wuichet
Department of Ecophysiology
Max Planck Institute for Terrestrial Microbiology
Marburg
Germany

kristin.wuichet@mpi-marburg.mpg.de

Zhaomin Yang
Department of Biological Sciences
Virginia Tech
Blacksburg, VA
USA

zmyang@vt.edu

David R. Zusman
Department of Molecular and Cell Biology
University of California
Berkeley, CA
USA

zusman@berkeley.edu

Current books of interest

Biofuels: From Microbes to Molecules	2014
Human Pathogenic Fungi: Molecular Biology and Pathogenic Mechanisms	2014
Applied RNAi: From Fundamental Research to Therapeutic Applications	2014
Halophiles: Genetics and Genomes	2014
Phage Therapy: Current Research and Applications	2014
Bioinformatics and Data Analysis in Microbiology	2014
The Cell Biology of Cyanobacteria	2014
Pathogenic *Escherichia coli*: Molecular and Cellular Microbiology	2014
Campylobacter Ecology and Evolution	2014
Burkholderia: From Genomes to Function	2014
Next-generation Sequencing: Current Technologies and Applications	2014
Omics in Soil Science	2014
Applications of Molecular Microbiological Methods	2014
Mollicutes: Molecular Biology and Pathogenesis	2014
Genome Analysis: Current Procedures and Applications	2014
Bacterial Toxins: Genetics, Cellular Biology and Practical Applications	2013
Bacterial Membranes: Structural and Molecular Biology	2014
Cold-Adapted Microorganisms	2013
Fusarium: Genomics, Molecular and Cellular Biology	2013
Prions: Current Progress in Advanced Research	2013
RNA Editing: Current Research and Future Trends	2013
Real-Time PCR: Advanced Technologies and Applications	2013
Microbial Efflux Pumps: Current Research	2013
Cytomegaloviruses: From Molecular Pathogenesis to Intervention	2013
Oral Microbial Ecology: Current Research and New Perspectives	2013
Bionanotechnology: Biological Self-assembly and its Applications	2013
Real-Time PCR in Food Science: Current Technology and Applications	2013
Bacterial Gene Regulation and Transcriptional Networks	2013
Bioremediation of Mercury: Current Research and Industrial Applications	2013
Neurospora: Genomics and Molecular Biology	2013
Rhabdoviruses	2012
Horizontal Gene Transfer in Microorganisms	2012
Microbial Ecological Theory: Current Perspectives	2012
Two-Component Systems in Bacteria	2012

Full details at www.caister.com

Preface

For well over a century, the myxobacteria have been the premiere example of prokaryotes with fascinating 'eukaryotic-like' behaviours because of their social lifestyle. During vegetative or active growth, myxobacteria spread or move over solid surfaces as multicellular swarms. This allows them to hunt and feed on prey microorganisms cooperatively as they share secreted antimicrobials and hydrolytic enzymes. When faced with nutrient limitation, myxobacteria may enter a developmental programme with the formation of dormant spores within beautiful and macroscopic multicellular fruiting bodies. The complex multicellular behaviours of these supposedly unicellular microorganisms have had biologists spellbound for many generations.

The earliest work on myxobacteria was pioneered by Roland Thaxter around 1900. He published elegant illustrations of the macroscopic fruiting bodies as well as careful description of microscopic morphologies of vegetative and differentiated cells of various myxobacteria. Like a spore, however, the field largely laid dormant with few signs of life for quite a period. A renaissance started in the 1960s, led by Martin Dworkin, David White, Eugene Rosenberg and Hans Reichenbach. Among many notable accomplishments, they attained new techniques of isolation and cultivation, and provided insights into the physiology and self-organization of these amazing bacteria. Development and application of tools in genetics and molecular biology around the 1980s sparked a booming era of scientific progress on *Myxococcus xanthus*, the reigning model for this bacterial group. Dale Kaiser and David Zusman deserve the most credit as leaders of this era and they rooted family trees which have produced many a myxobiologist in recent history. Some hallmarks of this period include the discovery of the two motility systems of *M. xanthus*, insights into the sensory regulation of gliding motility and the establishment of the molecular framework for fruiting body development. The genomics and post-genomics era was launched by the complete sequencing of *M. xanthus* and *Sorangium cellulosum* genomes which remain two of the largest among bacteria to this date. Bioinformatic analyses revealed that these genomes encoded record numbers of proteins for signal transduction, transcriptional regulations, secondary metabolite production and of unknown function. As a sign of the new age, we can now investigate myxobacteria at the level of systems biology instead of individual genes. In more recent years, tools in computational biology and cell biology including advanced imaging are being increasingly applied to myxobacteria to reveal their secrets. As the saying goes, the rest is history.

This book, which has a more exclusive focus on the model *M. xanthus*, comes at the most exciting time in the field. New discoveries and insights are streaming out in quickened pace and more depth. With tools in functional genomics and the increasing power of technology, it is now possible to study myxobacteria by comparative genomics as well as global transcriptomics and proteomics. Quantitative approaches and high-resolution imaging are providing insights into regulatory networks and promising to unravelling many mechanistic mysteries. All chapters in this book highlight recent advances in defining molecular mechanisms on their respective topics and provide perspectives on important outstanding questions. In addition, the authors structured their chapters such that this book can serve as a gateway for established scientists of other fields as well as advanced undergraduate and beginning graduate students who are interested in venturing into myxobacteria. Through this book, they will gain an understanding of our current state of knowledge and perspectives of unsolved scientific issues in the field.

This book is divided into four sections which may be read in any order or selectively. The first two chapters have an ecological and evolutionary theme: Chapter 1 discusses behaviours and evolution of myxobacteria in the context of their natural environment, while Chapter 2 focuses on comparative genomics and the evolution of the developmental regulatory programme. Chapters 3 through 5 take on cell biology and differentiation in *M. xanthus*: Chapter 3 reviews phenotypic and cellular heterogeneities and their critical roles in fruiting body development; Chapter 4 highlights recent advances in understanding cell cycle in this unusual bacterium; Chapter 5 provides expert insights into the mechanism of cell-to-cell protein transfer. The third section describes the complex regulatory capacity of *M. xanthus*: Chapter 6 reviews recent progress in defining the regulatory network of development; Chapter 7 covers recent advances in unravelling signalling proteins and mechanisms which include both those expected for a prokaryote and those more frequently associated with eukaryotes, a fascinating mix of seemingly mixed evolutionary origins; Chapter 8 introduces computational modelling of myxobacterial behaviours, which is at the forefront of biology at the present time.

The last four chapters are devoted to motility: Chapter 9 presents our current understanding and challenges of Adventurous (A) motility; Chapter 10 provides a discussion on social (S) motility which is powered by Type IV pilus (T4P) as bacterial twitching; Chapter 11 focuses on how *M. xanthus* uses sensory regulation to govern the two motility systems in action; and last but not least, Chapter 12 discusses the biophysics of motion in theory and practice.

The myxobacteria community deserves special thanks for this book coming to its fruition. Thank you, the authors, for taking your precious time to share your expertise and insights. Thank you, the reviewers, for providing excellent feedback on all chapters. A special acknowledgement goes to Annette Griffin at Horizon Press for her initiatives, encouragement and patience. We would like to especially acknowledge Dr Yue-zhong Li (Shandong University, China) for hosting the 40th International Meeting on the Myxobacteria in China in 2013.

We are indebted to our families. We thank our spouses, Jenny Wang and Stuart Huntley, for their sacrifices and their ever-necessary support in our editorial endeavour. A shout-out to our children, Jewel JuJu Yang and Storey Huntley, for bringing endless joy and unconditional love which allow us to endure.

Zhaomin Yang
Penelope I. Higgs

Whence Comes Social Diversity? Ecological and Evolutionary Analysis of the Myxobacteria

1

Gregory J. Velicer, Helena Mendes-Soares and Sébastien Wielgoss

Abstract

Recent discoveries have found the myxobacteria to be much more diverse – both across and within species – than previously known, from global to micrometre spatial scales. Evolutionary analysis of such extant diversity promises to reveal much about how myxobacteria have adapted to natural ecological habitats in the past and continue to evolve in the present, particularly with regard to their intriguing social phenotypes. Experimental populations propagated under defined laboratory conditions undergo very rapid evolution at cooperative traits in a manner that radically changes their social composition. Analysis of such experimentally evolved populations allows detailed characterization of social evolutionary dynamics in real time. Moreover, traditional genetic tools and new genome sequencing technologies together allow deep investigation of the molecular basis of adaptation by experimental populations to known ecological habitats, which in turn can lead to new discoveries regarding the molecular mechanisms governing social behaviour.

Introduction

Why do myxobacteria cooperate at all? What selective forces led to the evolution of coordinated group predation and fruiting body formation? Why do fruiting body morphologies vary so drastically across species? What is the spatial distribution of genetic and phenotypic diversity in this clade? Who cooperates with whom? What ecological and evolutionary forces and what molecular mechanisms lead to social divergence and isolation among myxobacterial lineages? Do myxobacteria prey upon one another? Why are most *Myxococcus xanthus* isolates yellow? Are there natural strains of myxobacteria that are maintained by natural selection because they 'cheat' socially on other strains? What roles do myxobacteria play in global carbon cycling? To what degree and in what manner are the compositional dynamics of terrestrial microbial communities influenced by myxobacteria?

The majority of myxobacteria research to date has sought to understand molecular interactions occurring within individual cells and between genetically identical cells that generate the unique group-level behaviours exhibited by the myxobacteria that have fascinated researchers for generations. However, the vast array of open questions regarding the evolutionary origins and maintenance and ecological contexts of myxobacterial sociality, including those listed above and many more, can readily fuel a healthy number of incipient research careers for the foreseeable future.

The myxobacteria constitute the monophyletic order *Myxococcales* in the δ-subdivision of the Gram-negative proteobacteria and are best known for the strikingly complex and beautiful social behaviours that many myxobacterial species display. For example, the model species *Myxococcus xanthus* and others engage in group-coordinated movement through soil habitats driven by two genetically distinct motility systems (Chapters 9–11) in search of growth substrates. Growth is often fuelled by other microbes, which myxobacteria prey upon by secreting a suite of compounds that kill and decompose prey cells (this chapter). Upon depletion of local resources, some myxobacterial species respond by initiating aggregative development of multicellular fruiting bodies, within which some cells convert into stress-resistant spores (Chapters 3 and 6–8). Spores then germinate and resume growth upon encountering favourable conditions. All of these social traits are highly variable both across and within species, but much remains unknown regarding the origins and maintenance of myxobacterial diversity.

The potential scope of a chapter on myxobacteria ecology and evolution effectively interfaces with all of myxobacterial biology, because populations of whole, integrated individuals evolve while interacting with a complex array of dynamic environmental features including other organisms, both conspecific and heterospecific. This chapter only touches on a small fraction of the potential themes in these fields, both due to space limitations and because these research areas remain in their infancy. This chapter will highlight recent developments rather than providing an encyclopaedic historical summary of relevant research. We first revisit long-standing questions regarding the basic benefits of myxobacterial social traits. We then describe recent investigations of marine myxobacteria that significantly expand the Myxococcales clade at macroevolutionary scales and also describe biogeographic and phenotypic patterns of diversity in the model species *Myxococcus xanthus* at the microevolutionary scale, including the intriguing finding of substantial diversity within individual fruiting bodies isolated from soil. The power of social environments to impose selection on evolving populations is highlighted by two evolution experiments, one of which generated strains capable of suppressing cheating behaviour and the other of which led to the mechanistic discovery of a small regulatory RNA that controls the onset of *Myxococcus* development. Finally, we overview recent advances in our understanding of myxobacterial predation, which is likely to affect the composition and dynamics of many soil communities. For additional background, readers are encouraged to see alternative reviews on the following themes: taxa descriptions (Shimkets *et al.*, 2006; Velicer and Hillesland, 2008), secondary metabolism diversity (Weissman and Müller, 2010), A-motility evolution (Chapter 9 and Luciano *et al.*, 2011), myxobacterial social evolution (Velicer and Hillesland, 2008; Velicer and Vos, 2009) and microbial social evolution more broadly (e.g. Foster *et al.*, 2007; Nadell *et al.*, 2008; Strassmann *et al.*, 2011; Strassmann and Queller, 2011; Travisano and Velicer, 2004; Velicer and Vos, 2009; West, 2006).

A fundamental unanswered question: why be social?

The complex social traits exhibited by the myxobacteria must be beneficial, on average, to the lineages that display them; otherwise their evolutionary loss would be expected due to the cost of expressing those traits. This intuition is supported by the observation that relaxation of selection favouring development or social motility leads to the rapid evolutionary loss of these traits in experimental populations (Velicer *et al.*, 1998). Nonetheless, the legitimacy of the assumption that genetically complex social traits are beneficial does not reveal what those benefits are.

The social trait for which we have the most evidence regarding its evolutionary benefits is Type-IV pili-based social ('S') motility (see Chapter 10). The fundamental benefit of motility *per se*, namely the ability of cells to migrate in search of conditions more favourable to growth and survival, is non-controversial. But considering the fact that *M. xanthus* also employs a second motility system, namely the relatively asocial A-motility system (Chapter 9), the question arises as to why two distinct genetically complex forms of motility are used rather than only one. Shi and Zusman (1993) found that under conditions of resource abundance, possession of both A- and S-motility allowed effective movement across a broader range of surfaces than did either system alone. Specifically, mutational disruption of A-motility greatly slowed group swarming on hard (1.5%) agar and disruption of S-motility both slowed swarming on hard agar and nearly eliminated it on soft (0.3%) agar. Hillesland and Velicer (2005) showed that the surface-type specific performance of these two motility systems is qualitatively resilient across a range of growth resource levels but also found that the relative contribution of each system to overall swarm movement could vary greatly as a function of nutrient concentration. An additional benefit of S-motility may be the tight packing of spores into fruiting bodies, which requires pili (Wu *et al.*, 1998). The Type IV pili employed by *M. xanthus* in S-motility have been implicated as necessary for entry into prey cells by *Bdellovibrio bacteriovorus* (Evans *et al.*, 2007) but do not appear to be similarly directly involved in *Myxococcus* predation beyond their role in motility (Pham *et al.*, 2005).

Why make fruiting bodies? The basic benefits of constructing fruiting bodies remain unclear. Some *Myxococcus* genotypes are able to form spores without constructing fruiting bodies (Velicer *et al.*, 1998) and some myxobacterial species do not appear to make fruiting bodies at all (Shimkets *et al.*, 2006), thus raising the question of why so many strains bother to engage in a genetically complex and energetically costly process to do so. Several plausible hypotheses regarding the benefits of fruiting body formation have been proposed but none have yet been demonstrated to be true. Spores in fruiting bodies may somehow be individually hardier and/or better protected from environmental stress or predation (Dahl *et al.*, 2011) due to group-level effects than non-fruiting body spores (Velicer and Hillesland, 2008). If growth rate increases with local cell density under some conditions (see section below), tight packing of spores in fruiting bodies may enhance germination and growth when fruiting bodies are exposed to sufficient nutrients (Kaiser, 2001; Rosenberg *et al.*, 1977). Enhanced dispersal of spores due to placement in fruiting bodies has been a commonly proposed benefit of fruiting body formation (e.g. Kaiser, 2001) but the veracity of this hypothesis has not been demonstrated.

Because the ability to form fruiting bodies is readily lost under some selective conditions, non-fruiting strains of *M. xanthus* and other species known for their ability to undergo development might be common (Jiang *et al.*, 2007; Velicer *et al.*, 2002). Indeed, natural strains of *M. xanthus* that produce few or no spores in pure culture under laboratory conditions have recently been isolated (albeit from within fully formed fruiting bodies; Kraemer and Velicer, 2011). Standardized isolation methods independent of development will need to be employed to rigorously test for the presence and frequencies of non-fruiting strains among fruiting species.

Density dependence

Why do myxobacteria tend to live in groups for much of their life cycle? Group living is prevalent throughout the spectrum of life (Krause and Ruxton, 2002): flocking birds, schooling fish, complex society forming hymenopterans with division of labour (e.g. bees and ants) and swarming bacteria. This prevalence indicates that social proximity is often beneficial overall and that the benefits of grouping more than compensate for its potential costs, which include within-group competition for limiting resources (Alexander, 1974), more extensive exposure to parasites (Arneberg *et al.*, 1998), and cheating (Velicer *et al.*, 2000). Net benefits of grouping imply an Allee effect (Allee, 1931, 1949; Stephens and Freckleton, 1999), or a positive relationship between density and fitness over some density ranges and several forms of Allee effect have been demonstrated or proposed for the myxobacteria (Kaplan and Plamann, 1996; Rosenberg *et al.* 1977; Velicer and Vos, 2009).

As might be expected, the initiation of myxobacterial fruiting body development is density dependent (Kaplan and Plamann, 1996), although the relationship between density and spore production can vary dramatically across strains of *M. xanthus* (Kadam and Velicer, 2006). It has also been suggested that reproduction rates should correlate positively with cell density within groups during growth on prey, as they might benefit from elevated local concentrations of secreted hydrolytic peptides (Rosenberg *et al.*, 1977). Rosenberg *et al.* (1977) demonstrated a beneficial effect of high cell density in well-mixed liquid cultures of *M. xanthus*. Populations grew faster in high cell-density cultures when feeding on long peptides that required hydrolysis prior to uptake than did low density populations. This effect was likely due to an elevated concentration of extracellular digestive enzymes in the high-density treatment, making more hydrolysed growth substrate available per *M. xanthus* cell, on average, than occurred in the low density treatment. The density-dependent nature of enzyme release was further demonstrated by the existence of a critical density threshold below which no growth was detected. In contrast, growth rate was independent of density when cultures were grown on pre-digested short peptides, since bacterial cells can readily take up these nutrients without releasing digestive enzymes. Density-dependent growth on prey has not yet been demonstrated.

The well-mixed liquid cultures used by Rosenberg *et al.* (1977) do not represent natural viscous habitats in which environmental features likely often inhibit rapid diffusion of secreted compounds in a manner that mitigates benefits of high cell density. Peitz and Velicer (unpublished) have tested for positive effects of high density on bacterial vegetative growth, development and survival in structured environments with varying degrees of pH stress. In a first assay, they spotted *M. xanthus* cells onto solid agar medium containing pre-hydrolysed growth substrate and varied both cell density and pH (from 5.5 to 9.0). They found that maximum growth rates correlated positively with cell density at all pH levels, including levels nearly optimal for growth (near pH 7.0). However, the degree of this correlation increased strongly with pH stress (acid stress in particular), suggesting that high cell density may buffer social groups in the soil from some forms of environmental stress. The degree to which the ability to form fruiting bodies depends on density also increased markedly with acid stress. Why growth rate would correlate with density in a structured habitat even at nearly optimal pH during growth on pre-hydrolysed Casitone [as opposed to the non-hydrolysed casein used by Rosenberg *et al.* (1977)] is not clear.

Sociality and cheating

Cooperative behaviours benefit others at an individual cost and are thus theoretically susceptible to being undermined by selfish 'cheating' (Crespi, 2001; Krebs and Davies, 1997; Velicer, 2003). Cheating occurs when one individual or genotype does not fully cooperate during a social process by contributing proportionately to some social good, such as an inter-cellular developmental signal (e.g. the A- or C-signals in *Myxococcus* development), and thereby gains a short-term, selfish fitness advantage over cooperative genotypes that do contribute their 'fair share' of the social good (Travisano and Velicer, 2004; Velicer, 2003; Velicer *et al.*, 2000; West *et al.*, 2006). In principle, all myxobacterial traits involving the production of cell-surface or diffusible molecules that are beneficial to other cells are potentially susceptible to cheating, including S-motility, predation, secondary metabolite production and development.

Myxococcus strains that show strong cheating phenotypes (very poor pure-culture sporulation associated with large fitness advantages over developmentally proficient strains in mixed culture sporulation) have readily been generated by both mutagenesis and experimental evolution (Velicer *et al.*, 2000). However, the cheating advantage of *M. xanthus* developmental cheaters is highly frequency dependent and in some cases is lost completely at high cheater frequencies, thus hindering cheater genotypes from fully displacing cooperators (Fiegna and Velicer, 2003; Smith *et al.*, 2010; Velicer *et al.*, 2000; Velicer and Vos, 2009). At least partial social complementation of a defect in an experimentally evolved form of motility that involves social interaction has been demonstrated (Velicer and Yu, 2003), but outright cheating in which a socially defective strain has a relative fitness advantage over a socially proficient strain in mixed culture has yet to be demonstrated for motility, predation or secondary metabolite production in the myxobacteria [but see Xavier *et al.* (2011) for an example of cheating during *Pseudomonas aeruginosa* swarming].

All evolutionarily stable forms of biological cooperation are buttressed by mechanisms that limit the frequency of non-cooperative genotypes. The reader can see Velicer and Vos (2009) for a survey of such mechanisms and how they apply to the myxobacteria. More broadly, investigations into how microbes successfully cooperate and how microbial cheating is correspondingly limited have been burgeoning in recent years (e.g. Gilbert *et al.*, 2007; Harrison and Buckling, 2009; Kümmerli *et al.*, 2009; Rainey and Rainey, 2003; Sachs *et al.*, 2004; Simms *et al.*, 2006). Comparison of how social cooperation and conflict evolve across a wide variety of microbial species can be expected to yield both important empirically based general principles and insights into features of cooperation and conflict that are specific to particular biological systems.

Empirical methods of evolutionary analysis

Comparative analysis has been the primary empirical mode of addressing evolutionary questions since Darwin and Wallace (Harvey and Pagel, 1991). In this method, patterns of natural biological variation (e.g. genetic, biochemical, phenotypic) are documented and inferences regarding phylogenetic relationships, evolutionary forces (e.g. selection vs. drift) and temporal evolutionary dynamics are drawn from such patterns. In this book, the comparative method is used in Chapter 2 to address basic questions about the origins and divergence of the genetic programmes that underlie myxobacterial development across the entire myxobacterial clade. It is also used in this chapter to broadly summarize our current understanding of the phylogeny of myxobacteria and to gain early insights into the evolutionary processes that generate natural social diversity within a single myxobacterial species – *Myxococcus xanthus.*

Alternatively, many evolutionary questions can be addressed directly with experiments in which populations are propagated in manipulated environments that allow evolutionary effects of specific variables to be rigorously tested (Hindré *et al.*, 2012; Kawecki *et al.*, 2012). Large microbial populations generate substantial novel variation by mutation upon which selection can act and the rapid growth of microbes allows for extensive evolutionary change to be tracked and characterized over short periods of time. In one example with *M. xanthus,* selection was imposed for improved fitness during colony growth on soft agar by a mutant that was defective at S-motility (Chapter 10) due to deletion of the *pilA* gene from a 'wild-type' (WT) lab reference strain (Velicer and Yu, 2003). This selection led to dramatic evolutionary improvements in socially mediated swarming that were attained by a novel interaction between the A-motility system and the extracellular matrix (Fig. 1.1). Moreover, the genetic manipulability of many microbial species combined with recent advances in genome-sequencing technology allows the molecular basis of evolutionary change to be characterized with greater rigor than ever before (Blount *et al.*, 2012; Hindré *et al.*, 2012; Kawecki *et al.*, 2012; Wielgoss *et al.*, 2013; Yu *et al.*, 2010).

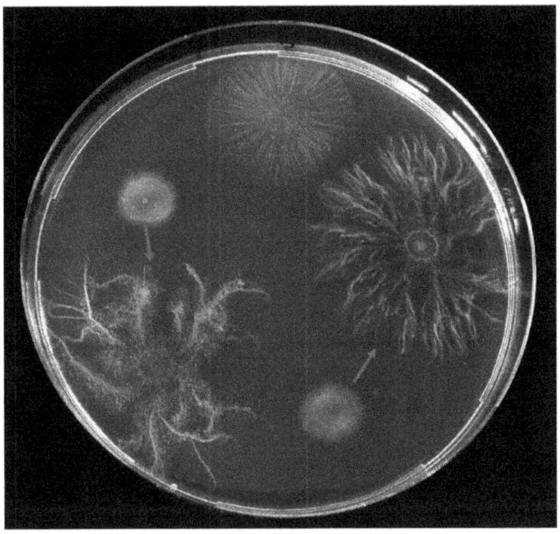

Figure 1.1 Major phenotypic change during experimental evolution. Colony phenotypes on soft agar: WT parent (top position), two *ΔpilA* mutants (second and fourth positions counterclockwise from top) derived from WT (blue arrows) and their lab-evolved descendants E7 and E8 (third and fifth positions counterclockwise from top, respectively, red arrows). Adapted and reprinted with permission from Yu and Velicer (2003). A colour version of this figure is located in the plate section at the back of the book.

Comparative evolutionary analysis thus seeks to infer how natural populations actually have evolved while lacking much information about their complex selective and genetic histories, whereas experimental evolutionary analysis seeks to explore how populations evolve in defined selective regimes. These two approaches are now also being complemented by real time observations of how natural populations evolve over time in their native habitats (Barrett, 2010; Barrett *et al.*, 2010) and evolution experiments conducted in complex habitats that approximate natural environments. Analogously, empirical ecological research can be largely subdivided into careful observation and interpretation of natural ecological interactions and correlations and experimental testing of specific ecological hypotheses in controlled, simplified environments (Bohannan and Lenski, 2000; Elena and Lenski, 2003; Fukami *et al.*, 2007; Jessup *et al.*, 2004; Kassen and Rainey, 2004).

Classification and phylogenetic analysis

The myxobacteria are composed of all species within the monophyletic order *Myxococcales* (Reichenbach, 2005) (Fig. 1.2). Their closest relatives are Gram-negative sulphate and sulphur reducers and the bacterivorous *Bdellovibrionales*, which together group with *Myxococcales* to form the delta-subclass of the *Proteobacteria* (Oyaizu and Woese, 1985; Reichenbach, 2005; Stackebrandt *et al.*, 1988; Woese *et al.*, 1985). Myxobacteria have been isolated from a wide range of terrestrial soils all over the world and appear to be most abundant in nutrient-rich soils of tropical and temperate environments (Dawid, 2000; Reichenbach, 1993, 2005; Watve *et al.*, 1999). They are also found in organic waste such as animal dung or decaying plant material and in

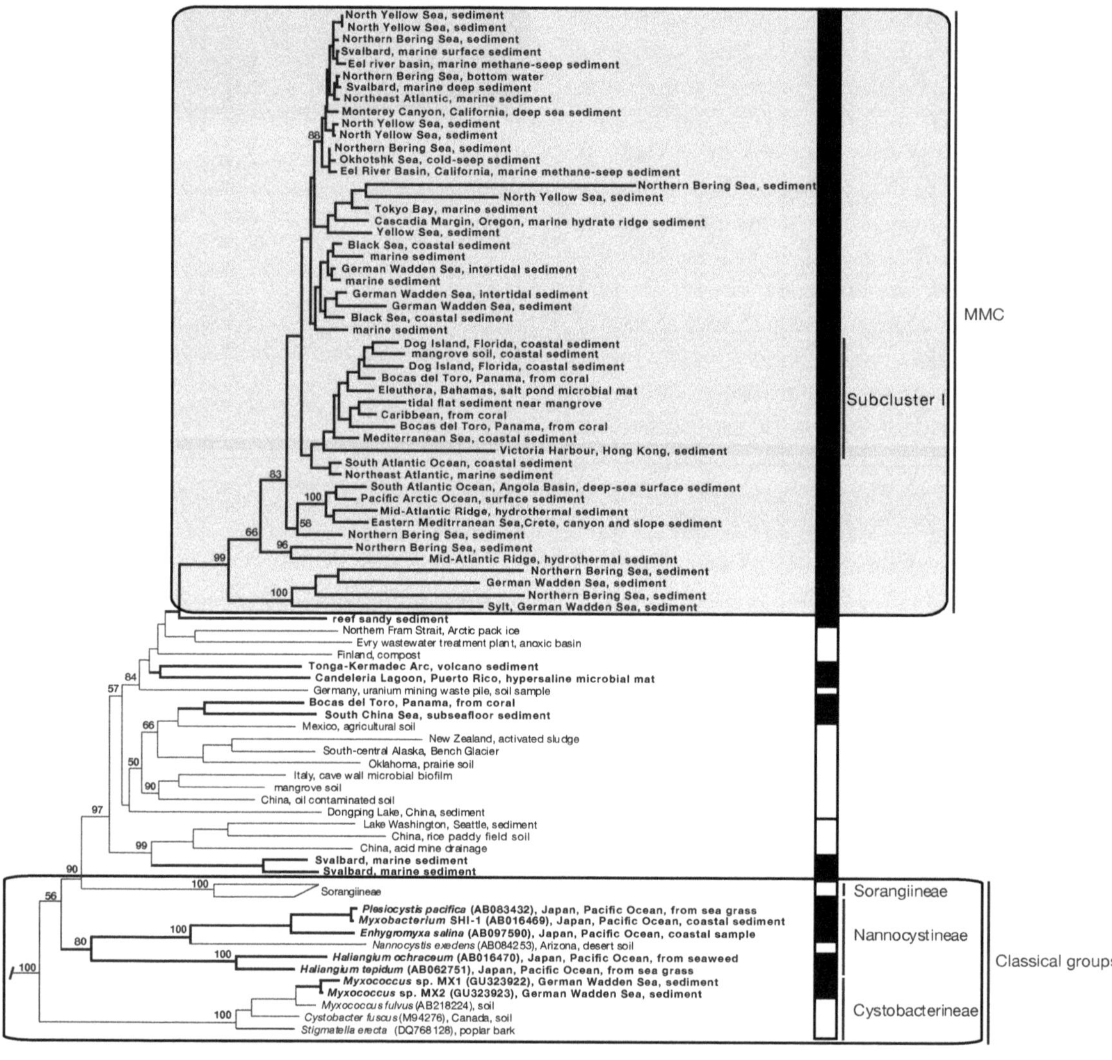

Figure 1.2 Neighbour-joining tree depicting a modified 16S rRNA phylogeny of the *Myxococcales* order by Brinkhoff *et al.* (2012) (reproduced with permission), with special focus on sequences of marine origin (bold text) and a monophyletic subclade composed of only marine sequences (MMC, shaded in grey). Members of the cyanobacteria were used to root the tree (not shown). The three classical groups *Sorangiineae*, *Nannocystineae* and *Cystobacterineae* are bordered without shading. Sequences of terrestrial origin (including glaciers, freshwater, soil and waste water treatment plants) are given in non-bold text. Bootstrap values >50% are shown and the scale bar length corresponds to 0.10 substitutions per nucleotide position.

seemingly less hospitable locations such as on rock surfaces and in sand (Reichenbach, 1999). Most phylogenetic studies have almost exclusively analysed culture collections from terrestrial environments (Garcia *et al.*, 2010; Huntley *et al.*, 2011; Ludwig *et al.*, 1983; Shimkets and Woese, 1992; Spröer *et al.*, 1999). These studies have consistently divided *Myxococcales* into either two or three suborders, the more basal *Cystobacterineae* and the sister taxa *Sorangiineae* and *Nannocystineae* which have recently been distinguished as distinct suborders (Shimkets *et al.*, 2006). The topologies of phylogenetic trees based on 16S rRNA gene sequences have been found to agree well with traditional classifications based on phenotypic criteria at higher taxonomic levels, including the shape, size and colour of fruiting bodies (Reichenbach, 1993; Shimkets *et al.*, 2006; Spröer *et al.*, 1999). Importantly, more recent phylogenetic studies have begun to shift the focus from solely terrestrial soil samples towards inclusion of myxobacterial cultures and/or DNA sequences isolated from marine environments and have thus vastly expanded our understanding of the habitat range occupied by myxobacterial taxa (Brinkhoff *et al.*, 2012; Jiang *et al.*, 2010).

Marine-inclusive phylogenies

Jiang *et al.* (2010) screened DNA collected from four deep-sea sediments ranging in depth from 853 to 4675 m and one hydrothermal vent (at 204 m depth) for myxobacteria-like 16S rRNA. This study provided the first evidence that marine myxobacteria can be phylogenetically distinct from terrestrial species. They appear to be ubiquitous in marine sediments, as they were present at all sampled locations and depths. This pervasiveness of myxobacterial DNA across ocean sediment sites was interpreted as evidence against the hypothesis that terrestrial myxobacteria present in marine environments have merely dispersed there by accident and are not evolutionarily adapted to these habitats (Jiang *et al.*, 2010). However, despite the prevalence of myxobacterial sequences, Jiang *et al.* (2010) were largely unable to cultivate myxobacteria from their sediment samples and inferred that this was due to a lack of suitable isolation protocols that do not require fruiting body formation.

Brinkhoff *et al.* (2012) have conducted the most comprehensive survey of myxobacterial sequences from ocean floor sediments and marine water columns at a global scale to date (Fig. 1.2). They found that the structure of phylogenetic relationships between myxobacterial sequences of marine vs. terrestrial origin was intermediate between its two possible extremes, as Jiang *et al.* (2010) had found previously as well. Specifically, the marine sequences neither clustered into a single clade separate from all terrestrial sequences, nor did they pervasively intersperse with sequences of terrestrial origin among terminal branches. Rather, marine samples tended to cluster into multi-tip clades dominated by marine sequences that are distinct from, yet phylogenetically interspersed with, multi-tip clades dominated by predominantly terrestrial isolates. However, in contrast to Jiang *et al.* (2010), Brinkhoff *et al.* (2012) found a deep-branching, exclusively marine myxobacteria cluster (MMC) within the *Myxococcales* that was monophyletic and entirely distinct from the three established suborders. The MMC was mainly composed of sequences from sediments and included samples from coral surfaces and microbial mats. The tree topology of Brinkhoff *et al.* (2012) (Fig. 1.2) places the *Cystobacterineae* clade at the basal position within the *Myxococcales*, with the *Nannocystineae*, *Sorangiineae* and MMC clades later branching out successively.

Overall, the relative phylogenetic patterns of marine versus terrestrial myxobacteria sequences suggest that at least several, and possibly many, successful evolutionary invasions across these major habitat categories have occurred since the origin of the myxobacteria. The interspersed marine and terrestrial subclades appear to vary greatly in their phylogenetic breadth and degree of within-subclade heterogeneity (with respect to marine versus terrestrial origin) and some subclades may represent monophyletic adaptive radiations within only one habitat or the other. Further research is needed in order to thoroughly characterize the phylogeographical patterns of marine vs. terrestrial myxobacteria and infer the evolutionary processes that have generated those patterns.

Of particular interest is understanding how readily myxobacteria residing in one habitat can successfully invade and become evolutionarily established in the other as well as the strength of cross-migration barriers among habitats, especially along natural salinity gradients. While cultures of both marine halotolerant and halophilic myxobacteria are able to develop fruiting bodies in the presence of pure or slightly diluted seawater (Wang *et al.*, 2007; Zhang *et al.*, 2005), proper morphogenesis of these structures in terrestrial myxobacteria typically requires much lower salt concentrations, at a level closer to freshwater. Since habitat switching appears to have happened several times independently in all major lineages of myxobacteria, it seems reasonable to study this phenomenon using an experimental evolution approach. Replaying evolution in the lab has led to significant insights into how organisms adapt to novel environments or induced stress over time, including nutrient depletion (Elena and Lenski, 2003). For example, one might allow a relatively halo-intolerant species such as *Myxococcus xanthus* to evolve while gradually increasing levels of salinity in a structured habitat and thus select for increased halo-tolerance over time and assess if evolved populations fully retain their social phenotypes.

Biogeography: patterns and their causes

It has long been postulated that bacteria are capable of dispersing freely over extensive distances so that 'everything is everywhere' (Baas-Becking, 1934). In fact, most bacteria and archaea are very broadly distributed at the level of domains (DeLong and Pace, 2001), bacterial classes (e.g. beta-Proteobacteria, Actinobacteria, Flavobacteria, Cyanobacteria), and genera (e.g. *Bacillus*, *Pseudomonas*) and even individual species (e.g. *Bacillus subtilis*, *Escherichia coli* and *M. xanthus*) are often globally distributed. However, recent studies have provided strong evidence against a universal 'everything is everywhere' model of prokaryotic biogeography at the level of populations. Several studies have shown that endemism (genetic variation specific to a local area) and isolation-by-distance are readily detectable in an increasing number of bacterial species. Such biogeographic patterns are most evident in the case of extremophiles, such as *Sulfolobus* and *Pyrococcus* species (Escobar-Páramo *et al.*, 2005; Whitaker *et al.*, 2003), that occupy 'island'-like habitats and barriers to dispersal are clear-cut (MacArthur and Wilson, 1967). However, only few studies have focused on the structure of widely distributed and free-living soil microbes (e.g. Vos and Velicer, 2006, 2008a). Below we review recent efforts to begin characterizing the biogeography of marine myxobacteria at the inter-specific scale and then proceed to summarize our knowledge of the biogeography of a single myxobacterial species.

Marine biogeography

Halo-tolerant myxobacteria and myxobacteria-like gene sequences sampled from seawater have been known for some time (Iizuka *et al.*, 1998; Li *et al.*, 2002; Zhang *et al.*, 2005). However, as for terrestrial myxobacteria, our understanding of the roles that marine myxobacteria play in the ecology and evolution of marine sediment communities is extremely limited. For example, we know effectively nothing regarding the degree to which marine myxobacteria prey upon other microbes as a carbon source. However, myxobacteria appear to account for a substantial fraction of all bacteria present in many sediment samples (often several per cent), suggesting that they are ecologically important in these communities (Brinkhoff *et al.*, 2012).

Iizuka *et al.* (1998) investigated samples derived from coastal marine water and detected halophilic myxobacterial swarms in 6 out of 90 isolates, which were later identified as novel taxa (Fudou *et al.*, 2002; Iizuka *et al.*, 2003a; Iizuka *et al.*, 2003b). Myxobacteria-like sequences were also present at all four deep–sea sites and a hydrothermal vent sampled by Jiang *et al.* (2010), although only very few were found at the two deepest locations (2961 and 4675 m). No evidence of phylogeographic structure among the sequences of marine origin in this study was detected. It is possible that ocean currents and marine animals promote greater dispersal of marine sediment bacteria than terrestrial animals and wind disperse soil bacteria, but the results of Jiang *et al.* (2010) do not imply the absence of structure among sediment populations at some geographic and phylogenetic scales. Such structure might be found within narrower phylogenetic ranges across large spatial scales, as has been found among terrestrial *Myxococcus* populations (Vos and Velicer, 2008a).

In fact, the phylogenetic survey by Brinkhoff *et al.* (2012) suggests that certain myxobacterial taxa are structured at finer sampling scales. By sampling a transect through a region of the North Sea they identified a novel group of marine myxobacteria (the aforementioned MMC) that are clearly phylogenetically isolated from members of the three 'classical' suborders. While MMC bacteria are rather ubiquitously distributed on a global scale, this marine clade contains a locally isolated monophyletic group of eight samples that are derived from shallow marine sediments in the Caribbean (part of subcluster I; Fig. 1.2). However, it remains to be tested if this apparent structure is due to limited dispersal or strong local selection in the absence of dispersal barriers.

Marine habitats

The phylogenetically distinct marine myxobacteria cluster (MMC) has been shown to be ubiquitously distributed both locally in a North-to-South transect across the North Sea, as well as in 72 global samples taken from marine sediments around the world (Brinkhoff *et al.*, 2012). Using a real-time PCR approach, the MMC were also shown to be a prominent component of the bacterial community associated with marine sediments, with much higher abundances in the North Sea (0.8–13.1% per sampling spot) compared to the global samples from other regions (0.01–0.71%). Importantly, members of the MMC appear to be limited to salinity ranging from 6 to 60 PSUs (practical salinity units), suggesting that they are restricted to marine and brackish water, and should be absent in both freshwater and hyperhaline environments.

MMC members appear to thrive under anoxic conditions and were detected in sediment horizons up to ~2 m below the ocean floor surface, composing up to ~3% of the entire microbial community. This finding is consistent with the discovery of anaerobic myxobacteria in terrestrial soils (genus *Anaeromyxobacter*), which demonstrated the potential of some myxobacteria to grow under oxygen-limited conditions (Cole *et al.*, 1994; Sanford *et al.*, 2002). Using specific primers, Brinkhoff *et al.* (2012) found that marine myxobacteria appear to be predominantly located in marine floor sediments (0–0.5 cm horizon) irrespective of the absolute water depth (27–260 m) rather than in the open water column. Importantly, the probability of detecting MMC bacteria increased the closer the samples were taken to the ocean floor (starting from around 3 m above the floor surface). Moreover, particle samples (>5 μm) taken in a time series at one station near Helgoland during a tidal cycle demonstrated the consistent presence of MMC in the nepheloid layer (which contains a high concentration of suspended particles) from 5 m depth to the bottom. Additionally, Jiang *et al.* (2010) found evidence for myxobacterial life in deep-sea sediments (up to 4675 m) and at a hydrothermal vent.

A major impediment to understanding the ecology and evolution of marine myxobacteria is the difficulty of cultivating them and studying them under ecologically relevant conditions. With the exception of a few halo-tolerant strains that have been isolated from coastal regions (Brinkhoff *et al.*, 2012; Iizuka *et al.*, 1998, 2003a,b; Wang *et al.*, 2007) and characterized with respect to social phenotypes, successful isolation of live myxobacterial cultures from oceanic samples has been rare (Iizuka *et al.*, 1998, 2003a,b), indicating that marine-specific isolation protocols require further development.

Intraspecific diversity in *Myxococcus xanthus*

One of the primary goals of evolutionary biology is to explain the origin and maintenance of natural biological diversity, but such diversity must be documented and characterized before it can be explained. While characterization of natural variation among bacterial species has long been commonplace, until recently, evolutionary analysis of intraspecific variation has received less attention (Habets *et al.*, 2012; Heath *et al.*, 2012; Williams and Wernegreen, 2012). Phylogenetic branching events that ultimately lead to full-fledged speciation initially represent the divergence of conspecific lineages (i.e. lineages belonging to the same species). Thus, understanding the evolutionary processes and ecological forces driving intraspecific divergence, as well as the genetic basis and spatial dynamics of that divergence, represent keys to ultimately understanding processes of speciation in the myxobacteria.

To date and to our knowledge, natural isolates of *M. xanthus* have been found to vary heritably at every quantifiable phenotype that has been measured across many isolates (Fig. 1.3). These variable traits include predatory performance profiles across a wide range of prey (Mendes-Soares and Velicer, 2012; Morgan *et al.*, 2010), swarming rates and phenotypes (Kraemer and Velicer, 2011; Vos and Velicer, 2008b), group merger compatibility during swarming (Vos and Velicer, 2009), pure culture spore productivity (Kraemer *et al.*, 2010; Kraemer and Velicer, 2011), competitiveness in chimeric groups during development (Fiegna and Velicer, 2005; Vos and Velicer, 2009), individual fruiting body morphology, the size and spatial distribution of fruiting bodies on lab media (Fiegna and Velicer, 2005; Kraemer *et al.*, 2010), population-level rates of development (Kraemer *et al.*, 2010), the density-dependence of spore production (Kadam and Velicer, 2006), secondary metabolite production profiles (Krug *et al.*, 2008) and even the presence versus absence of regulatory elements required for development in some strains (G. Lippert-Viera, unpublished data). Importantly, many of these traits vary not only between highly divergent *M. xanthus* strains isolated from distant locations, but also among closely related strains isolated from a cm-scale patch of soil (Krug *et al.*, 2008; Morgan *et al.*, 2010; Vos and Velicer, 2006, 2008a, 2009) and even among extremely genetically similar isolates sampled from within the same fruiting body (Kraemer and Velicer, 2011).

Terrestrial biogeography of intraspecific variation

A fundamental characteristic of animal and plant populations is that their genetic and phenotypic diversity is often spatially structured. Such structure can strongly affect evolutionary processes and is therefore a necessary component of population genetics theory that is relevant to real biological populations (Whitlock, 2003). Microbes can disperse great distances (Smith *et al.*, 2012; Vos and Velicer, 2008a) owing to their small size and such migration was long emphasized in speculations regarding the spatial dynamics of microbial populations (Baas-Becking, 1934; Finlay, 2002). However, a number of recent empirical studies have demonstrated that natural populations of microbes are also highly structured (Ramette and Tiedje, 2007; Reno *et al.*, 2009; Sul *et al.*, 2013), including populations of free-living, spore-forming species such as *Myxococcus xanthus* (Vos and Velicer, 2008a). As for plants and animals, spatial structure is a critical feature of microbial population genetics (Hawlena *et al.*, 2010a,b) and rigorous characterization of that structure is necessary for understanding the spatio-temporal dynamics of evolutionary processes that drive microbial evolution, especially for species that engage in cooperative behaviours discriminately across conspecific genotypes.

Large-scale spatial structure

Soil populations of *M. xanthus* are highly structured at both broad and fine spatial scales. Initial studies testing for such structure examined single nucleotide polymorphisms at several relatively conserved loci (*clpX*, *csgA*, *fibA*, *icd*, and *sglK*) for multi-locus sequence type (MLST) analysis of >150 *M. xanthus* strains isolated from soil in a nested design with sampling scales ranging from centimetres to thousands of kilometres (Vos and Velicer, 2008a). Nucleotide diversity was analysed across seven sampling scales graded at roughly order-of-magnitude increments with sample sites centred on a 16 × 16 cm patch of soil from which cm-scale isolates were obtained. Diversity at these MLST loci was found to increase substantially from cm to metre scales, level off across intermediate scales (metres to hundreds of kilometres) and then increase dramatically again from hundreds to thousands of kilometres (Fig. 1.3E). In a distinct analysis, MLST diversity was examined within vs. across several sets of metre-scale samples collected from ten globally distributed sites. Pairwise comparisons of these population sets revealed not only that most metre-scale *M. xanthus* populations separated by large distances are genetically differentiated from one another but also that the

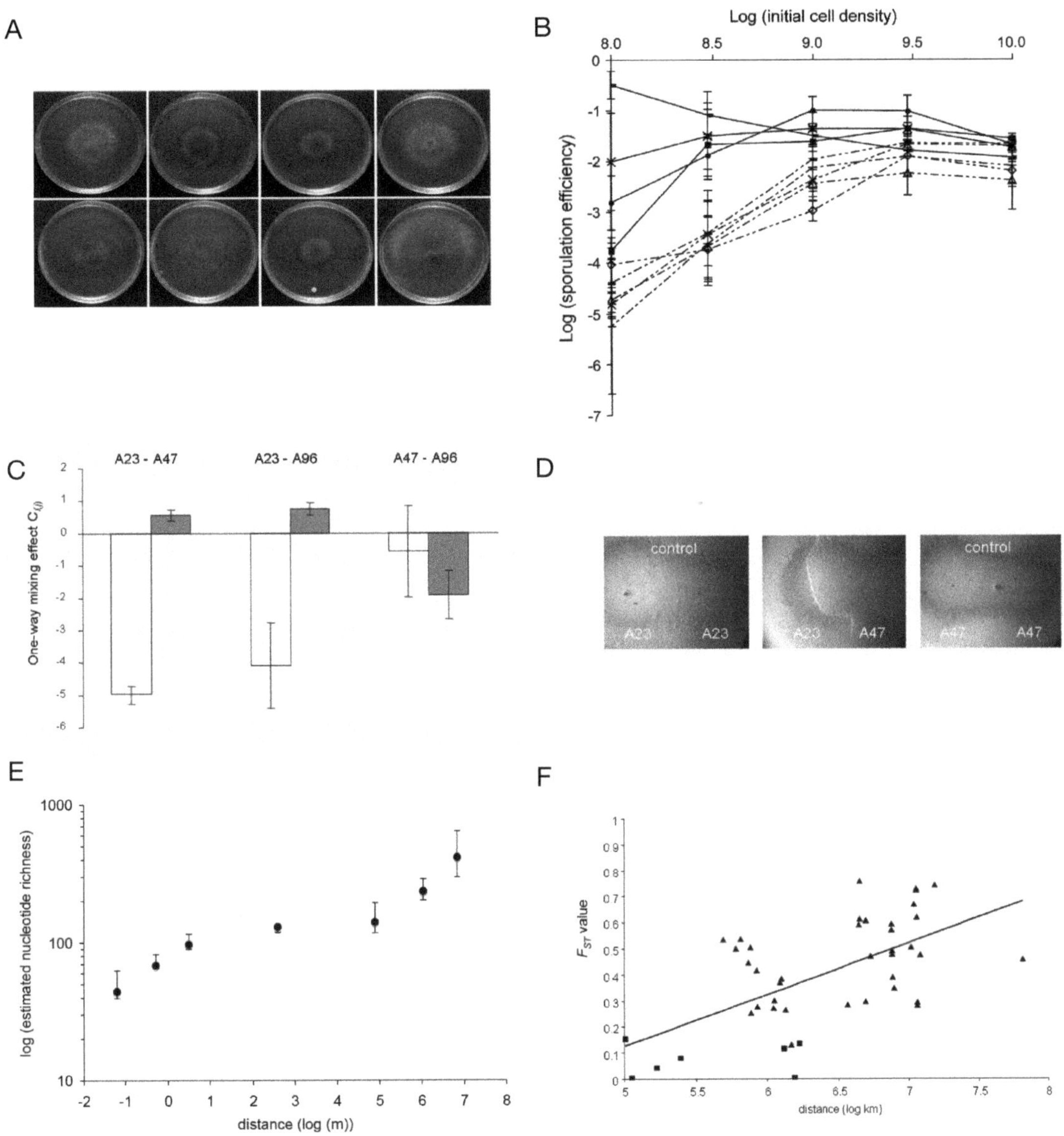

Figure 1.3 Heritable intraspecific variation at several social traits and genetic diversity in *M. xanthus*. (A) Variable swarming phenotypes of distinct clones isolated from the same fruiting body on soft agar. Reprinted with permission from Kraemer and Velicer 2011. (B) Variable relationships between density and spore production among isolates from globally distributed origins. Reprinted with permission from Kadam and Velicer 2006. (C) Variable competitiveness during development in pair-wise mixes of three genetically similar cm-scale isolates. The y-axis shows the effect of mixing on the sporulation efficiency of individual competitors. The one-way mixing effect, $Ci(j)$, shows the sporulation efficiency of a strain when mixed with a competitor relative to its performance in pure culture (on a $\log_{10}$ scale). (D) Variable social compatibility during group swarming among cm-scale isolates. The controls show swarms of the same genotype that have merged with no line of demarcation. Panels c and d reprinted with permission from Vos and Velicer 2009. (E) Nucleotide richness estimates representing the estimated number of polymorphisms in a multi-locus sequence typing (MLST) concatemer for *M. xanthus* isolate sets across a wide range of spatial scales. (F) F_{ST} values among 10 globally distributed metre-scale populations plotted against distance between populations. F_{ST} values represent the degree of genetic differentiation between paired populations. Significantly differentiated population pairs are indicated by triangles; non-significantly differentiated population pairs are indicated by squares. (E) and (F) adapted and reprinted with permission from Vos and Velicer (2008). A colour version of this figure is located in the plate section at the back of the book.

degree of differentiation tends to increase with inter-population distance (Fig. 1.3F).

The observed correlation between distance and population differentiation might be explained by selective forces (and resulting genetic signatures of local adaptation) diverging increasingly with inter-sample site distance, differential patterns of genetic drift and/or hitchhiking (Barton *et al.*, 2013) of non-adaptive mutations across isolated populations or some combination of these processes. The genetic drift hypothesis is supported by the finding that synonymous sites in the *pilA* gene – which should be either neutral or under relatively weak selection (Zhou *et al.*, 2010) – reveal a correlation between inter-population distance and differentiation of similar magnitude to that revealed by non-synonymous *pilA* sites (Vos and Velicer, 2006). If genetic population differentiation reflected divergent adaptation at *pilA*, no correlation between inter-population distance and differentiation at synonymous sites would be expected. The finding of such a relationship suggests that migration of *M. xanthus* cells across distances of hundreds to thousands of kilometres is insufficient to eliminate population differentiation generated by the random forces of mutation and genetic drift.

Migration of *M. xanthus* cells, spores or lineages across very large distances does appear to occur to a limited degree, as a small number of isolates differ greatly from most of their local neighbours and share identical or extremely similar MLST genotypes with isolates from a very distant locale (Vos and Velicer, 2009). However, long-distance migration is limited relative to rates of differential local evolution. The inference of isolation by distance (population differentiation due to limited dispersal) does not exclude the possibility that populations at these scales also adapt locally to distinct selective habits, but it does suggest that natural selection alone does not fully explain the spatial structure of *M. xanthus* genetic diversity at global scales. Comparative genomic analyses employing advanced sequencing technologies will allow further phylogentically broad comparison of myxobacterial genomes and biogeography at the inter-specific level (Goldman *et al.*, 2006; Huntley *et al.*, 2011, 2012; Ivanova *et al.*, 2010; Li *et al.*, 2011; Schneiker *et al.*, 2007; Thomas *et al.*, 2008). Importantly, however, such analyses will also allow extremely fine-scale resolution of myxobacterial biogeography within species and corresponding insights into how the key forces of evolutionary change – mutation, recombination, selection, genetic drift and migration – have shaped the social divergence of the myxobacteria.

Fine-scale spatial structure

At the smallest spatial scale that microbial populations occupy, the micrometre scale, spatial structure must exist at least transiently in viscous habitats due to the locality of colony growth by binary fission. Within colonies, cells are surrounded primarily by genetically identical clone-mates, with the exceptions of new spontaneous mutants (which still remain identical at almost all genome sites) and any divergent migrant cells that have penetrated colony space. Thus, one of the primary criteria for the evolution of cooperation in kin selection models, namely sufficiently high relatedness among interactants (Queller, 2011), may be met at the within-colony scale by most microbes that dwell in highly viscous habitats, even those that exhibit fewer and less complex cooperative traits than the myxobacteria. High average within-group relatedness is likely to have been a necessary but not sufficient condition for the evolution of forms of cooperation unique to the myxobacteria.

How large are clonal patches of myxobacteria in the soil? Are they sufficiently large that most cells aggregating into a fruiting body tend to be genetically identical, or nearly so, due to recent common descent from a colony founder? Or do cells of fruiting species migrate sufficiently (via motility or vector-enhanced dispersal) between fruiting body cycles to cause genetically distinct individuals from separate natal colonies to merge together into chimeric fruiting bodies?

Although a hint of structure was suggested by the distribution of MLST genotypes in the centimetre-scale patch of *M. xanthus* isolates examined by Vos and Velicer (2006), this structure was marginally non-significant, indicating that the genotypes sampled migrate extensively at the centimetre scale. Given that the relatively conserved loci used in that study nearly showed significant structure, it is likely that structure would have been detected if more highly variable loci had been used. In a subsequent study, Kraemer (2011) genotyped multiple *M. xanthus* isolates from each of several fruiting bodies that emerged in a millimetre-scale cluster on the same soil particle for variation at several loci showing the highest degree of polymorphism between several of the genetically similar centimetre-scale isolates of Vos and Velicer (2006) based on whole genome sequencing (Kraemer and Velicer, 2011). She found that most neighbouring fruiting bodies share the same dominant genotype, indicating that *M. xanthus* clonal patches are sufficiently large to support the formation of multiple fruiting bodies, at least at one sampling location. Taken together, the studies by Vos and Velicer (2006, 2008a) and Kraemer and Velicer (2011) suggest that *M. xanthus* clonal patch sizes fall in the millimetre to centimetre range. However, additional similar studies of local diversity at more sample sites and utilizing the same genetic loci are needed to test the accuracy and generality of this inference.

If neighbouring fruiting bodies in the soil often share the same dominant genotype at loci that are highly polymorphic at broader sampling scales (e.g. ≥ the centimetre scale), most fruiting bodies might be expected to be largely genetically homogeneous, like individual plants and metazoans. Kraemer and Velicer (2011) investigated the genetic composition of *M. xanthus* fruiting bodies that emerged from soil collected at three woodland sites in southern Indiana. They found that 8 of the 10 fruiting bodies examined contained cells that differed reproducibly in one or more phenotypic trait (e.g. spore production, swarming rate and phenotype (Fig. 1.3A), colony colour, adhesiveness, etc.) and/or in MLST genotype from other isolates from the same fruiting body. Instances of highly reproducible phenotypic variation are considered to reflect underlying genetic variation that may not yet have been identified at the level of DNA sequence. Despite the fact that a majority of fruiting bodies harboured significant variation, strains isolated from the same fruiting body were found to be much more genetically similar to one another than to clones from other fruiting bodies sampled just centimetres to metres apart. In fact, zero genetic polymorphisms were found within four of the fruiting bodies for five rapidly diversifying loci examined by Kraemer and Velicer (2012) (MXAN_0128, MXAN_0176, MXAN_0396, MXAN_0533 and MXAN_4405) (Goldman *et al.*, 2006) or *pilA* and only one fruiting body harboured polymorphism among the five loci other than *pilA*.

The finding that a majority of fruiting bodies harbour significant but limited genetic variation (either inferred indirectly from variable phenotypes or detected directly by sequencing) indicates that most such variation is likely to be of recent endemic origin and raises a number of intriguing questions regarding the evolutionary forces responsible for it. Why does apparently endemic phenotypic variation exist within such a large proportion of fruiting bodies examined and at such high frequencies within fruiting bodies? Such variation might merely reflect genetic drift of neutral or mildly deleterious mutations (Hershberg *et al.*, 2008), but its high frequency (variants at frequencies greater than 0.02 in most fruiting bodies) suggests that selection may be at play. Detected variation might also represent temporal snapshots of selective sweeps in progress (Majewski and Cohan, 1999) or recently diverged lineages that are being maintained by balancing selection (Zhang *et al.*, 2012). Finally, in principle the reproducible phenotypic variation might be of epigenetic origin, but the fact that the variable phenotypes are heritably maintained between lineages even after multi-generational pre-conditioning under common, high-nutrient culture environments strongly suggests that much of the observed variation is caused by true genetic polymorphisms.

To begin testing between adaptive and non-adaptive explanations for within fruiting-body variation, Kraemer (2011) asked whether clones that produce few spores in pure culture might be maintained in natural populations by cheating on other high-sporulating strains within the same fruiting body during development. Intriguingly, Kraemer (2011) found that although developmentally defective clones were partially rescued by high-sporulating strains from the same fruiting body, such rescue did not confer a relative fitness advantage over the cooperative strains and the defective strains decreased in frequency over several rounds of growth and development in mixed-culture competitions rather than increasing or being maintained about an equilibrium frequency. It is possible that such developmentally defective strains do, in fact, cheat during development under natural conditions but, for an unknown reason, do not do so under laboratory conditions. Alternatively, naturally developmentally defective strains might have a selective advantage over their developmentally proficient counterparts in some component of overall fitness other than development (e.g. motility and/or predation) that counter-balances their developmental defects. Finally, such defective mutants may not be positively maintained by selection at all but rather may have reached high frequency within some fruiting body-forming groups via genetic drift. Extensive further work is required to understand the evolutionary forces, genetic polymorphisms and molecular mechanisms responsible for the high levels of endemic phenotypic diversity found to exist within myxobacterial social groups.

Evolution of social fragmentation

The distinct genotypes found together within fruiting bodies by Kraemer and Velicer (2011) can obviously co-aggregate. However, genetically similar clones from the Tübingen cm-scale patch of Vos and Velicer (2006, 2008a) were found to have evolved social barriers that prevent the merger of swarming colonies that occurs when spatially separate colonies of the same genotype grow into one another on agar plates (Vos and Velicer, 2009). Remarkably, among the 78 clones isolated from the centimetre-scale patch, 45 distinct 'allo-recognition' types – representatives of which showed reduced or no group swarm merger across types – were present, despite the fact that distinct swarm compatibility types often had very similar or even identical MLST profiles (Fig. 1.4). These swarming incompatibilities are likely to represent incipient forms of social isolation that might be reinforced over evolutionary time to ultimately prevent any form of cooperation between distinct types. Indeed, initial results suggest that distinct swarm compatibility types among this collection of isolates show a reduced propensity to co-aggregate into common fruiting bodies at the interface of oncoming swarms on CF agar (Olaya Rendueles-Garcia, unpublished). Complete social isolation, perhaps defined as the absence of any propensity of distinct genotypes to merge into common cooperative groups from distinct points

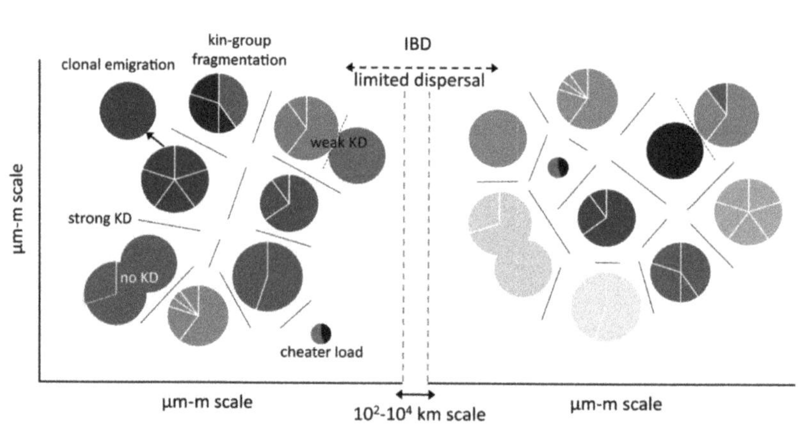

Figure 1.4 Simple hypothetical model of natural *Myxococcus* population biology. Circles represent social groups within which individuals directly interact. Sectors represent genetically distinct within-group variants. Distinctly coloured circles represent among-group genetic differentiation, with lines separating kin discrimination (KD) units. Overlapping circles represent lack of KD between highly similar, but nonetheless genetically distinct, social groups. Multicolored circles represent kin-group fragmentation (see text). Small circles with a black sector represent cheater-infected groups burdened by cheating load (Travisano and Velicer, 2004). Distinct colour ranges across left vs. right panels represent population differentiation across large spatial scales due to isolation by distance (IBD). The arrow in the left panel represents establishment of a new clonal group by clonal emigration. Reprinted with permission from Kraemer and Velicer 2011. A colour version of this figure is located in the plate section at the back of the book.

of spatial origin under any ecological conditions, might be worth considering as a criterion for species or subspecies distinction in the myxobacteria that has a more intuitive biological basis than some traditional criteria (Vos, 2011). However, further work is required to identify the scale of genetic divergence at which such complete social isolation is reached (if it is reached at all).

It remains unclear what drives the evolution of social fragmentation within local populations of *M. xanthus*. Do new adaptive mutations often confer a fitness advantage to their bearers *because* they reduce social interactions with individuals of the parental genotype? Or might mutations that are adaptive at some other trait often reduce social compatibility between parental and mutant genotypes as a pleiotropic side effect?

It is important to distinguish between primary and secondary mechanisms of social incompatibility or kin discrimination (here used interchangeably for simplicity). Primary forms of social incompatibility are the mechanisms that actually function to prevent or reduce cooperative interactions between two genotypes under natural conditions. Secondary forms of social incompatibility derive from differential independent evolution of co-evolving social gene networks between lineages that are already socially isolated by primary mechanisms of incompatibility. For example, two distinct lineages might first evolve mechanisms of non-self swarm exclusion as a primary mechanism of social incompatibility that would prevent aggregation into a common group if swarms of those two strains ever met and then later evolve secondary (i.e. post-isolation) social incompatibilities such as dysfunctionally divergent developmental signalling (Chapter 7) or an inability to engage in extracellular protein transfer (Chapter 5). In such a scenario, secondary forms of social incompatibility would not be considered natural forms of kin discrimination because primary mechanisms of social incompatibility would largely or completely prevent them from ever being manifested under natural conditions. This distinction between primary and secondary forms of social incompatibility is analogous to the distinction in sexual eukaryotes between functional mechanisms of reproductive isolation that lead to full speciation and post-speciation divergence due to differential selection, adaptation trajectories, and/or genetic drift.

Socially mediated selection

The finding that local social groups (in this case fruiting bodies) are composed of closely related individuals that are nonetheless phenotypically diverse (Kraemer and Velicer, 2011) is consistent with the possibility of social co-evolution among lineages that are clustered in space and time. Diverse interacting *Myxococcus* lineages may tend to remain statistically clustered over many generations within social groups engaging in cooperative behaviours, but the degree to which they do so remains an open question. Clustering over evolutionarily substantial timescales would align the evolutionary interests of associated lineages at the scale of group-level selection and thus might lead to synergistic co-evolution that enhances long-term group productivity by reducing within-group conflict (Sachs and Bull, 2005). Such synergistic co-evolution could take the negative form of mitigating group-level fitness costs of phenotypic chimerism (chimeric load) generated by negative interactions among genotypes maintained by balancing selection (e.g. frequency-dependent fitness rank reversals; Velicer *et al.*, 2000) or a positive form of generating novel complementary phenotypes among distinct lineages that mutually enhance long-term group productivity. Synergistic co-evolution is less expected if within-group lineage combinations frequently reassort due to extensive inter-group migration that de-couples the evolutionary interests of interacting lineages.

Evolution of selfish policing

Social environments can impose strong selective pressure (Queller, 2011), as two evolution experiments with *M. xanthus* exemplify in the myxobacteria. In these experiments, evolving populations of an *M. xanthus* developmental cooperator strain (a genotype that is highly proficient at fruiting body formation and sporulation) repeatedly interacted with a developmentally defective cheater. In one case, the cooperator subpopulation evolved rapidly (Manhes and Velicer, 2011), and, in the other, the cheater subpopulation evolved (Fiegna *et al.*, 2006).

Manhes and Velicer (2011) tested whether *M. xanthus* populations initially composed of a single socially proficient genotype could evolve to reduce or eliminate the ability of a developmentally defective strain (in this case a *csgA*⁻ mutant) to cheat during development. The *csgA*⁻ mutant (hereafter called 'CH') had previously been demonstrated to have a large relative fitness advantage over a socially proficient lab reference strain (hereafter called 'WT') in mixed groups during development despite the fact that CH sporulates very poorly in pure culture (Manhes and Velicer, 2011; Velicer *et al.*, 2000). WT was marked with an antibiotic resistance not shared by the *csgA*⁻ mutant CH.

Replicate populations of WT were allowed to evolve while repeatedly encountering the same non-evolving CH genotype during development. Briefly, WT cells were mixed at a 50:50 ratio with CH and the chimeric populations were subjected to starvation, after which viable spores were selected for by heating and sonication. The total spore population from each replicate was then inoculated into medium containing kanamycin, which allowed WT-derived spores, but not spores of the CH mutant, to germinate and divide. After the WT-derived subpopulations grew to high density, they were mixed again with a fresh culture of CH and the resulting mixed populations were starved. This process was repeated for 20 cycles of development, thus potentially allowing spontaneous advantageous mutants of WT to arise and increase to detectable frequencies. In this experiment, CH cells were an environmental feature that might affect the relative fitness of distinct WT-derived genotypes, and they were replaced with the same genotype at each cycle of development.

WT cooperators rapidly evolved to suppress cheating by CH (Fig. 1.5). After 20 cycles of evolution, all four replicate WT-derived populations (hereafter 'EV' for *Ev*olved) had higher fitness than CH in 50:50 mixed cultures, thus reversing the fitness advantage of CH over the ancestral WT. Intriguingly, the EV populations not only suppressed CH sporulation to their own advantage in two-party mixes, but also benefited the absolute fitness of the ancestral WT in three-party mixes (25:25:50 EV:WT:CH) relative to WT performance in two-party mixes with CH (50:50 WT: CH). This beneficial effect of EV on WT was specific to the presence of the cheater, because

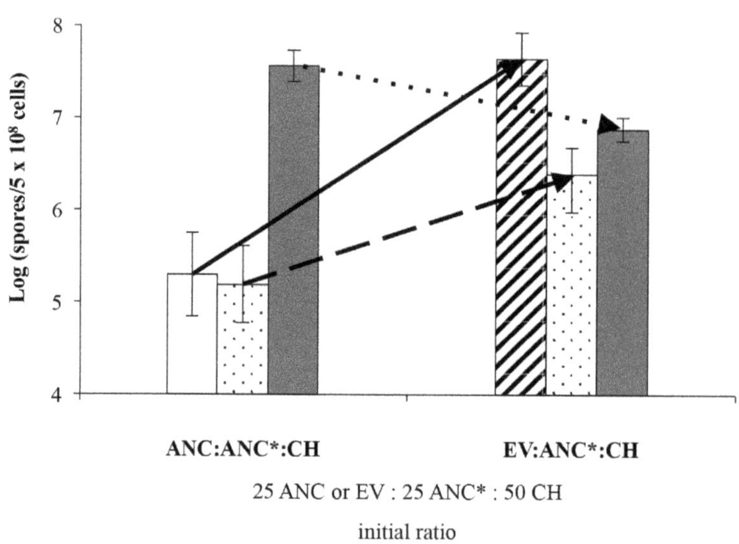

Figure 1.5 Evolution of cheater suppression. Spore production in three-party competitions with ANC:ANC*:CH (white:speckled:grey) and EV:ANC*:CH (hatched:speckled:grey). ANC, developmentally proficient ancestor; ANC*, genetically marked variant of ANC; CH, non-evolved cheater; EV, evolved populations (data show average for four independently evolved populations). Arrows highlight changes after evolution over 20 cycles of development from ANC to EV. Reprinted with permission from Manhes and Velicer (2011).

in two-party mixes in the absence of CH the EV populations reduced WT fitness (Manhes and Velicer, 2011).

The third-party benefit of cheater suppression demonstrated in this study is analogous to effects of policing behaviour in some social animals (Zanette *et al.*, 2012). However, cheater suppression was nonetheless directly beneficial to EV populations (i.e. 'selfish' in evolutionary terms rather than altruistic) not only in two-party mixes, but also in three-party mixes of EV, WT and CH. In three-party mixes, EV fitness was higher than that of WT even though EV populations benefited WT relative to its low spore production in 50:50 mixes with CH. Importantly, the advantage of EV over WT was largely specific to the presence of the cheater. In three-party mixes, all four EV populations had higher fitness than WT, but in the absence of the cheater [considering both 50:50 WT:EV mixes and pure cultures (100:0 and 0:100 WT:EV)] the EV populations had lower overall fitness than WT.

Socially mediated re-evolution of developmental proficiency

In another experiment, a wild-type cooperator (WT) and an experimentally evolved, developmentally defective cheater (OC) were mixed at a 99:1 ratio and allowed to compete over six successive cycles of starvation-induced development and intermittent growth in liquid medium (Fiegna and Velicer, 2003). Only spores were allowed to survive each round of development for inoculation into the next round of growth medium. Due to the extremely large relative fitness advantage of OC over WT when OC is rare (i.e. its strong cheating ability) (Fiegna and Velicer, 2003; Fiegna *et al.*, 2006; Velicer *et al.*, 2002), OC rapidly rose to high frequency in the mixed population. However, due to its own developmental defect, at high frequency OC caused the entire mixed population to 'crash', or produce extremely few spores, in the fourth development cycle of the competition.

Intriguingly, the social rescue of OC by WT allowed a spontaneous mutant of OC with an adaptive advantage over both OC and WT to arise and survive the fourth-cycle crash caused by OC. This mutant, named 'PX' (Phoenix), had surprisingly regained the ability to sporulate at high levels in pure culture (even higher than WT) that had been lost by OC and went to apparent fixation in the mixed population (Fiegna *et al.*, 2006). Not only did PX exhibit higher pure culture sporulation than both of its ancestors, but in paired competitions PX had a relative fitness advantage over both WT and OC at all tested frequencies. This result implies that whatever mechanism restored developmental proficiency to PX did so in a manner that simultaneously conferred resistance to being cheated on by its immediate ancestor OC (which, as stated previously, cheats very effectively against WT). Subsequent sequencing of the PX genome (Velicer *et al.*, 2006) revealed a single nucleotide mutation that conferred the superior PX phenotype and ultimately led to the discovery of a major regulatory gateway controlling the initiation of *Myxococcus* development (Yu *et al.*, 2010; see also 'From experimental evolution to mechanistic discovery', below, for further details).

The two instances of laboratory evolution highlighted above examined the evolution of one social partner in the presence of distinct social partners rather than actual co-evolution among multiple interacting lineages. Nonetheless, these examples highlight the selective power of microbial social environments and illustrate how rapidly microbes can evolve in response to presence of particular social partners. The rapidity of evolution in response to socially mediated selection in laboratory experiments (Fiegna *et al.*, 2006; Manhes and Velicer, 2011; Zanette *et al.*, 2012), together with observed patterns of diversity within and across natural *Myxococcus* social groups (Kraemer and Velicer, 2011) suggest that spatiotemporally clustered lineages in the wild may remain associated and interact over sufficient periods of time to undergo true co-evolution, whether synergistic or antagonistic. Evolution experiments in which spatial structure is maintained across generations will facilitate investigations into the potential for such spatially localized co-evolution.

Evolution of development (Myxo Evo-Devo)

Comparative macro-evolutionary studies of animal and plant ontogeny have dominated the 'evo-devo' field (de Bruijn *et al.*, 2012; Mallarino and Abzhanov, 2012). However, microbial fruiting body development shares many features with multicellular development from a single-cell zygote, including cell–cell communication, cell differentiation, multicellular structure formation and multicellular-level life history traits (e.g. rate of development). Thus, fruiting microbes such as the myxobacteria and cellular slime moulds offer excellent opportunities to address fundamental questions about the evolution of developmental pathways and morphologies that are general to both social (i.e. microbial) and zygotic forms of multicellular development. Indeed, even relatively asocial microbial species are now being employed to address development themes traditionally associated with plant and animal research, for example ageing (Ackermann *et al.*, 2003) and pattern formation (Liu *et al.*, 2011). Recently, a call has been made to explicitly include microbial species in the definition and development of model organisms for development biology (Love and Travisano, 2013).

Major questions central to the evolution of development field have already been broached by studies with myxobacteria. For example, how variable can the genetic programme underlying a common developmental phenomenon within a defined clade be? At the macroevolutionary scale, Huntley *et al.* (2011) conducted a genome-wide comparative analysis of five myxobacteria species (representing all three classical subdivisions) to discern the degree to which loci essential for development in the model species *M. xanthus* are present across the myxobacteria. They found that many of these essential *M. xanthus* development genes are not present in other species that also form robust fruiting bodies, thus suggesting that a wide range of evolutionarily malleable gene regulation programmes can underlie the general developmental phenomenon of fruiting body morphogenesis. This finding is consistent with an earlier genomic study of the *M. xanthus* genome which found that major loci required for fruiting body formation were apparently acquired by horizontal gene transfer after divergence from the most recent common ancestor of all fruiting species (Goldman *et al.*, 2007).

If the genetic basis of myxobacterial development has diverged so greatly across species that a common genomic signature of proficiency at fruiting body morphogenesis is difficult to discern, then just how easily can myxobacterial developmental programmes be 'rewired' by evolution? Both evolution experiments and intraspecific variation analysis indicate that such rewiring can occur at very fine phylogenetic scales. For example, the PX strain that re-evolved developmental proficiency from the developmentally defective ancestor OC did so by mutationally eliminating the function of a small RNA molecule (*pxr*) that was found to negatively regulate fruiting body formation in the wild-type in the presence of abundant nutrients (Yu *et al.*, 2010). This single mutation generated a sporulation proficient strain (PX) that differs dramatically in its developmental gene expression profile from the sporulation-proficient WT ancestor of OC (Kadam *et al.*, 2008).

Moreover, *M. xanthus* natural isolates also vary in what genetic elements are required for proficient development. The lab strain DK1622 requires the *dev* regulatory region for normal development (Viswanathan *et al.*, 2007). However, a high proportion of developmentally proficient natural isolates have been found to entirely lack a large section of this region requisite for normal DK1622 development (G. Lippert-Vieira and G. Velicer, unpublished), implying a significant natural rewiring of the *M. xanthus* developmental regulatory network. Clearly, extensive genomic comparisons of many *M. xanthus* isolates and associated experimental analysis will be necessary to fully grasp just how variable the genetic programmes that underlie myxobacterial development can be within a single species.

Of course, a fundamental requirement for the evolutionary rewiring of development is variation in developmental genes and their associated phenotypes. While the instances of the PX mutation in a laboratory lineage and the *dev* region polymorphism among natural isolates represent rather dramatic instances of such variation, subtler instances exist as well. Kraemer *et al.* (2010) demonstrated that *M. xanthus* natural isolates vary continuously in the rate at which they form fruiting bodies and spores. It is easy to imagine scenarios in which the rate of development is under selective pressure, either positively or negatively. For example, fast development might confer a competitive advantage if strains with distinct intrinsic development rates co-aggregate into common fruiting bodies (Buttery *et al.*, 2012; Kraemer *et al.*, 2010). Alternatively, one could imagine a tradeoff between development rate and spore quantity and/or quality that could lead to selection for slower development under some conditions.

Another classic evo-devo theme touched upon by myxobacterial research is the question of how much the evolution of morphological features is driven by mutations that regulate protein expression as opposed to changes in protein sequences per se (Carroll, 2008). The evolutionary restoration of development in strain PX was wrought by a mutation in a *trans* acting regulatory gene, namely a small RNA gene, by debilitating its negative regulatory function (Yu *et al.*, 2010). But how might more subtle regulatory mutations affect development? Escalante *et al.* (2012) examined the potential for selection to act on variation in *trans*-acting developmental gene regulation in the myxobacteria by testing for pleiotropic effects of altering expression of the key developmental regulatory gene *fruA*. [See Chapter 6 and Mittal and Kroos, 2009(a,b) for a summary of the role of *fruA* in *Myxococcus* development.] To accomplish this, four distinct trans-acting regulators of *fruA* were deleted, resulting in variable degrees of increased *fruA* expression, as anticipated. Increased *fruA* expression was found to correlate very strongly with rate of fruiting body development, as well as affecting several other developmental parameters, including the number of fruiting bodies produced (and variation in that number), viable spore count and degree of spore viability. The study by Escalante *et al.* (2012) is important for demonstrating that predictable quantitative variation in a suite of developmental traits (that are likely exposed to selection in some environments) can be generated as a function of defined mutations in a well-characterized regulatory pathway. It will be of interest to investigate the degree to which natural variation in development rates may be caused by regulatory variation in the *fruA* pathway network versus loci outside that network.

From experimental evolution to mechanistic discovery

As noted above, the mutation that restored developmental proficiency to strain PX found by Fiegna *et al.* (2006) occurred in a gene encoding a previously undiscovered small regulatory RNA, Pxr, which is predicted to fold into a triple stem–loop structure similar to many regulatory sRNAs found in other species such as *E. coli* (Gottesman, 2005). Pxr was shown to block development in OC, because both the PX mutation itself and deletion of the entire *pxr* locus restore developmental proficiency (Fiegna *et al.*, 2006; Yu *et al.*, 2010). Pxr was subsequently shown to negatively regulate the initiation of development in the wild-type (WT, a derivative of DK1622) in a nutrient-dependent manner. WT does not initiate development when growth substrate is abundant, but a mutant of WT from which *pxr* was deleted forms robust fruiting bodies and many spores even at high nutrient levels (Fig. 1.6). Thus, Pxr appears to serve as a major gatekeeper to prevent the initiation of development under conditions in which continued growth is a more beneficial strategy. Two RNA products that vary in length result from an intact *pxr* locus, one long (Pxr-L) and one short (Pxr-S). The short form is likely to be the active negative regulator, as Pxr-S disappears quickly after the onset of development in WT, whereas Pxr-L remains present at high levels throughout early development (Yu *et al.*, 2010). Pxr as a regulatory element appears to have arisen since the origin of the *Myxococcales* order, as the *pxr* gene can only be detected in a monophyletic subclade of the mxyobacteria that includes the Cystobacteraceae and Myxococcaceae families but not more deeply branching lineages (I-C.K. Chen *et al.*, unpublished). The discovery of the Pxr regulatory systems represents an example of how molecular-level analysis of experimental evolution outcomes can generate novel mechanistic discoveries that were not uncovered by traditional molecular-genetic methods.

Predation

Microbes have long been viewed primarily as decomposers, but recent research has shown that a variety of phylogenetically disparate bacterial

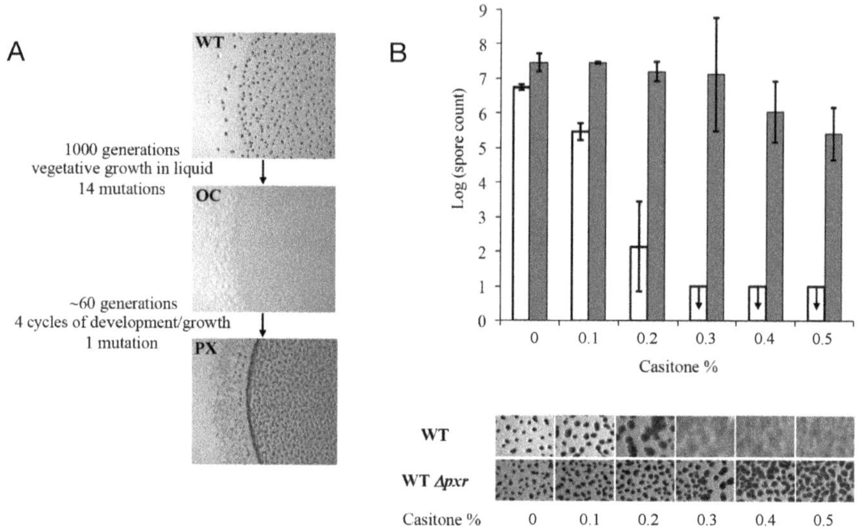

Figure 1.6 Evolutionary restoration of sporulation to a developmentally defective cheater. (A) The lineage from wild type (WT) to the obligate cheater (OC) accumulated 14 mutations during 1000 generations of evolution in liquid and lost developmental proficiency. OC evolved into the developmentally proficient strain Phoenix (PX) by a single mutation during four cycles of competition with a marked variant of WT. Adapted and reprinted with permission from Velicer *et al.* (2006). (B) *pxr* inhibits development in the presence of nutrients. The graph and images show spore production and developmental phenotypes, respectively, at different casitone concentrations. WT=GJV1, a derivative DK1622 (Yu *et al.* 2010). Arrows indicate zero spores at the limit of detection. Reprinted with permission from Yu *et al.* (2010).

species prey on a plethora of other microbial species (Jurkevitch, 2007). Among these predators are the myxobacteria.

Several distinct microbial predatory strategies exist. While bacteria such as *Bdellovibrio* and *Daptobacter* feed on other bacterial species by periplasmic and cytoplasmic invasion, respectively (Guerrero *et al.*, 1987; Mark, 2002; Martin and Bull, 2002; Rogosky *et al.*, 2006), myxobacteria have an extracellular predatory strategy, popularized as 'wolfpack' predation because it is hypothesized to be cooperative in a manner that goes beyond the contribution of social motility to predation (Martin and Bull, 2002; Rosenberg *et al.*, 1977). In tightly packed groups, *M. xanthus* cells use their motility mechanisms to search for the prey, possibly by the use of elasticotaxis (Fontes and Kaiser, 1999) and/or chemotaxis (Kearns and Shimkets, 1998), but at the very least by undirected swarming. The predator cells produce a cocktail of antibiotics and lytic enzymes that kill and lyse prey cells (Mendes-Soares and Velicer, 2013; Reichenbach and Höfle, 1993), thereby releasing their molecular components to the environment. Predator cells then uptake these components that are further degraded and integrated into the metabolism of the myxobacterium (Berleman and Kirby, 2009; Reichenbach, 1999). Myxobacteria have been shown to feed on a wide variety of soil microbes, including both pathogens and commensals of macroorganisms (Berleman *et al.*, 2006; Bull *et al.*, 2002; Martin and Bull, 2002; Mendes-Soares and Velicer, 2013; Morgan *et al.*, 2010; Reichenbach, 1999). Thus, due to the range of prey that they can consume, predatory microbes such as *M. xanthus* may have deep impacts on the dynamics of complex ecosystems.

Stigmatella aurantiaca, Sorangium cellulosum, and *M. xanthus* are all myxobacteria species that are able to utilize nutrients made available from enzymatic digestion of other soil inhabitants, including both bacteria and eukaryotic organisms such as yeast. Even though in some species such as *Myxococcus xanthus,* cells are also able to feed on other organisms individually (McBride and Zusman, 1996), it is often hypothesized that they are more efficient predators when in groups (e.g. Berleman and Kirby, 2009), because a higher concentration of degradative enzymes is reached sooner (Rosenberg *et al.*, 1977). However, a synergistic effect of high cell density on predatory growth has yet to be demonstrated directly. Predation appears to be a costly trait to maintain, since the predatory ability of *M. xanthus* has been found to degrade during evolution in the absence of selection for its maintenance, both in structured and unstructured environments (Velicer and Stredwick, 2002).

While the developmental stage of the myxobacterial life cycle has been extensively studied for decades, research on the predatory behaviours of this group has been relatively limited until recently (Berleman and Kirby, 2009; Hillesland *et al.*, 2007; Hillesland *et al.*, 2009; Mendes-Soares and Velicer, 2013; Morgan *et al.*, 2010). Researchers use several methods to assess the predatory efficiency of myxobacterial populations, including the common assay of simply measuring the rate of outward expansion of a colony growing and swarming across a lawn of prey or a set of prey patches (Hillesland *et al.*, 2007, 2009; Morgan *et al.*, 2010; Pham *et al.*, 2005). However, other parameters might also be used as predatory fitness proxies (e.g. killing rate, killing efficiency and predator yield (Mendes-Soares and Velicer, 2013). In nature, swarming rate through soil during predation may be a reflection of how well a strain is able to feed and move to another patch once prey are depleted, and thus spread more efficiently to new, more favourable, environments. A poor swarmer would be unable to do so, and may have to enter development sooner or else suffer greater cell death. The predatory ability of a given myxobacterial genotype is determined both by its genetic makeup and the surrounding environment, both physical and biological (Morgan *et al.*, 2010).

M. xanthus has a complex life cycle composed of two distinct stages: one of growth when nutrients are sufficiently abundant, and one of development into fruiting bodies when nutrients are scarce (Dworkin, 1996; Shimkets, 1990; Wireman and Dworkin, 1975). Because the overall fitness of *M. xanthus* genotypes in nature will emerge from their collective performance across both life-cycle stages, extensive regulatory co-evolution among the genes involved in these different life phases is expected (Guillaume and Otto, 2012), and mutations affecting one *M. xanthus* social trait (e.g. development) will often affect others (e.g. predation; Berleman and Kirby, 2009). This is indeed the case. Mutations in the *frz* chemotaxis-like system not only affect motility and development but also reduce predatory ability (Berleman *et al.*, 2006; Pham *et al.*, 2005). Similarly, mutations affecting developmental signalling also affect predation. For example, mutations in three A-signalling genes, *asgA*, *asgC* and *asgE*, reduced predatory ability 64–80% relative to the wild-type). A significant effect, although not as marked, was also observed for strains with mutations in the early developmental genes *sdeK* and *csgA* (36% and 33% reduction relative to the wild-type, respectively; Pham *et al.*, 2005). Finally, laboratory populations of *M. xanthus* evolving in liquid culture accumulated mutations that negatively affected not only development and social motility (Velicer *et al.*, 1998), but predation as well (Hillesland, 2005) and it is likely that at least some of these negative mutations simultaneously affected multiple traits.

While these results were obtained with mutants constructed or evolved in the laboratory, other work has demonstrated phenotypic variation in predatory swarming among *M. xanthus* natural isolates (Morgan *et al.*, 2010). Although the genetic basis of these differences remains to be identified, it is possible that some of the polymorphisms responsible for this variation may occur at the loci examined by Pham *et al.* (2005). Strains with different swarming abilities may be maintained in nature as a result of complex social interactions between genetically distinct populations. For instance, a good predator may benefit a poor predator by producing predatory compounds that the latter can exploit. Such complementation of the poor predator's disadvantage may be partial, with the good predator retaining a relative fitness advantage despite the absolute benefit to the poor predator from the interaction. Alternatively, the poor predator may cheat, or gain both an absolute benefit and a relative advantage over the good predator when they interact in a shared social environment (Velicer and Vos, 2009).

Environmental features other than social context, including both biotic and abiotic factors, can also significantly affect the predatory efficiency of *M. xanthus* populations. For example, the niches that *M. xanthus* occupy in nature should be strongly shaped by its microbial prey. Indeed, *M. xanthus* predatory efficiency has been found to vary greatly depending on its microbial prey (Hillesland *et al.*, 2007; Mendes-Soares and Velicer, 2013; Morgan *et al.*, 2010) and this should affect the speed at which the predator's local population is able to grow and disperse and how rapidly it enters into a physiologically quiescent stage as spores.

Hillesland *et al.* (2007) showed that surface type, prey type and the spatial distribution of prey patches influence the swarming rate of *M. xanthus*, and thus its predatory ability. Even though no influence of prey species was found on the rate of prey killing within a prey patch in that study, (an observation later also supported for the species used by Mendes-Soares and Velicer (2013)), the rate of prey-patch encounter was affected by prey species (Hillesland *et al.*, 2007). The effect was stronger in densely distributed patches on hard-agar surfaces, compared with soft-agar surfaces or arrangements with low-density patches. To test whether the searching ability of *M. xanthus* populations could be improved under conditions where prey patches were hard to find, as predicted by optimal foraging theory (Holling, 1959), Hillesland *et al.* (2009) set up a one-year evolution experiment on agar plates comprised of 16 populations that evolved while swarming across agar arenas with either a high or low density of prey (*E. coli*) patches. Equating between-patch swarming to 'searching' and within-patch swarming to 'handling' during predation (Holling, 1959, 1961), Hillesland *et al.* (2009) found that *M. xanthus* searching rate increased to a greater degree in the low prey-patch density populations than those evolving on high patch-density plates. There was no significant change in handling, as reflected by the lack of evolution on the rate of swarming within patches.

Other studies have demonstrated significant variation in the swarming rate of different *M. xanthus* strains on different prey species (Fig. 1.7a; Mendes-Soares and Velicer, 2013; Morgan *et al.*, 2010). By analyzing the swarming ability of natural isolates of *M. xanthus* on diverse prey arenas, this work established that genotype-by-environment interactions matter for predatory efficiency. For example, some strains that are relatively good swarmers on a particular prey (compared to other strains) can be relatively poor swarmers on an alternative prey type.

Furthermore, recent work by Mendes-Soares and Velicer (2013) showed that several predation-related parameters other than swarming ability also vary across prey species. In this study, populations of *M. xanthus* encountered nine phylogenetically diverse bacterial prey species and the extent of correlation between swarming rate on each prey type with each of four additional parameters measured in the absence of swarming – population growth rate, maximum yield, prey kill rate, and maximum prey kill – were assessed. Swarming rate was found to correlate significantly with all four of the other measured parameters, and all the parameters except maximum yield were found to vary significantly across prey species. Moreover, all prey species tested supported predator growth, including one species (*Curtobacterium citreum*) that had previously been hypothesized to be non-consumable by *M. xanthus* because it prevents territorial expansion of *M. xanthus* colonies. It was also found that there is substantial variation in the relative dynamics of predator growth and prey death across prey species. Overall, these results shed new light on the population dynamics stemming from microbial predator-prey interactions. The strength of such correlations can have an impact on the evolutionary trajectories of each parameter and may impose constraints on the evolution of an optimal predatory efficiency for this species (Heineman and Bull, 2007).

The efficiency of predation may also be affected by a strain's social encounters in nature. As social entities, *M. xanthus* cells interact with one another throughout their life cycle. Even though intraspecific (and inter-specific) interactions during development have been intensively studied (e.g. Velicer *et al.*, 2000; Wireman and Dworkin, 1975), interactions that occur during predation are not well understood. For example, mixing of genotypes during development can result in both increased (Velicer *et al.*, 2000; Vos and Velicer,

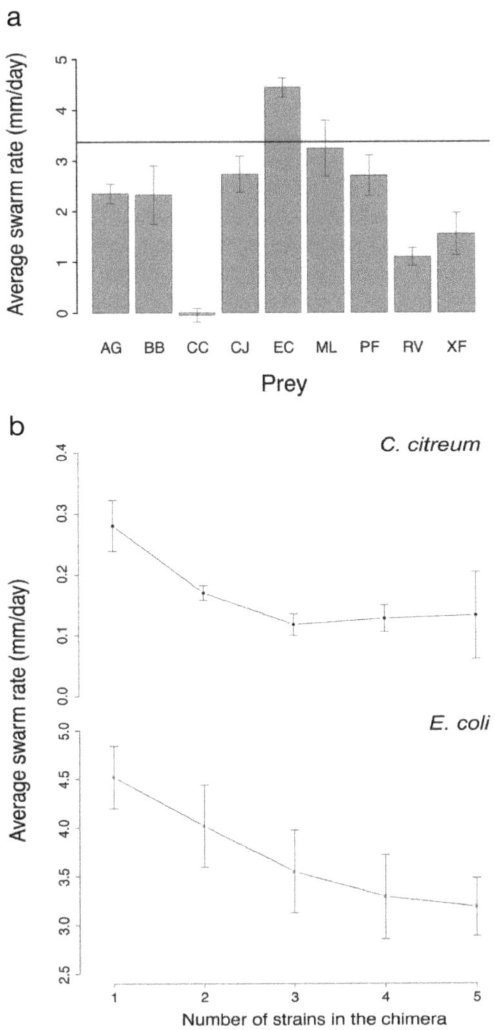

Figure 1.7 Predation. (a) Predatory performance across distinct prey species by WT strain GJV1 (a derivative of DK1622). Values represent prey-specific swarming rates across lawns of prey (i.e. the difference between absolute swarming rate on a given prey and swarming on nutrient-free buffered agar). Reprinted with permission from Mendes-Soares and Velicer (2013). (b) Effects of chimerism on predatory performance. *M. xanthus* swarming rate while preying upon *E. coli* and *Curtobacterium citreum* decreases as a function of diversity within chimeric swarms. Reprinted from Mendes-Soares (2012).

2009) and decreased group productivity (Fiegna and Velicer, 2005; Velicer *et al.*, 2000; Vos and Velicer, 2009). In a recent study (Mendes-Soares, 2012; Mendes-Soares *et al.*, submitted) the number of closely related natural isolates (Vos and Velicer, 2006) mixed in equal proportions prior to growth on prey was varied, thereby forming chimeric groups of different genotypes. More diverse chimeric groups were found to show significantly decreased group swarming rates on the same prey arena (Fig. 1.7b). This phenomenon was termed chimeric load (Kraemer and Velicer, 2011) and was specific to predatory swarming, as it was absent on control plates rich in pre-hydrolysed nutrients. It is noteworthy that all strains used in this study were isolated from a small patch of soil (Vos and Velicer, 2006) and were proficient swarmers. The impact of chimeric load on groups of strains that are closely related may lead to or reinforce the maintenance of sympatric but socially isolated kin-groups, ultimately contributing to their diversification (Fig. 1.4; Kraemer and Velicer, 2011). Thus, the potential for chimeric load during predation may be an important force that shapes genetic diversity and population structure in natural populations of myxobacteria (Mendes-Soares *et al.* submitted).

The detailed mechanisms of myxobacterial predation are not well understood. While bacterial predators appear to share some genetic features (Pasternak *et al.*, 2012), different species may have adaptations to their particular strategy of predation. Myxobacteria produce and secrete a plethora of secondary metabolites, i.e. organic compounds that are not directly implicated in cell growth, development or reproduction. Approximately 5% of all known microbial secondary metabolites are of myxobacterial origin, and they exhibit a wide array of biological activities that may prove to be of significant clinical value (Wenzel and Müller, 2009). The genome of *M. xanthus*, for instance, is highly enriched for genes encoding enzymes that produce secondary metabolites, such as polyketides, non-ribosomally synthesized peptides, or hybrids of the two structural classes (Wenzel and Müller, 2009).

It has long been suggested that secondary metabolites play a major role in predation (Dworkin, 1966). For example, Rosenberg *et al.* (1973) showed that the antibiotic TA produced by *M. xanthus* is effective at killing *E. coli* and a mutant defective at TA production was recently found to show reduced growth while utilizing *E. coli* as its sole carbon source (Xiao *et al.*, 2011). However, the actual evolutionary *raison d'etre* of the vast majority of myxobacterial secondary metabolites (including the antibiotic TA) remains unknown. Do they primarily function as anti-competitor compounds, predatory compounds, simultaneously (or alternately) as both anti-competitor (Fiegna and Velicer, 2005) and predatory compounds or some other function? Lytic enzymes have also been implicated in myxobacterial predation (Rosenberg and Varon, 1984), but their identities and roles in predation relative to secondary metabolites largely remain to be characterized.

Regardless of their actual identities, the mode by which predatory compounds come into contact with prey to kill and degrade them has large implications for the social evolution of predation. Predatory compounds that diffuse freely away from the producing cell effectively become 'public goods' and should be more susceptible to being exploited than any predatory molecules that remain attached to the producing cell (Goldman *et al.*, 2006) by cheaters that reap nutrients from prey carcasses without a corresponding cooperative contribution to the cost of predation (Velicer and Mendes-Soares, 2009; Velicer and Vos, 2009).

We can only fully grasp the ecology and evolution of *M. xanthus* populations if we take into account the multifaceted characteristics of myxobacterial predation. Genetic variation leads to differential predatory ability, which in turns is expected to be under intense selection in natural populations. It remains to be determined if phenotypically distinct neighbours from the same group tend to vary in predation ability as well as other phenotypes (Kraemer and Velicer, 2011). Local variations in prey environments are expected to be a major component of fine-scale spatial variation in selective conditions at the micrometre- to centimetre-scale and greatly affect the evolutionary dynamics of *M. xanthus* populations.

References

Ackermann, M., Stearns, S.C., and Jenal, U. (2003). Senescence in a bacterium with asymmetric division. Science *300*, 1920.

Alexander, R.D. (1974). The evolution of social behavior. Annu. Rev. Ecol. Syst. *5*, 325–383.

Allee, W.C. (1931). Animal aggregations, a study in general sociology (Chicago: University of Chicago Press).

Allee, W.C. (1949). Principles of animal ecology (Philadelphia: Saunders Co.).

Arneberg, P., Skorping, A., Grenfell, B., and Read, A.F. (1998). Host densities as determinants of abundance in parasite communities. Proc. R Soc. Lond, Ser. B: Biol. Sci. *265*, 1283–1289.

Baas-Becking, L.G.M. (1934). Geobiologie of inleiding tot de milieukunde (The Hague, Netherlands: Van Stockkum & Zoon).

Barrett, R.D.H. (2010). Adaptive evolution of lateral plates in three-spined stickleback *Gasterosteus aculeatus*: a case study in functional analysis of natural variation. J. Fish Biol. *77*, 311–328.

Barrett, R.D.H., Paccard, A., Healy, T.M., Bergek, S., Schulte, P.M., Schluter, D., and Rogers, S.M. (2010). Rapid evolution of cold tolerance in stickleback. Proc. R Soc. Lond, Ser. B: Biol. Sci. *278*, 233–238.

Barton, N.H., Etheridge, A.M., Kelleher, J., and Veber, A. (2013). Genetic hitchhiking in spatially extended populations. Theor. Popul. Biol. *85*, 75–89.

Berleman, J.E., and Kirby, J.R. (2009). Deciphering the hunting strategy of a bacterial wolfpack. FEMS Microbiol. Rev. *33*, 942–957.

Berleman, J.E., Chumley, T., Cheung, P., and Kirby, J.R. (2006). Rippling is a predatory behavior in *Myxococcus xanthus*. J. Bacteriol. *188*, 5888–5895.

Blount, Z.D., Barrick, J.E., Davidson, C.J., and Lenski, R.E. (2012). Genomic analysis of a key innovation in an experimental *Escherichia coli* population. Nature *489*, 513–518.

Bohannan, B.J.M., and Lenski, R.E. (2000). Linking genetic change to community evolution: insights from studies of bacteria and bacteriophage. Ecol. Lett. *3*, 362–377.

Brinkhoff, T., Fischer, D., Vollmers, J., Voget, S., Beardsley, C., Thole, S., Mussmann, M., Kunze, B., Wagner-Dobler, I., Daniel, R., *et al.* (2012). Biogeography and phylogenetic diversity of a cluster of exclusively marine myxobacteria. ISME J. *6*, 1260–1272.

de Bruijn, S., Angenent, G.C., and Kaufmann, K. (2012). Plant 'evo-devo' goes genomic: from candidate genes to regulatory networks. Trends Plant Sci. *17*, 441–447.

Bull, C.T., Shetty, K.G., and Subbarao, K.V. (2002). Interactions between myxobacteria, plant pathogenic fungi, and biocontrol agents. Plant Dis. *86*, 889–896.

Buttery, N.J., Jack, C.N., Adu-Oppong, B., Snyder, K.T., Thompson, C.R.L., Queller, D.C., and Strassmann, J.E. (2012). Structured growth and genetic drift raise relatedness in the social amoeba *Dictyostelium discoideum*. Biol. Lett. *8*, 794–797.

Carroll, S.B. (2008). Evo-Devo and an expanding evolutionary synthesis: a genetic theory of morphological evolution. Cell *134*, 25–36.

Cole, J.R., Cascarelli, A.L., Mohn, W.W., and Tiedje, J.M. (1994). Isolation and characterization of a novel bacterium growing via reductive dehalogenation of 2-chlorophenol. Appl. Environ. Microbiol. *60*, 3536–3542.

Crespi, B.J. (2001). The evolution of social behavior in microorganisms. Trends Ecol. Evol. *16*, 178–183.

Dahl, J.L., Ulrich, C.H., and Kroft, T.L. (2011). Role of phase variation in the resistance of *Myxococcus xanthus* fruiting bodies to *Caenorhabditis elegans* predation. J. Bacteriol. *193*, 5081–5089.

Dawid, W. (2000). Biology and global distribution of myxobacteria in soils. FEMS Microbiol. Rev. *24*, 403–427.

DeLong, E.F., and Pace, N.R. (2001). Environmental diversity of bacteria and archaea. Syst. Biol. *50*, 470–478.

Dworkin, M. (1966). Biology of the myxobacteria. Annu. Rev. Microbiol. *20*, 75–106.

Dworkin, M. (1996). Recent advances in the social and developmental biology of the myxobacteria. Microbiol. Rev. *60*, 70–102.

Elena, S.F., and Lenski, R.E. (2003). Evolution experiments with microorganisms: the dynamics and genetic bases of adaptation. Nature Rev. Genet. *4*, 457–469.

Escalante, A.E., Inouye, S., and Travisano, M. (2012). A spectrum of pleiotropic consequences in development due to changes in a regulatory pathway. PLoS One *7*, e43413.

Escobar-Páramo, P., Ghosh, S., and DiRuggiero, J. (2005). Evidence for genetic drift in the diversification of a geographically isolated population of the hyperthermophilic archaeon *Pyrococcus*. Mol. Biol. Evol. *22*, 2297–2303.

Evans, K.J., Lambert, C., and Sockett, R.E. (2007). Predation by *Bdellovibrio bacteriovorus* HD100 requires type IV pili. J. Bacteriol. *189*, 4850–4859.

Fiegna, F., and Velicer, G.J. (2003). Competitive fates of bacterial social parasites: persistence and self-induced extinction of *Myxococcus xanthus* cheaters. Proc. R Soc. Lond, Ser. B: Biol. Sci. *270*, 1527–1534.

Fiegna, F., and Velicer, G.J. (2005). Exploitative and hierarchical antagonism in a cooperative bacterium. PLoS Biol. *3*, 1980–1987.

Fiegna, F., Yu, Y.T., Kadam, S.V., and Velicer, G.J. (2006). Evolution of an obligate social cheater to a superior cooperator. Nature *441*, 310–314.

Finlay, B.J. (2002). Global dispersal of free-living microbial eukaryote species. Science *296*, 1061–1063.

Fontes, M., and Kaiser, D. (1999). Myxococcus cells respond to elastic forces in their substrate. Proc. Natl. Acad. Sci. U.S.A. *96*, 8052–8057.

Foster, K.R., Parkinson, K., and Thompson, C.R. (2007). What can microbial genetics teach sociobiology? Trends Genet. *23*, 74–80.

Fudou, R., Jojima, Y., Iizuka, T., and Yamanaka, S. (2002). *Haliangium ochraceum* gen. nov., sp. nov. and *Haliangium tepidum* sp. nov.: novel moderately halophilic myxobacteria isolated from coastal saline environments. J. Gen. Appl. Microbiol. *48*, 109–116.

Fukami, T., Beaumont, H.J.E., Zhang, X.-X., and Rainey, P.B. (2007). Immigration history controls diversification in experimental adaptive radiation. Nature *446*, 436–439.

Garcia, R., Gerth, K., Stadler, M., Dogma Jr, I.J., and Müller, R. (2010). Expanded phylogeny of myxobacteria and evidence for cultivation of the 'unculturables'. Mol. Phylogen. Evol. *57*, 878–887.

Gilbert, O.M., Foster, K.R., Mehdiabadi, N.J., Strassmann, J.E., and Queller, D.C. (2007). High relatedness maintains multicellular cooperation in a social amoeba by controlling cheater mutants. Proc. Natl. Acad. Sci. U.S.A. *104*, 8913–8917.

Goldman, B.S., Nierman, W.C., Kaiser, D., Slater, S.C., Durkin, A.S., Eisen, J.A., Ronning, C.M., Barbazuk, W.B., Blanchard, M., Field, C., *et al.* (2006). Evolution of sensory complexity recorded in a myxobacterial genome. Proc. Natl. Acad. Sci. U.S.A. *103*, 15200–15205.

Goldman, B., Bhat, S., and Shimkets, L.J. (2007). Genome evolution and the emergence of fruiting body development in *Myxococcus xanthus*. PLoS ONE *2*, e1329.

Gottesman, S. (2005). Micros for microbes: non-coding regulatory RNAs in bacteria. Trends Genet. *21*, 399–404.

Guerrero, R., Esteve, I., Pedros-Alio, C., and Gaju, N. (1987). Predatory bacteria in prokaryotic communities. Ann. N. Y. Acad. Sci. *503*, 238–250.

Guillaume, F., and Otto, S.P. (2012). Gene functional trade-offs and the evolution of pleiotropy. Genetics *192*, 1389–1409.

Habets, M.G., Rozen, D.E., and Brockhurst, M.A. (2012). Variation in *Streptococcus pneumoniae* susceptibility to human antimicrobial peptides may mediate intraspecific competition. Proc. R Soc. Lond, Ser. B: Biol. Sci. *279*, 3803–3811.

Harrison, F., and Buckling, A. (2009). Siderophore production and biofilm formation as linked social traits. ISME J. *3*, 632–634.

Harvey, P.H., and Pagel, M.D. (1991). The comparative method in evolutionary biology, Vol 1 (Oxford: Oxford University Press).

Hawlena, H., Bashey, F., and Lively, C.M. (2010a). The evolution of spite: population structure and bacteriocin-mediated antagonism in two natural populations of *Xenorhabdus* bacteria. Evolution *64*, 3198–3204.

Hawlena, H., Bashey, F., Mendes-Soares, H., and Lively, C.M. (2010b). Spiteful interactions in a natural population of the bacterium *Xenorhabdus bovienii*. Am. Nat. *175*, 374–381.

Heath, T.A., Holder, M.T., and Huelsenbeck, J.P. (2012). A dirichlet process prior for estimating lineage-specific substitution rates. Mol. Biol. Evol. *29*, 939–955.

Heineman, R.H., and Bull, J.J. (2007). Testing optimality with experimental evolution: lysis time in a bacteriophage. Evolution *61*, 1695–1709.

Hershberg, R., Lipatov, M., Small, P.M., Sheffer, H., Niemann, S., Homolka, S., Roach, J.C., Kremer, K., Petrov, D.A., Feldman, M.W., *et al.* (2008). High functional diversity in *Mycobacterium tuberculosis* driven by genetic drift and human demography. PLoS Biol. *6*, e311.

Hillesland, K.L. (2005). Evolutionary ecology of predation by the soil bacterium, *Myxococcus xanthus* (East Lansing: Michigan State University).

Hillesland, K.L., and Velicer, G.J. (2005). Resource level affects relative performance of the two motility systems of *Myxococcus xanthus*. Microb. Ecol. *49*, 558–566.

Hillesland, K.L., Lenski, R.E., and Velicer, G.J. (2007). Ecological variables affecting predatory success in *Myxococcus xanthus*. Microb. Ecol. *53*, 571–578.

Hillesland, K.L., Velicer, G.J., and Lenski, R.E. (2009). Experimental evolution of a microbial predator's ability to find prey. Proc. R Soc. Lond, Ser. B: Biol. Sci. *276*, 459–467.

Hindré, T., Knibbe, C., Beslon, G., and Schneider, D. (2012). New insights into bacterial adaptation through *in vivo* and in silico experimental evolution. Nat. Rev. Microbiol. *10*, 352–365.

Holling, C.S. (1959). Some characteristics of simple types of predation and parasitism. Canad. Entomol. *91*, 385–398.

Holling, C.S. (1961). Principles of insect predation. Annu. Rev. Entomol. *6*, 163–182.

Huntley, S., Hamann, N., Wegener-Feldbrügge, S., Treuner-Lange, A., Kube, M., Reinhardt, R., Klages, S., Müller, R., Ronning, C.M., Nierman, W.C., *et al.* (2011). Comparative genomic analysis of fruiting body formation in *Myxococcales*. Mol. Biol. Evol. *28*, 1083–1097.

Huntley, S., Zhang, Y., Treuner-Lange, A., Kneip, S., Sensen, C.W., and Sogaard-Andersen, L. (2012). Complete genome sequence of the fruiting myxobacterium *Corallococcus coralloides* DSM 2259. J. Bacteriol. *194*, 3012–3013.

Iizuka, T., Jojima, Y., Fudou, R., and Yamanaka, S. (1998). Isolation of myxobacteria from the marine environment. FEMS Microbiol. Lett. *169*, 317–322.

Iizuka, T., Jojima, Y., Fudou, R., Hiraishi, A., Ahn, J.W., and Yamanaka, S. (2003a). *Plesiocystis pacifica* gen. nov., sp. nov., a marine myxobacterium that contains dihydrogenated menaquinone, isolated from the Pacific coasts of Japan. Int. J. Syst. Evol. Microbiol. *53*, 189–195.

Iizuka, T., Jojima, Y., Fudou, R., Tokura, M., Hiraishi, A., and Yamanaka, S. (2003b). *Enhygromyxa salina* gen. nov., sp. nov., a slightly halophilic myxobacterium isolated from the coastal areas of Japan. Syst. Appl. Microbiol. *26*, 189–196.

Ivanova, N., Daum, C., Lang, E., Abt, B., Kopitz, M., Saunders, E., Lapidus, A., Lucas, S., Glavina Del Rio, T., Nolan, M., *et al.* (2010). Complete genome sequence of *Haliangium ochraceum* type strain (SMP-2). Standards in Genomic Sciences 2, 96–106.

Jessup, C.M., Kassen, R., Forde, S.E., Kerr, B., Buckling, A., Rainey, P.B., and Bohannan, B.J.M. (2004). Big questions, small worlds: microbial model systems in ecology. Trends Ecol. Evol. *19*, 189–197.

Jiang, D.M., Wu, Z.H., Zhao, J.Y., and Li, Y.Z. (2007). Fruiting and non-fruiting myxobacteria: a phylogenetic perspective of cultured and uncultured members of this group. Mol. Phylogenet. Evol. *44*, 545–552.

Jiang, D.-M., Kato, C., Zhou, X.-W., Wu, Z.-H., Sato, T., and Li, Y.-Z. (2010). Phylogeographic separation of marine and soil myxobacteria at high levels of classification. ISME J. *4*, 1520–1530.

Jurkevitch, E. (2007). Predatory Prokaryotes: Biology, Ecology and Evolution (Berlin: Springer).

Kadam, S.V., and Velicer, G.J. (2006). Variable patterns of density-dependent survival in social bacteria. Behav. Ecol. *17*, 833–838.

Kadam, S.V., Wegener-Feldbrugge, S., Sogaard-Andersen, L., and Velicer, G.J. (2008). Novel transcriptome patterns accompany evolutionary restoration of defective social development in the bacterium *Myxococcus xanthus*. Mol. Biol. Evol. *25*, 1274–1281.

Kaiser, D. (2001). Building a multicellular organism. Annu. Rev. Genet. *35*, 103–123.

Kaplan, H.B., and Plamann, L. (1996). A *Myxococcus xanthus* cell density-sensing system required for multicellular development. FEMS Microbiol. Lett. *139*, 89–95.

Kassen, R., and Rainey, P.B. (2004). The ecology and genetics of microbial diversity. Annu. Rev. Microbiol. *58*, 207–231.

Kawecki, T.J., Lenski, R.E., Ebert, D., Hollis, B., Olivieri, I., and Whitlock, M.C. (2012). Experimental evolution. Trends Ecol. Evol. *27*, 547–560.

Kearns, D.B., and Shimkets, L.J. (1998). Chemotaxis in a gliding bacterium. Proc. Natl. Acad. Sci. U.S.A. *95*, 11957–11962.

Kraemer, S.A. (2011). The structure of social and genetic diversity in local soil populations of the social bacterium *Myxococcus xanthus* (Bloomington, IN: Indiana University).

Kraemer, S.A., and Velicer, G.J. (2011). Endemic social diversity within natural kin groups of a cooperative bacterium. Proc. Natl. Acad. Sci. U.S.A. *108*, 10823–10830.

Kraemer, S.A., Toups, M.A., and Velicer, G.J. (2010). Natural variation in developmental life-history traits of the bacterium *Myxococcus xanthus*. FEMS Microbiol. Ecol. *73*, 226–233.

Krause, J., and Ruxton, G.D. (2002). Living in groups (Oxford, UK: Oxford University Press).

Krebs, J.R., and Davies, N.B. (1997). Behavioural ecology: an evolutionary approach, 4th edn (Oxford, UK: Wiley-Blackwell).

Krug, D., Zurek, G., Revermann, O., Vos, M., Velicer, G.J., and Müller, R. (2008). Discovering the hidden secondary metabolome of *Myxococcus xanthus*: a study of intraspecific diversity. Appl. Environ. Microbiol. *74*, 3058–3068.

Kümmerli, R., Griffin, A.S., West, S.A., Buckling, A., and Harrison, F. (2009). Viscous medium promotes cooperation in the pathogenic bacterium *Pseudomonas aeruginosa*. Proc. R Soc. Lond, Ser. B: Biol. Sci. *276*, 3531–3538.

Li, Y.Z., Hu, W., Zhang, Y.Q., Qiu, Z., Zhang, Y., and Wu, B.H. (2002). A simple method to isolate salt-tolerant myxobacteria from marine samples. J. Microbiol. Methods *50*, 205–209.

Li, Z.F., Li, X., Liu, H., Liu, X., Han, K., Wu, Z.H., Hu, W., Li, F.F., and Li, Y.Z. (2011). Genome sequence of the halotolerant marine bacterium *Myxococcus fulvus* HW-1. J. Bacteriol. *193*, 5015–5016.

Liu, C., Fu, X., Liu, L., Ren, X., Chau, C.K.L., Li, S., Xiang, L., Zeng, H., Chen, G., Tang, L.-H., *et al.* (2011).

Sequential establishment of stripe patterns in an expanding cell population. Science *334*, 238–241.

Love, A.C., and Travisano, M. (2013). Microbes modeling ontogeny. Biol. Philos. *28*, 161–188.

Luciano, J., Agrebi, R., Le Gall, A.V., Wartel, M., Fiegna, F., Ducret, A., Brochier-Armanet, C., and Mignot, T. (2011). Emergence and modular evolution of a novel motility machinery in bacteria. PLoS Genet. *7*, e1002268.

Ludwig, W., Schleifer, K.H., Reichenbach, H., and Stackebrandt, E. (1983). A phylogenetic analysis of the myxobacteria *Myxococcus fulvus, Stigmatella aurantiaca, Cystobacter fuscus, Sorangium cellulosum* and *Nannocystis exedens*. Arch. Microbiol. *135*, 58–62.

MacArthur, R.H., and Wilson, E.O. (1967). The theory of island biogeography (Princeton, NJ: Princeton University Press).

McBride, M.J., and Zusman, D.R. (1996). Behavioral analysis of single cells of *Myxococcus xanthus* in response to prey cells of *Escherichia coli*. FEMS Microbiol. Lett. *137*, 227–231.

Majewski, J., and Cohan, F.M. (1999). Adapt globally, act locally: the effect of selective sweeps on bacterial sequence diversity. Genetics *152*, 1459–1474.

Mallarino, R., and Abzhanov, A. (2012). Paths less traveled: evo-devo approaches to investigating animal morphological evolution. Annu. Rev. Cell. Dev. Biol. *28*, 743–763.

Manhes, P., and Velicer, G.J. (2011). Experimental evolution of selfish policing in social bacteria. Proc. Natl. Acad. Sci. U.S.A. *108*, 8357–8362.

Mark, M.O. (2002). Predatory prokaryotes: an emerging research opportunity. J. Mol. Microbiol. Biotechnol. *4*, 467–477.

Martin, F.N., and Bull, C.T. (2002). Biological approaches for control of root pathogens of strawberry. Phytopathol. *92*, 1356–1362.

Mendes-Soares, H. (2012). The ecology and evolution of microbial predation by *Myxococcus xanthus* across heterogeneous prey environments (Bloomington, IN: Indiana University).

Mendes-Soares, H., and Velicer, G.J. (2013). Decomposing predation: testing for parameters that correlate with predatory performance by a social bacterium. Microb. Ecol. *65*, 415–423.

Mendes-Soares, H., Chen, I.-C.K., Fitzpatrick, K., and Velicer, G.J. Chimeric load among sympatric social bacteria increases with genotype richness. Submitted for publication.

Mittal, S., and Kroos, L. (2009a). A combination of unusual transcription factors binds cooperatively to control *Myxococcus xanthus* developmental gene expression. Proc. Natl. Acad. Sci. U.S.A. *106*, 1965–1970.

Mittal, S., and Kroos, L. (2009b). Combinatorial regulation by a novel arrangement of FruA and MrpC2 transcription factors during *Myxococcus xanthus* development. J. Bacteriol. *191*, 2753–2763.

Morgan, A.D., MacLean, R.C., Hillesland, K.L., and Velicer, G.J. (2010). Comparative analysis of *Myxococcus* predation on soil bacteria. Appl. Environ. Microbiol. *76*, 6920–6927.

Nadell, C., Bassler, B., and Levin, S. (2008). Observing bacteria through the lens of social evolution. J. Biol. *7*, 27.

Oyaizu, H., and Woese, C.R. (1985). Phylogenetic relationships among the sulfate respiring bacteria, Myxobacteria and Purple Bacteria. Syst. Appl. Microbiol. *6*, 257–263.

Pasternak, Z., Pietrokovski, S., Rotem, O., Gophna, U., Lurie-Weinberger, M.N., and Jurkevitch, E. (2013). By their genes ye shall know them: genomic signatures of predatory bacteria. ISME J. *7*, 756–769.

Pham, V.D., Shebelut, C.W., Diodati, M.E., Bull, C.T., and Singer, M. (2005). Mutations affecting predation ability of the soil bacterium *Myxococcus xanthus*. Microbiology *151*, 1865–1874.

Queller, D.C. (2011). Expanded social fitness and Hamilton's rule for kin, kith, and kind. Proc. Natl. Acad. Sci. U.S.A. *108 (Suppl 2)* 10792–10799.

Rainey, P.B., and Rainey, K. (2003). Evolution of cooperation and conflict in experimental bacterial populations. Nature *425*, 72–74.

Ramette, A., and Tiedje, J.M. (2007). Biogeography: an emerging cornerstone for understanding prokaryotic diversity, ecology, and evolution. Microb. Ecol. *53*, 197–207.

Reichenbach, H. (1993). Biology of the Myxobacteria: ecology and taxonomy. In Myxobacteria II, M. Dworkin, and D. Kaiser, eds. (Washington, DC: American Society for Microbiology).

Reichenbach, H. (1999). The ecology of the myxobacteria. Environ. Microbiol. *1*, 15–21.

Reichenbach, H. (2005). Order VIII. Myxococcales Tchan, Pochon and Prévot 1948, 398AL. In Bergey's manual of systematic bacteriology, D.J. Brenner, N.R. Krieg, J.T. Staley, and G.M. Garrity, eds. (New York: Springer), pp. 1059–1072.

Reichenbach, H., and Höfle, G. (1993). Biologically active secondary metabolites from myxobacteria. Biotechnol. Adv. *11*, 219–277.

Reno, M.L., Held, N.L., Fields, C.J., Burke, P.V., and Whitaker, R.J. (2009). Biogeography of the *Sulfolobus islandicus* pan-genome. Proc. Natl. Acad. Sci. U.S.A. *106*, 8605–8610.

Rogosky, A., Moak, P., and Emmert, E.B. (2006). Differential predation by *Bdellovibrio bacteriovorus* 109J. Curr. Microbiol. *52*, 81–85.

Rosenberg, E., and Varon, M. (1984). Antibiotics and lytic enzymes. In Myxobacteria, development and cell interaction, E. Rosenberg, ed. (New York: Springer-Verlag), pp. 109–125.

Rosenberg, E., Vaks, B., and Zuckerberg, A. (1973). Bactericidal action of an antibiotic produced by *Myxococcus xanthus*. Antimicrob. Agents Ch. *4*, 507–513.

Rosenberg, E., Keller, K., and Dworkin, M. (1977). Cell density-dependent growth of *Myxococcus xanthus* on casein. J. Bacteriol. *129*, 770–777.

Sachs, J.L., and Bull, J.J. (2005). Experimental evolution of conflict mediation between genomes. Proc. Natl. Acad. Sci. U.S.A. *102*, 390–395.

Sachs, J.L., Mueller, U.G., Wilcox, T.P., and Bull, J.J. (2004). The evolution of cooperation. Q. Rev. Biol. *79*, 135–160.

Sanford, R.A., Cole, J.R., and Tiedje, J.M. (2002). Characterization and description of *Anaeromyxobacter dehalogenans* gen. nov., sp. nov., an aryl-halorespiring facultative anaerobic myxobacterium. Appl. Environ. Microbiol. *68*, 893–900.

Schneiker, S., Perlova, O., Kaiser, O., Gerth, K., Alici, A., Altmeyer, M.O., Bartels, D., Bekel, T., Beyer, S., Bode, E., *et al.* (2007). Complete genome sequence of the myxobacterium *Sorangium cellulosum*. Nat. Biotechnol. *25*, 1281–1289.

Shi, W., and Zusman, D.R. (1993). The two motility systems of *Myxococcus xanthus* show different selective advantages on various surfaces. Proc. Natl. Acad. Sci. U.S.A. *90*, 3378–3382.

Shimkets, L.J. (1990). The *Myxococcus xanthus* Fpra Protein causes increased flavin biosynthesis in *Escherichia coli*. J. Bacteriol. 172, 24–30.

Shimkets, L., and Woese, C.R. (1992). A phylogenetic analysis of the myxobacteria: basis for their classification. Proc. Natl. Acad. Sci. U.S.A. *89*, 9459–9463.

Shimkets, L.J., Reichenbach, H., and Dworkin, M. (2006). The Myxobacteria. In The Prokaryotes, M. Dworkin, ed. (New York: Springer), pp. 31–115.

Simms, E.L., Taylor, D.L., Povich, J., Shefferson, R.P., Sachs, J.L., Urbina, M., and Tausczik, Y. (2006). An empirical test of partner choice mechanisms in a wild legume–rhizobium interaction. Proc. R Soc. Lond, Ser. B: Biol. Sci. *273*, 77–81.

Smith, D.J., Timonen, H.J., Jaffe, D.A., Griffin, D.W., Birmele, M.N., Perry, K.D., Ward, P.D., and Roberts, M.S. (2012). Intercontinental dispersal of bacteria and archaea in transpacific winds. Appl. Environ. Microbiol. *79*, 1134–1139.

Smith, J., Van Dyken, J.D., and Zee, P.C. (2010). A generalization of Hamilton's rule for the evolution of microbial cooperation. Science *328*, 1700–1703.

Spröer, C., Reichenbach, H., and Stackebrandt, E. (1999). The correlation between morphological and phylogenetic classification of myxobacteria. Int. J. Syst. Bacteriol. *49*, 1255–1262.

Stackebrandt, E., Murray, R.G.E., and Trüper, H.G. (1988). *Proteobacteria classis* nov., a name for the phylogenetic taxon that includes the 'purple bacteria and their relatives'. Int. J. Syst. Bacteriol. *38*, 321–325.

Stephens, P.A., and Freckleton, R.P. (1999). What is the Allee effect? Oikos *87*, 185–190.

Strassmann, J.E., and Queller, D.C. (2011). Evolution of cooperation and control of cheating in a social microbe. Proc. Natl. Acad. Sci. U.S.A. *108*, 10855–10862.

Strassmann, J.E., Gilbert, O.M., and Queller, D.C. (2011). Kin discrimination and cooperation in microbes. Annu. Rev. Microbiol. *65*, 349–367.

Sul, W.J., Oliver, T.A., Ducklow, H.W., Amaral-Zettler, L.A., and Sogin, M.L. (2013). Marine bacteria exhibit a bipolar distribution. Proc. Natl. Acad. Sci. U.S.A. *110*, 2342–2347.

Thomas, S.H., Wagner, R.D., Arakaki, A.K., Skolnick, J., Kirby, J.R., Shimkets, L.J., Sanford, R.A., and Loffler, F.E. (2008). The mosaic genome of *Anaeromyxobacter dehalogenans* strain 2CP-C suggests an aerobic common ancestor to the delta-proteobacteria. PLoS ONE *3*, e2103.

Travisano, M., and Velicer, G.J. (2004). Strategies of microbial cheater control. Trends Microbiol. *12*, 72–78.

Velicer, G.J. (2003). Social strife in the microbial world. Trends Microbiol. *11*, 330–337.

Velicer, G.J., and Hillesland, K.L. (2008). Why cooperate? The ecology and evolution of the myxobacteria. In Myxobacteria: Multicellularity and Differentiation, D.E. Whitworth, ed. (Washington, DC: American Society for Microbiology Press), pp. 17–40.

Velicer, G.J., and Mendes-Soares, H. (2009). Bacterial predators. Curr. Biol. *19*, R55-R56.

Velicer, G.J., and Stredwick, K.L. (2002). Experimental social evolution with *Myxococcus xanthus*. Anton. Leeuw. Int. J. G. *81*, 155–164.

Velicer, G.J., and Vos, M. (2009). Sociobiology of the myxobacteria. Annu. Rev. of Microbiol. *63*, 599–623.

Velicer, G.J., and Yu, Y.T. (2003). Evolution of novel cooperative swarming in the bacterium *Myxococcus xanthus*. Nature *425*, 75–78.

Velicer, G.J., Kroos, L., and Lenski, R.E. (1998). Loss of social behaviors by *Myxococcus xanthus* during evolution in an unstructured habitat. Proc. Natl. Acad. Sci. U.S.A. *95*, 12376–12380.

Velicer, G.J., Kroos, L., and Lenski, R.E. (2000). Developmental cheating in the social bacterium *Myxococcus xanthus*. Nature *404*, 598–601.

Velicer, G.J., Lenski, R.E., and Kroos, L. (2002). Rescue of social motility lost during evolution of *Myxococcus xanthus* in an asocial environment. J. Bacteriol. *184*, 2719–2727.

Velicer, G.J., Raddatz, G., Keller, H., Deiss, S., Lanz, C., Dinkelacker, I., and Schuster, S.C. (2006). Comprehensive mutation identification in an evolved bacterial cooperator and its cheating ancestor. Proc. Natl. Acad. Sci. U.S.A. *103*, 8107–8112.

Viswanathan, P., Ueki, T., Inouye, S., and Kroos, L. (2007). Combinatorial regulation of genes essential for *Myxococcus xanthus* development involves a response regulator and a LysR-type regulator. Proc. Natl. Acad. Sci. U.S.A. *104*, 7969–7974.

Vos, M. (2011). A species concept for bacteria based on adaptive divergence. Trends Microbiol. *19*, 1–7.

Vos, M., and Velicer, G.J. (2006). Genetic population structure of the soil bacterium *Myxococcus xanthus* at the centimeter scale. Appl. Environ. Microbiol. *72*, 3615–3625.

Vos, M., and Velicer, G.J. (2008a). Isolation by distance in the spore-forming soil bacterium *Myxococcus xanthus*. Curr. Biol. *18*, 386–391.

Vos, M., and Velicer, G.J. (2008b). Natural variation of gliding motility in a centimetre-scale population of *Myxococcus xanthus*. FEMS Microbiol. Ecol. *64*, 343–350.

Vos, M., and Velicer, G.J. (2009). Social conflict in centimeter and global-scale populations of the bacterium *Myxococcus xanthus*. Curr. Biol. *19*, 1763–1767.

Wang, B., Hu, W., Liu, H., Zhang, C.Y., Zhao, J.Y., Jiang, D.M., Wu, Z.H., and Li, Y.Z. (2007). Adaptation of salt-tolerant *Myxococcus* strains and their motility systems to the ocean conditions. Microb. Ecol. *54*, 43–51.

Watve, M.G., Shete, A.M., Jadhav, N., Wagh, S.A., Shelar, S.P., Chakraborti, S.S., Botre, A.P., and Kulkarni, A.A. (1999). Myxobacterial diversity of Indian soils – how many species do we have?. Curr. Sci. *77*, 1089–1095.

Weissman, K.J., and Müller, R. (2010). Myxobacterial secondary metabolites: bioactivities and modes-of-action. Nat. Prod. Rep. *27*, 1276–1295.

Wenzel, S.C., and Müller, R. (2009). The biosynthetic potential of myxobacteria and their impact in drug discovery. Curr. Opin. Drug Discov. Devel. *12*, 220–230.

West, S.A. (2006). Cooperation and the scale of competition in humans. Curr. Biol. *16*, 1103–1106.

West, S.A., Griffin, A.S., Gardner, A., and Diggle, S.P. (2006). Social evolution theory for microbes. Nature Rev. Microbiol. *4*, 597–607.

Whitaker, R.J., Grogan, D.W., and Taylor, J.W. (2003). Geographic barriers isolate endemic populations of hyperthermophilic *Archaea*. Science *301*, 976–978.

Whitlock, M.C. (2003). Fixation probability and time in subdivided populations. Genetics *164*, 767–779.

Wielgoss, S., Barrick, J.E., Tenaillon, O., Wiser, M.J., Dittmar, W.J., Cruveiller, S., Chane-Woon-Ming, B., Medigue, C., Lenski, R.E., and Schneider, D. (2013). Mutation rate dynamics in a bacterial population reflect tension between adaptation and genetic load. Proc. Natl. Acad. Sci. U.S.A. *110*, 222–227.

Williams, L.E., and Wernegreen, J.J. (2012). Purifying selection, sequence composition, and context-specific indel mutations shape intraspecific variation in a bacterial endosymbiont. Genome Biol. Evol. *4*, 44–51.

Wireman, J., and Dworkin, M. (1975). Morphogenesis and developmental interactions in myxobacteria. Science *189*, 516–523.

Woese, C.R., Stackebrandt, E., Macke, T.J., and Fox, G.E. (1985). A phylogenetic definition of the major eubacterial taxa. Syst. Appl. Microbiol. *6*, 143–151.

Wu, S.S., Wu, J., Cheng, Y.L., and Kaiser, D. (1998). The *pilH* gene encodes an ABC transporter homologue required for type IV pilus biogenesis and social gliding motility in *Myxococcus xanthus*. Mol. Microbiol. *29*, 1249–1261.

Xavier, J.B., Kim, W., and Foster, K.R. (2011). A molecular mechanism that stabilizes cooperative secretions in *Pseudomonas aeruginosa*. Mol. Microbiol. *79*, 166–179.

Xiao, Y., Wei, X., Ebright, R., and Wall, D. (2011). Antibiotic production by myxobacteria plays a role in predation. J. Bacteriol. *193*, 4626–4633.

Yu, Y.T., Yuan, X., and Velicer, G.J. (2010). Adaptive evolution of an sRNA that controls *Myxococcus* development. Science *328*, 993.

Zanette, L.R.S., Miller, S.D.L., Faria, C.M.A., Almond, E.J., Huggins, T.J., Jordan, W.C., and Bourke, A.F.G. (2012). Reproductive conflict in bumblebees and the evolution of worker policing. Evolution *66*, 3765–3777.

Zhang, L., Thomas, J.C., Didelot, X., and Robinson, D.A. (2012). Molecular signatures identify a candidate target of balancing selection in an *arcD*-like gene of *Staphylococcus epidermidis*. J. Mol. Evol. 75, 43–54.

Zhang, Y.Q., Li, Y.Z., Wang, B., Wu, Z.H., Zhang, C.Y., Gong, X., Qiu, Z.J., and Zhang, Y. (2005). Characteristics and living patterns of marine myxobacterial isolates. Appl. Environ. Microbiol. *71*, 3331–3336.

Zhou, T., Gu, W., and Wilke, C.O. (2010). Detecting positive and purifying selection at synonymous sites in yeast and worm. Mol. Biol. Evol. *27*, 1912–1922.

Genome Evolution and Content in the Myxobacteria

2

Stuart Huntley, Kristin Wuichet and Lotte Søgaard-Andersen

Abstract

Nearly 2000 microbial genomes have been completely sequenced since the first bacterial genome sequence was released in 1995. Since then, comparative and functional genomics in combination with advances in sequencing techniques have significantly changed the way (micro) biological research is done. Intra- and inter-genome comparisons are now common practice to understand possible evolutionary trajectories and for identifying genes of interest. Similarly, functional genomics approaches such as transcriptome and proteome analyses, which rely on genome sequences, have been developed in many model organisms including *Myxococcus xanthus*. Here, we present a summary of the myxobacteria genome sequences available to date (July 2012) as well as an overall comparison of these genomes. Most members of the myxobacteria have large genome with sizes of approximately 10 Mb or even larger. We explore hypotheses for myxobacteria genome evolution, including genome size, genome organization with conserved synteny and genetic content with a special emphasis on genes for signal transduction proteins. Moreover, we discuss the level of shared genetic content of the hallmark characteristic of the myxobacteria, i.e. fruiting body formation. Finally, we look at what is on the near horizon for the future of myxobacteria genomics.

Myxobacteria genome sequencing history

Determining the DNA sequence of an entire genome of an organism first became a reality in 1995, when The Institute for Genomic Research (TIGR) successfully determined the complete nucleotide sequence of the 1.8 Mb *Haemophilus influenzae* RD KW20 chromosome (Fleischmann *et al.*, 1995). By utilizing an innovative combination of whole genome shotgun sequencing and new computational methods of assembling the relatively short sequences generated by this method, TIGR was able to identify and assemble the entire genome sequence of *H. influenzae* in a fraction of the time it was taking contemporary genome sequencing projects. At that time, most large sequencing projects, such as that which produced the *Escherichia coli* MG1655 complete genome sequence (Blattner *et al.*, 1997), involved creating libraries of vectors bearing 10–40 kb target sequence clones. The clones were then sheared and subcloned, these subclones sequenced and the generated sequences were assembled to produce the sequence of the DNA insert in the original clone. This process was then repeated for the entire library, which usually consisted of hundreds of clones.

The power of the TIGR sequence assembly method using short sequence reads allowed them to bypass the arduous and time consuming cloning, mapping, and subcloning steps. Instead, the brunt of their sequencing was performed on a whole genome shotgun library, in which the vectors contained 2 kb randomly sheared genomic DNA inserts. The over 24,000 sequences generated from this library were assembled into less than 150 sets of contiguous, overlapping DNA sequences (contigs), which were then ordered and orientated relative to each other. Gaps between contigs were filled using a combination of methods, including PCR, Southern analysis and end sequencing of lambda clones (Fleischmann *et al.*, 1995). While still requiring significant effort to 'close' the assembled sequence, this method lent itself to automation of a large fraction of the effort, and manual work was focused on resolving problem areas. This somewhat revolutionary method of sequencing whole genomes lead to an exponential growth in the number of complete and nearly complete bacterial genome sequences. As sequencing technology and techniques improved, costs plummeted, and for researchers around the world, thoughts of sequencing an entire bacterial genome moved from fantastic to feasible.

Eleven years after the release of the *H. influenzae* genome sequence, the first complete genome sequence of a myxobacterium was made publicly available in 2006 (and described in a publication in 2008) with the release of the 5 Mb chromosome of *Anaeromyxobacter dehalogenans* 2CP-C by the DOE Joint Genome Institute (JGI) (Thomas *et al.*, 2008). The myxobacteria generally prefer an aerobic environment, possess relatively large genomes averaging 9–10 Mb and with an average GC content of 70% (Table 2.1), move by gliding motility, and under starvation conditions, produce fruiting bodies containing thousands of environmentally resistant, quiescent myxospores (Reichenbach, 1999). While *A. dehalogenans* shares several of these traits, its genome size of 5.0 Mb is half that of typical myxobacteria and is in fact more similar in size to the genomes of other Deltaproteobacteria. Further, it prefers a microaerophilic environment and does not seem to form fruiting bodies. These differences make *A. dehalogenans* a potentially interesting species with which to contrast at a whole genome level, when comparing gene content among the fruiting myxobacteria (discussed further below). Subsequent to the release of the *A. dehalogenans* 2CP-C, three additional *Anaeromyxobacter* genomes have been completed and released by the JGI, allowing detailed comparisons of these closely related species.

Shortly after the release of the *A. dehalogenans* genome sequence, the genome sequence of *Myxococcus xanthus* DK1622 was released to the public by TIGR and a community-based consortium (Goldman *et al.*, 2006). *M. xanthus* more typically represents the myxobacteria in the above listed characteristics, possessing a 9.1 Mb chromosome (Table 2.1). Based on the sheer volume of publications, *M. xanthus* is the most studied of the myxobacteria. The release of the complete genome sequence of *M. xanthus* heralded a new age of research on myxobacteria, providing a tool for identification of genes of interest using comparative genomics and also for comparative analyses with other organisms within and outside the myxobacteria to understand genome evolution and the dynamic state of gene order and content. Similarly, functional genomics approaches such as transcriptome (Shi *et al.*, 2008; Müller *et al.*, 2010) and proteome analyses (Curtis *et al.*, 2007; Kahnt *et al.*, 2010), which rely on genome sequences, have been developed in *M. xanthus*.

Myxobacteria are members of the Deltaproteobacteria in the order Myxococcales. This order is deeply trifurcated, forming three distinct suborders: the Cystobacterineae, the Sorangiineae, and the Nannocystineae (Fig. 2.1A). While we have a picture of the broad landscape of bacterial diversity due to the ease of sequencing 16S rRNA from environmental samples, genome sequencing currently relies on cultured bacteria in order to get large enough quantities of pure DNA. Although recent studies have suggested that diverse myxobacteria may be more cultivatable than previously expected (Garcia *et al.*, 2010), only a minority of the myxobacteria have been cultured based on 16S rRNA sequences deposited in the Ribosomal Database Project (Cole *et al.*, 2009) (Fig. 2.1A).

The first two myxobacteria genomes (*A. dehalogens* 2CP-C and *M. xanthus* DK1622) were from species within the suborder Cystobacterineae. In 2007, the first genome sequence of a myxobacterium from outside this suborder, *Sorangium cellulosum* So ce 56 from the suborder Sorangiineae (Schneiker *et al.*, 2007), was released by Bielefeld University (Table 2.1). At 13 Mb, it is the largest bacterial single genome sequence released, to date. Also in 2007, the draft genomic sequence of *Plesiocystis pacifica* SIR-1 was released to the public by the Gordon and Betty Moore Foundation Microbial Genome sequencing Project and the J. Craig Venter Institute (JCVI), providing the first genome-scale data from the third suborder, the Nannocystineae. To date, the *P. pacifica*

Table 2.1 Characteristics of seven myxobacteria with completely sequenced genomes

Species	Native environment[a]	O_2 requirement[a]	Cell dimensions[a]	Spore shape and size[a]	Fruiting body morphology[a]	Genome size (bp)[b]	GC[b,c]	CDS[b,d]
M. xanthus	Terrestrial	Aerobic	Cylindrical rods with tapered ends, 3–6 × 0.8 µm	Spherical, 1.8–2.5 µm	Single orange/yellow slimy knobs, 80–400 µm diameter, constricted base	9,139,763	68.9%	7,316
M. fulvus	Marine	Aerobic	Cylindrical rods with tapered ends, 4–8 × 0.8 µm	Spherical, 1.2–1.8 µm	Single white/red/purple slimy knobs, 50–250 µm diameter, constricted base	9,003,593	70.6%	7,284
C. coralloides	Terrestrial	Aerobic	Cylindrical rods with tapered ends, 3–8 × 0.8 µm	Spherical, 1.5–1.9 µm	Single pink/red cartilaginous ridges or pustules 50–250 µm diameter, finger-like projections	10,080,619	69.9%	8,033
S. aurantiaca	Terrestrial	Aerobic	Boat-shaped rods with tapering ends, 4–8 × 0.8 µm	Short bent rods, 0.5 × 2 µm	5–15 orange/red spherical sporangioles, 35×50 µm diameter borne on a 60–140 µm (H) × 30–100 µm (W) stalk	10,260,756	67.5%	8,352
A dehalogenans	Terrestrial	Microaerophilic	Narrow rods with tapering ends, 4–8 × 0.3 µm	NA	NA	5,013,479	74.9%	4,346
H. ochraceum	Marine	Aerobic	Cylindrical rods with round, blunted ends, 4–8 × 0.5 µm	Spherical, 0.5–0.7 µm	One or more yellow/brown ovoid sporangioles 25–100 µm diameter in dense packs	9,446,314	69.0%	6,719
S. cellulosum	Terrestrial	Aerobic	Cylindrical rods with rounded ends, 3–8 × 1.0 µm	Short rods, 0.7 × 3 µm	12–40 round sporangioles, 20–25 µm diameter packed in sori 40–45 × 110–160 µm	13,033,779	71.4%	9,380

[a]Information obtained from (Reichenbach, 2005; Shimkets *et al.*, 2006).
[b]Data the National Centre for Biotechnology Information website (ftp://ftp.ncbi.nlm.nih.gov/genomes/Bacteria/).
[c]GC indicates the percentage of the genome composed of guanidine and cytosine nucleotides.
[d]CDS (coding sequence) indicates the total number of predicted protein-coding genes in a genome.

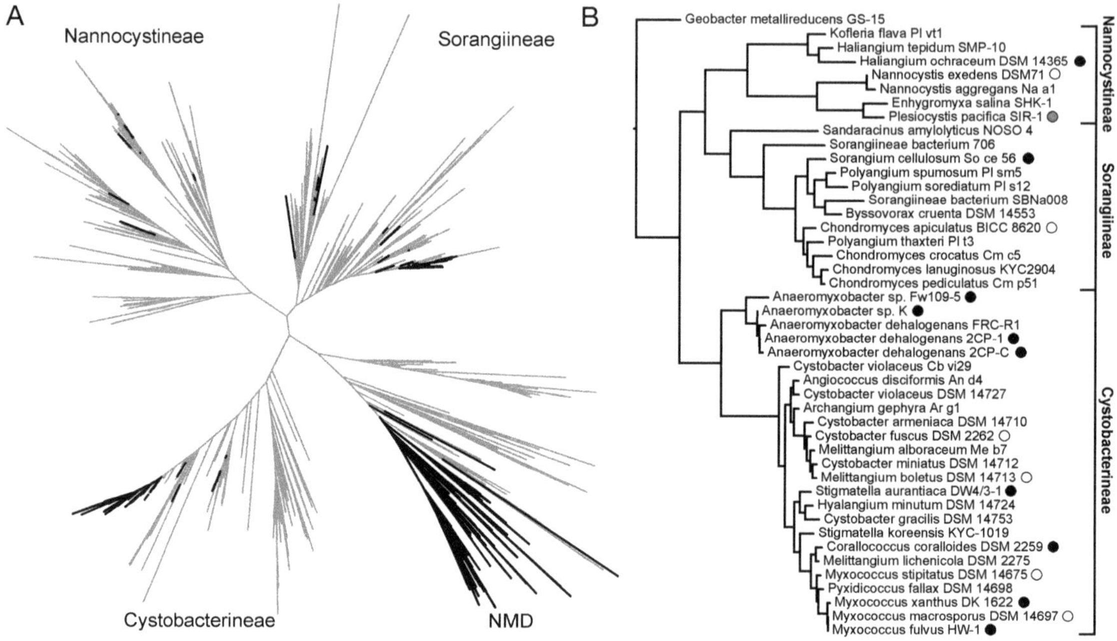

Figure 2.1 Phylogenetic diversity of myxobacteria. (A) A phylogenetic tree built from 16S rRNA sequences of myxobacteria and type strains of non-myxobacteria Deltaproteobacteria (NMD) retrieved from the Ribosomal Database Project (Cole *et al.*, 2009) reveals the clades that correspond to the three established suborders of myxobacteria. The tree includes myxobacteria sequences from isolates (black branches) and uncultured members (grey branches). A few sequences from uncultured members are more similar to sequences from NMD than other myxobacteria suggesting that they are misclassified. (B) A phylogenetic tree for a representative set of myxobacteria isolates and *Geobacter metallireducens* (as a NMD outgroup). Black circles indicate species for which the complete genome sequences is available, a grey circle indicates the incomplete *P. pacifica* genome, and open circles identify on-going genome projects.

genome data remains in a draft state, consisting of 237 unordered nucleotide contigs totalling 10.6 Mb. These contigs likely represent nearly the entire genome of this species. Two years later (2009), the first complete Nannocystineae genome was released. The 9.4 Mb genome of *Haliangium ochraceum* DSM 14365 was sequenced by the JGI (Natalia *et al.*, 2010) and so complete genomes from all three suborders became available (Table 2.1).

Since that time, three additional complete myxobacterial genome sequences have been released. *Stigmatella aurantiaca* DW4/3-1 was initially sequenced to a draft state as a part of the *M. xanthus* DK1622 sequencing project at TIGR, and released as 579 contigs totalling 10.3 Mb. In 2010, a sequencing consortium headed by the Max Planck Institute in Marburg (MPI-Marburg) released the complete 10.3 Mb genome sequence of the same *S. aurantiaca* strain (Huntley *et al.*, 2011) (Table 2.1). With this release, detailed, global comparative analyses within the fruiting Cystobacterineae were for the first time possible (Huntley *et al.*, 2011). Similarly, these comparative analyses could be extended to myxobacteria from the two other suborders. More recently, the complete 9.0 Mb genome sequences of the fruiting Cystobacterineae *Myxococcus fulvus* HW-1 was released by the College of Life Sciences, Shandong University (Li *et al.*, 2011) (Table 2.1) and the complete 10.1 Mb genome sequence of the fruiting Cystobacterineae *Corallococcus coralloides* DSM 2259 genome was released by MPI-Marburg (Huntley *et al.*, 2012). Thus, at the time of writing, a total of one draft and 10 complete myxobacteria genome sequences are available, representing 8 different genera (Fig. 2.1B).

Myxobacteria whole-genome comparisons: genome size

Throughout the remainder of this chapter, information is obtained from and comparisons are made amongst the following genome sequencing projects with the exception of underlined genomes when comparisons were made using information from the MiST2 database (Ulrich and Zhulin, 2010) as it did not contain these members at the time of the analysis (July 2012):

- Fruiting myxobacteria (six genomes): *Corallococcus coralloides* DSM2259 (uid157997), *Haliangium ochraceum* DSM14365 (uid41425), *Myxococcus fulvus* HW-1 (uid68443), *Myxococcus xanthus* DK1622 (uid58003), *Sorangium cellulosum* So-ce-56 (uid61629), *Stigmatella aurantiaca* DW4/3-1 (uid158509).
- All myxobacteria (seven genomes): all fruiting myxobacteria plus *Anaeromyxobacter dehalogenans* 2CP-C (uid58135).
- Non-myxobacteria Deltaproteobacteria (henceforth referred to as NMD, 36 genomes): *Bacteriovorax marinus* SJ (uid82341), *Bdellovibrio bacteriovorus* HD100 (uid61595), *Desulfarculus baarsii* DSM2075 (uid51371), *Desulfatibacillum alkenivorans* AK-01 (uid58913), *Desulfobacca acetoxidans* DSM11109 (uid65785), *Desulfobacterium autotrophicum* HRM2 (uid59061), *Desulfobulbus propionicus* DSM2032 (uid62265), *Desulfococcus oleovorans* Hxd3 (uid58777), *Desulfohalobium retbaense* DSM5692 (uid59183), *Desulfomicrobium baculatum* DSM4028 (uid59217), *Desulfotalea psychrophila* LSv54 (uid58153), *Desulfovibrio aespoeensis* Aspo-2 (uid42613), *Desulfovibrio africanus* Walvis-Bay (uid66847), *Desulfovibrio alaskensis* G20 (uid57941), *Desulfovibrio desulfuricans* ATCC27774 (uid59213), *Desulfovibrio desulfuricans* ND132 (uid63159), *Desulfovibrio magneticus* RS-1 (uid59309), *Desulfovibrio salexigens* DSM2638 (uid59223), *Desulfovibrio vulgaris* Miyazaki-F (uid59089), *Desulfovibrio vulgaris* DP4 (uid58679), *Desulfovibrio vulgaris* Hildenborough (uid57645), *Desulfurivibrio alkaliphilus* AHT2 (uid49487), *Geobacter bemidjiensis* Bem (uid58749), *Geobacter* sp FRC-32 (uid58543), *Geobacter lovleyi* SZ (uid58713), *Geobacter* sp M18 (uid55771), *Geobacter* sp M21 (uid59037), *Geobacter metallireducens* GS-15 (uid57731), *Geobacter sulfurreducens* PCA (uid57743), *Geobacter uraniireducens* Rf4 (uid58475), *Hippea maritima* DSM10411 (uid65267), *Lawsonia intracellularis* PHE-MN1-00 (uid61575), *Pelobacter carbinolicus* DSM2380 (uid58241), *Pelobacter propionicus* DSM2379 (uid58255), *Syntrophobacter fumaroxidans* MPOB (uid58177), *Syntrophus aciditrophicus* SB (uid58539)
- All Deltaproteobacteria (43 genomes): all myxobacteria plus all NMD.

In bacteria, the length of the genome is directly related to the size of the gene repertoire it possesses (Mira *et al.*, 2001; Kuo *et al.*, 2009) and to the complexity of the environment in which the bacterium lives (Koonin and Wolf, 2008). Species that tend to live in constant environments, such as intracellular parasites, generally possess genomes smaller than 2 Mb, while those that live in fluctuating environments, such as terrestrial habitats, generally possess larger genomes, that average ~5 Mb (Koonin and Wolf, 2008).

As previously stated, most myxobacterial genomes range in size from 9 to 13 Mb, but *A. dehalogenans* is roughly half this size and with the four complete *A. dehalogenans* genomes ranging in size from 5.0 to 5.3 Mb (Table 2.1) (Huntley *et al.*, 2011), which more closely matches the average size of the NMD genomes, 4 Mb. Based on the hypothesis that all Deltaproteobacteria are descendants of a common ancestor and because NMD have genome sizes ranging from 1.7–6.5 Mb, by parsimony it may be hypothesized that the common ancestral genome was around 5 Mb in size and that the larger size of most myxobacterial genomes is due to lineage-specific expansions, discussed further below (Goldman *et al.*, 2006, 2007). Following this line of thought, *A. dehalogenans* genome size could be due to two possibilities: (i) the genome size expansions seen in most myxobacteria did not occur in *A. dehalogenans*; or (ii) the expanded genome size was present in the *A. dehalogenans* ancestry, but was lost in that lineage by deletion of significant regions of the chromosome. Analysis of the *A.*

dehalogenans 2CP-C genome reveals that both possibilities may have played a part in the differences seen in genome size (Thomas *et al.*, 2008). First, based on the amount of 'foreign DNA' identified by determining local GC content fluctuations across the genome, it appears that *A. dehalogenans* contains less DNA obtained by horizontal gene transfer (HGT) compared to the others, which would support the first possibility. However, it was also found that while the origin and terminus of replication in most circular bacterial genomes are separated by approximately 180° (Mrazek and Karlin, 1998), the terminus appears to be offset 1.5 Mb in the *A. dehalogenans* 2CP-C genome, which would support the second possibility.

How did the large myxobacterial genomes reach their extant sizes? Possible reasons for the increased genome sizes among most myxobacteria in comparison to other Deltaproteobacteria include HGT, duplication of smaller genome segments, and whole genome duplication (WGD). Analysis of the *M. xanthus* genome suggests that nearly 25% of its annotated genes may have been acquired by HGT from outside the Deltaproteobacteria, based on best BlastP hits and GC content of the genes (Goldman *et al.*, 2006, 2007). Gene duplications have also been suggested as contributing to the large genome size in *M. xanthus* (Goldman *et al.*, 2006). As shown in Table 2.2, a reciprocal best BlastP hit analysis (Huntley *et al.*, 2011) indicates that approximately 35% of the genes in the large myxobacteria genomes have a paralogous gene.

The increase in genome size in the large genome myxobacteria has also been attributed to WGD [(Luciano *et al.*, 2011), citing Goldman *et al.* (2006)]. Originally described by Susumu Ohno (Ohno, 1970), WGD has been attributed to increases in genome size and repertoire in eukaryotes [recently reviewed in (Semon and Wolfe, 2007; Jaillon *et al.*, 2009)], but little has been published suggesting that WGD has been identified within the Bacteria. An analysis of genome duplication in *Saccharomyces cerevisiae* describes a method to detect an ancestral genome duplication event in a genome of a size similar to that of the large myxobacterial genomes (Kellis *et al.*, 2004). This method depends on a comparison between a genome of a species that diverged before the duplication and would represent the state of the myxobacterial genome prior to a WGD and the large myxobacterial genomes. If a WGD had occurred in the myxobacteria with large genomes, extended regions of the 'ancestral' genome should show conserved synteny to two regions of the large myxobacterial genomes. However, little conserved synteny is observed between the genomes of myxobacteria from different suborders (Fig. 2.2A) and even less to genomes from species outside the order (data not shown). It could be argued that this lack of conserved synteny is due to the fact that the myxobacteria diverged from

Table 2.2 Relative distances[a] between paralogous genes in myxobacterial genomes

Distance between paralogous genes	Per cent of pairs located within each range, for each species[b]						
	Mx	Mf	Cc	Sa	Ad	Ho	Sc
0–5%	18.9%	20.4%	22.4%	22.6%	18.5%	26.6%	18.7%
0–10%	28.7%	28.6%	32.4%	32.4%	29.3%	36.1%	28.1%
10–20%	18.5%	20.2%	18.0%	18.0%	16.7%	13.2%	16.3%
20–30%	19.1%	19.1%	16.7%	16.4%	15.6%	18.4%	18.6%
30–40%	14.8%	15.4%	15.3%	18.2%	19.6%	15.3%	18.6%
40–50%	19.0%	16.7%	17.6%	15.0%	18.8%	17.0%	18.4%
Total paralogues (% of all CDS)	2400 (33%)	2456 (34%)	2776 (35%)	2942 (35%)	1318 (30%)	2258 (34%)	3346 (36%)

[a]Distances measured as per cent of each genome's total CDS.
[b]Species are abbreviated as follows: Mx, *Myxococcus xanthus*; Mf, *Myxococcus fulvus*; Cc, *Corallococcus coralloides*; Sa, *Stigmatella aurantiaca*; Ad, *Anaeromyxobacter dehalogenans*; Ho, *Haliangium ochraceum*; and Sc, *Sorangium cellulosum*.

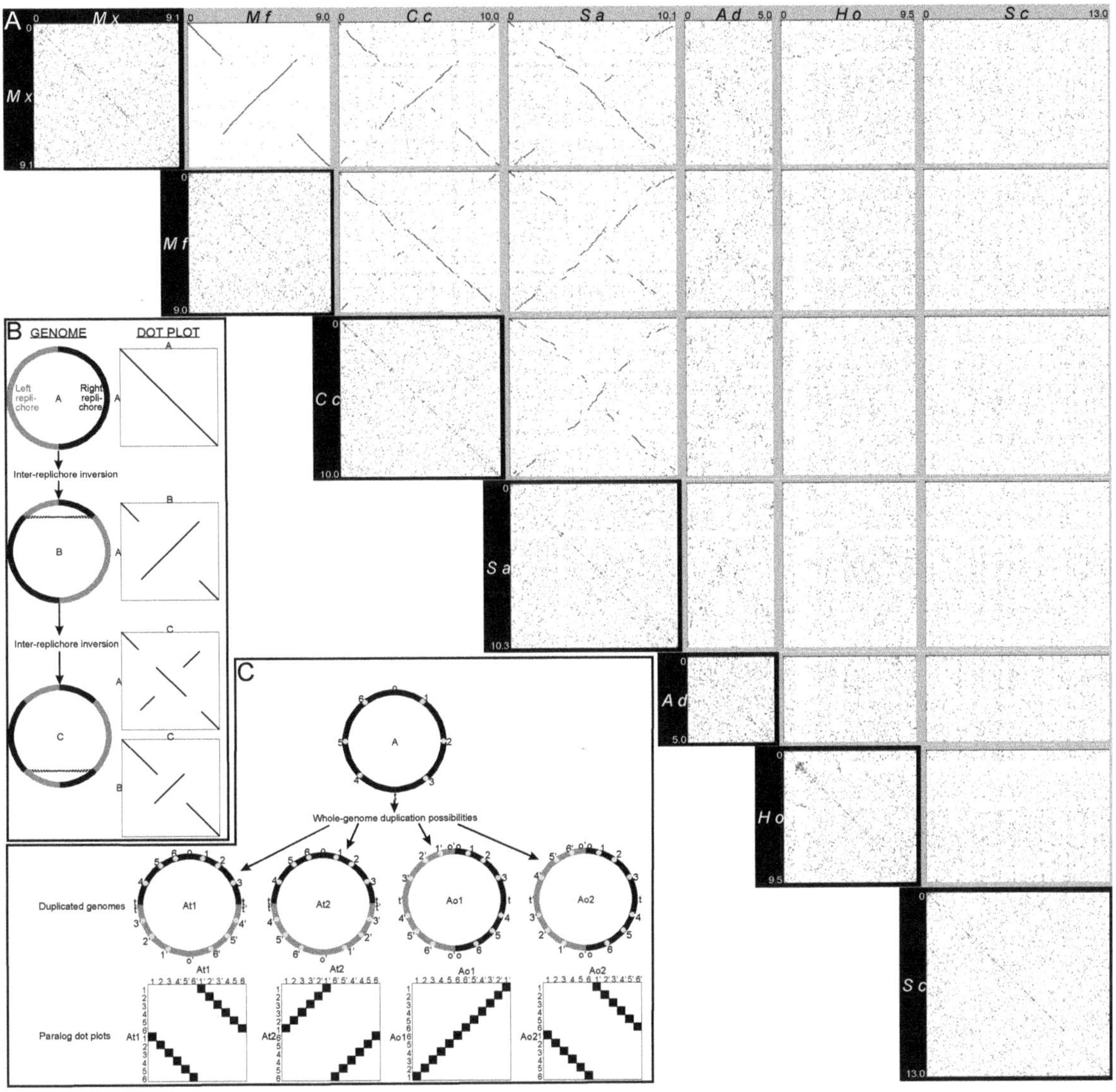

Figure 2.2 Conserved synteny in the myxobacterial genomes. Putative paralogues and orthologues were identified by reciprocal BLAST best hit analyses (Huntley *et al.*, 2011). Genomic context of the genes encoding identified pairs were then determined based on chromosomal gene order. This information was then used to generate a series of dot-plots indicating the relative position of each gene pair in the genome. (A) Synteny plot for paralogous genes and orthologous genes in pairwise combinations of genomes of myxobacteria. For each row of dot-plots, the first plot represents the relative positions of paralogous genes for a given species. In these plots, the diagonal indicates paralogous genes that lie relatively close to each other, whereas regions showing a random distribution indicate paralogous genes that are located on distant parts of the chromosome. The other plots in each row indicate the genetic context of orthologous genes for each species pair. Conserved synteny between species appears as contiguous lines which vary in length based on the number of conserved genes found in the segment. The 'X-pattern' seen in several of these plots is likely due to symmetric inter-replichore inversions, illustrated in (B). (B) Schematic illustrating connection between inter-replichore inversions and dot-blot patterns. A single inversion (such as that between genome A and B or B and C) would result in a pattern similar to that seen between *M. xanthus* and *M. fulvus*. Further inversions, at different symmetric positions on the chromosome, produce dot-plots such as that between A and C. From this, it is readily apparent how real plots, such as between *M. xanthus* and *S. aurantiaca* produce the prominent 'X-pattern'. (C) Schematic illustrating connection between various whole genome duplications and dot-blot patterns. Four possible duplications are shown, with alternate orientations of the duplication occurring at either the origin or terminus of replication. In none of the cases, the dot-plot pattern of duplicated genes (paralogues) is not seen when looking at the real paralogue dot-plots (A).

the other Deltaproteobacteria an estimated ca. 2 Gyr ago (Hedges *et al.*, 2006). Therefore, this analysis does not rule out that a WGD happened.

A different approach to test for a WGD relies on the assumption that WGD-derived paralogous genes should, at least initially, be located at a significant distance from the original genes and in a block of conserved synteny. Approximately 35% of the genes in the large myxobacterial genomes have paralogous genes (Table 2.2). When the genomic context of paralogous gene pairs is visualized in a dot-plot, paralogues produce a marked line running from the upper left to lower right corners of the graph (Fig. 2.2A) in all the large genomes. This pattern indicates that the majority of paralogues are, at the scale of these plots, located close to each other on the genome. Indeed we find roughly 30% of the paralogues are separated by no more than 10% of the total genome size (Table 2.2). The relative closeness of paralogues suggests that many independent gene duplication events resulted in segmental duplications of parts of the genomes (Rocha, 2004). The remaining paralogues appear distributed randomly over the rest of the genome, and no enrichment is seen of paralogues in diametrically opposed positions as might be expected as a result of WGD (Fig. 2.2C). In any case, while WGD cannot be fully discounted as a possible source contributing to the expansion of the large myxobacterial genomes, no evidence has been presented which supports such an event. Rather HGT and segmental genome duplications alone are adequate to explain the observed genome size differences observed between the large myxobacterial genomes and those of NMD.

Myxobacteria whole-genome comparisons: genome organization and conserved synteny

Synteny refers to the order of genes in a genome. Changes in synteny can be caused by gene gain or loss and genomic rearrangements, and can occur quite rapidly (Koonin and Wolf, 2008). Comparisons of synteny of conserved genes in bacterial genomes reveal that genomic rearrangements often occur as symmetric inversions relative to the origin of replication of the chromosome (Eisen *et al.*, 2000; Koonin and Wolf, 2008). It has been hypothesized that the observed symmetry of the inversions is due to illegitimate recombination of DNA between replication complexes, which are located approximately equidistant from the origin of replication, one on each replichore, during the process of chromosome replication (Tillier and Collins, 2000). Other forms of illegitimate recombination are associated with major changes in the genetic content of an organism, and include gene acquisition by HGT, segmental duplications, and gene deletions (Rocha, 2003). Recombination between repetitive DNA sequences also appears to be a major source of changes in synteny (Rocha, 2004). These repetitive elements, which include ribosomal RNA operons and other genes and regulatory elements, as well as regions of mobile DNA such as insertion sequences are often present in large numbers (100s to 1000s) in genomic sequences (Achaz *et al.*, 2003; Rocha, 2003).

Conserved synteny in myxobacterial genomes was identified by first identifying putative orthologues by reciprocal best BlastP hit analysis (Huntley *et al.*, 2011). Subsequently, the genomic context of orthologous genes was compared between species using simple visual methods such as dot-plots (Fig. 2.2A). These analyses show, as expected, that the genomes of more closely related species phylogenetically, based on 16S rRNA gene sequences, share more conserved synteny than species that are more distantly related. In the dot-plots, a distinct 'X pattern' is often observed. This pattern is indicative of symmetric, inter-replichore inversions having occurred in one of the lineages since the two species diverged from a common ancestor (Fig. 2.2B). In the dot-plots comparing more and more distantly related species (e.g. *M. xanthus*/*M. fulvus* to *M. xanthus*/*S. aurantiaca*), it is evident that these inversions have occurred repeatedly, and can significantly alter the overall synteny of a genome without significantly changing local gene order. Comparisons of even more distantly related species (e.g. *M. xanthus*/*A. dehalogenans*) reveals that a vague 'X pattern' can still be seen suggesting that symmetrical inversions combined with other recombination events have nearly obliterated any recognizable conserved synteny. Finally, for very distantly related myxobacteria (e.g. *M. xanthus*/*S. cellulosum*) essentially no conserved synteny is observed even though more of the *M. xanthus* genes are conserved in *S. cellulosum* than in *A. dehalogenans* (see Table 2.4).

Myxobacteria whole-genome comparisons: genetic content

Genetic content, or the gene repertoire of a genome, can be considered a snapshot of the sequenced organism's potential functionality. Gene density is another aspect of genetic content which can provide insight into the rapidity of change in the genome because organisms undergoing genome reductions may have a lower gene density due to the presence of higher than average numbers of pseudogenes (Kuo *et al.*, 2009). CDS genes constitute a slightly higher percentage of the myxobacteria genomes compared to that of the NMD (Table 2.3). A different means of looking at 'gene density' is determining the average number of CDS genes encoded per Kb of DNA. Relative to NMD, *A. dehalogenans* is more gene-dense, while *H. ochraceum* and *S. cellulosum* are less dense; and overall the myxobacterial genomes appear less gene-dense than NMD. These numbers most likely fluctuate due to the number of large genes (>5–10 kb) for secondary metabolite production in the myxobacteria, which would increase the percent of the genome dedicated to CDS genes, while lowering the overall number of genes and thus the gene density. Differences in the observed gene density can also be artificially influenced by the fidelity of the annotation performed on the genome: a more relaxed allowance for hypothetical CDS in an annotation may inflate the gene density above reality and conversely an overly stringent annotation may fail to include real genes and thus deflate the gene density for a genome. An analysis of the *E. coli* MG1655 genome revealed that over 200 (~5%) of its annotated CDS sequences were likely pseudogenes based on comparative analysis of genomes from close relatives (Ochman and Davalos, 2006), which if true for the large myxobacterial genomes, could mean >350 CDS genes from each species may not be functional. Regardless, the myxobacteria gene density overall appears similar to that of most bacterial genomes (Koonin and Wolf, 2008; Kuo *et al.*, 2009), which suggests the myxobacteria do not possess extraordinary numbers of pseudogenes, and are not undergoing genomic changes any more rapidly than an average bacterium.

Within the myxobacteria, the number of rRNA operons, which contain the 5S, 16S and 23S rRNA genes, varies from 2–4 per genome and does not correlate with genome size (Table 2.3). In the NMD the number of rRNA operons varies from 1 to 7 operons per genome and with an average of 3 (Table 2.3). Interestingly, the number of rRNA operons in the myxobacteria pales in comparison to the number found in certain Firmicutes species also found in the soil: species of the genera *Bacillus* (Loughney *et al.*, 1983) and *Clostridium* (Rainey *et al.*, 1996) can possess up to 15 rRNA operons. Like the myxobacteria, these organisms are capable of producing environmentally resistant spores. It has been hypothesized that the correlation between the large number of rRNA operons and the rapidity of growth observed when they are introduced to a nutritionally rich environment reflects a difference in competitive strategy compared to more slowly growing soil organisms (Klappenbach *et al.*, 2000). Specifically, the rapidly growing species possess more robust translational power, but at the cost of translational yield or amount of functional protein synthesized per unit energy consumed. We hypothesize that limited translational robustness and thus slower growth may favour the myxobacteria's ecological strategy of not necessarily competing with other soil bacteria for nutrients, but preying on bacteria growing within their niche. The number of tRNA-encoding genes also varies significantly among the myxobacteria, with an apparent positive correlation between rRNA operons and tRNA gene number (Table 2.3). This correlation has previously been described as being linked with translational power versus translational yield (Dethlefsen and Schmidt, 2007) and as discussed this could be related to the predatory myxobacterial lifestyle.

The assignment of CDS genes from each species to 22 functional categories (Huntley *et al.*, 2011) allows a comparison of functional gene repertoires in the myxobacteria (Table 2.3). From these analyses, several lineage-specific expansions and contractions in genetic content are evident. The functional gene repertoires (represented as

Table 2.3 Comparison of genetic content of seven complete myxobacterial genomes

Feature[a]	Number or percentage of total genes for each species[b]							Mean percentage for[c]			
	Mx	Mf	Cc	Sa	Ad	Ho	Sc	All Myxobacteria	Fruiting Myxobacteria	NMD	All Delta-proteobacteria
CDS encoding per cent of genome	91.5	88.1	91.8	91.2	91.1	88.0	86.8	89.8	89.6	87.8	88.1
Gene density (genes/kb DNA)	0.8	0.8	0.8	0.8	0.9	0.7	0.7	0.8	0.8	0.9	0.9
rRNA operons	4	3	3	2	2	2	4	3	3	3	3
tRNA genes	65	69	59	49	49	46	61	57	58	54	54
Functional category[d]											
Energy production and conversion	3.5%	3.4%	3.4%	3.1%	6.7%	4.0%	3.6%	3.8%	3.4%	7.2%	6.6%
Cell cycle control, cell division, chromosome partitioning	0.8%	0.9%	0.8%	0.9%	0.9%	1.1%	0.9%	0.9%	0.9%	1.0%	1.0%
Amino acid transport and metabolism	4.3%	4.2%	4.2%	4.0%	5.2%	4.4%	3.9%	4.3%	4.1%	5.3%	5.1%
Nucleotide transport and metabolism	1.3%	1.3%	1.4%	1.3%	1.7%	1.5%	1.2%	1.4%	1.3%	1.8%	1.7%
Carbohydrate transport and metabolism	2.7%	2.8%	3.3%	4.1%	3.5%	3.0%	4.4%	3.3%	3.2%	3.1%	3.1%
Coenzyme transport and metabolism	2.5%	2.4%	2.3%	2.5%	3.3%	2.8%	2.8%	2.6%	2.5%	4.1%	3.9%
Lipid transport and metabolism	2.8%	2.7%	2.6%	2.3%	2.9%	2.6%	2.0%	2.5%	2.5%	2.0%	2.1%
Translation and ribosome structure	2.9%	2.8%	2.6%	2.7%	4.3%	2.8%	2.2%	2.8%	2.6%	4.8%	4.5%
Transcription	4.2%	4.1%	4.7%	4.7%	4.2%	3.6%	4.5%	4.3%	4.4%	3.3%	3.5%
Replication, recombination, and DNA repair	3.7%	4.1%	2.8%	4.1%	3.8%	4.0%	3.0%	3.6%	3.5%	4.6%	4.4%
Cell wall, membrane, envelope biogenesis	5.0%	5.0%	4.9%	4.5%	6.0%	4.3%	4.0%	4.8%	4.6%	6.2%	5.9%

Cell motility	1.1%	1.0%	0.9%	0.8%	1.6%	0.5%	0.5%	0.9%	0.8%	1.9%	1.7%
Post-translation modification, prot. turnover, chaperones	3.5%	3.5%	3.2%	3.2%	4.2%	3.2%	3.1%	3.4%	3.3%	3.5%	3.5%
Inorganic ion transport and metabolism	2.9%	2.7%	2.6%	2.7%	3.3%	2.9%	3.1%	2.9%	2.8%	3.6%	3.5%
Secondary metabolite biosynthesis, transport, catabolism	2.9%	2.8%	3.0%	3.2%	1.6%	2.8%	2.7%	2.8%	3.0%	1.2%	1.5%
Signal transduction	8.6%	9.1%	8.6%	9.9%	8.6%	9.7%	10.0%	9.0%	9.1%	8.3%	8.5%
Intracellular traffic, secretion, and vesicular transport	1.5%	1.5%	1.3%	1.3%	1.5%	1.1%	0.8%	1.3%	1.3%	1.4%	1.4%
Defence mechanisms	1.9%	1.8%	1.8%	1.7%	2.2%	2.0%	1.6%	1.9%	1.8%	2.2%	2.1%
Other[E]	0.1%	0.1%	0.1%	0.1%	0.1%	0.1%	0.1%	0.1%	0.1%	0.1%	0.1%
Hypothetical[F]	44.0%	44.1%	45.3%	43.1%	34.4%	43.5%	45.8%	43.6%	44.9%	34.4%	36.1%

[a]All predicted genes from each genome were tabulated based on gene product type (protein, rRNA or tRNA).
[b]Individual myxobacteria abbreviated as follows: Mx, *Myxococcus xanthus*; Mf, *Myxococcus fulvus*; Cc, *Corallococcus coralloides*; Sa, *Stigmatella aurantiaca*; Ad, *Anaeromyxobacter dehalogenans*; Ho, *Haliangium ochraceum*; and Sc, *Sorangium cellulosum*. Black shaded cells indicate that the value is greater than one standard deviation above the mean for the seven species. Grey shaded cells indicate the value is lower than one standard deviations below the mean for the seven species.
[c]Four taxonomic groups described in the text. Black shaded cells indicate the value is greater than one standard deviation above the mean for all groups. Grey shaded cells indicate the value is greater than one standard deviation below the mean for all groups.
[d]Based on the Clusters of Orthologous Groups (COG) functional annotation (Tatusov *et al*., 2000), modified as described in Huntley *et al*. (2011).
[e]The categories RNA processing and modification, chromatin structure and dynamics and cytoskeleton contained few genes (less than 0.1% of each genomes repertoire) and were combined into the category labelled ‘Other’.
[f]The hypothetical category contains all genes for which only a very general or no function could be predicted (COG categories R and S, and genes with no COG assignment).

Table 2.4 Conservation of *M. xanthus* genes involved in development

Mx repertoire	*Mx* total	Per cent of *Mx* genes conserved in species[a]						
		Mf	Cc	Sa	Ad	Ho	Sc	NMD
Total genes	7316	80.7%	69.5%	63.5%	30.4%	30.9%	35.2%	50.4%
Known developers[b]	95	90.5%	82.1%	80.0%	27.4%	26.3%	37.9%	43.2%
Up-regulation during development[c]	408	89.7%	69.6%	68.9%	25.0%	29.4	31.6%	47.3%
Down-regulation during development[Cc]	423	88.9%	86.1%	80.1%	53.4%	51.5%	54.4%	65.5%

[a]Species are abbreviated as follows: Mx, *Myxococcus xanthus*; Mf, *Myxococcus fulvus*; Cc, *Corallococcus coralloides*; Sa, *Stigmatella aurantiaca*; Ad, *Anaeromyxobacter dehalogenans*; Ho, *Haliangium ochraceum*; and Sc, *Sorangium cellulosum*. Per cent of putative *M. xanthus* orthologues identified by reciprocal BLAST best hit analyses for each myxobacteria species, or in at least one non-myxobacteria Deltaproteobacteria (NMD). Black shaded cells indicate that percentage is higher than the overall conservation seen between *M. xanthus* and the given species.
[b]Set of genes published as being directly involved in *M. xanthus* developmental programme (Huntley *et al.*, 2011).
[c]Genes found to be significantly up- or down-regulated during a developmental time-course in *M. xanthus* (Shi *et al.*, 2008).

per cent of CDS genes in each category) are for the most part similar, with the clear exception of *A. dehalogenans*. *A. dehalogenans* possesses a smaller fraction of both secondary metabolite and hypothetical genes compared to the other myxobacteria. Loss of genes in these two categories may represent loss of non-critical genes that no longer provided an advantage after a change in the lineage's lifestyle. The small genome size results in the illusion of an enrichment of many other categories involved in 'core' metabolic activities (energy production, amino acid and nucleotide metabolism, coenzyme transport, cell wall biogenesis, motility, and protein modification/degradation) in *A. dehalogenans* compared to the myxobacteria with large genomes. In fact, the *A. dehalogenans* gene repertoire looks similar to that of the NMD average, and so may represent the Deltaproteobacteria base repertoire. As discussed, the fruiting myxobacteria may have added so many 'non-core' genes (i.e. secondary metabolite production and unknown function) that the composition of their repertoires relative to the NMDs is skewed.

Conserved genetic content in myxobacterial genomes can be identified by reciprocal best BlastP hit analysis to classify putative orthologues (Huntley *et al.*, 2011). This type of analysis previously revealed that 63.5% of *M. xanthus* genes have an orthologous gene in *S. aurantiaca* while only 30–35% of the *M. xanthus* genes have an orthologous gene in the more distantly related *H. ochraceum* and *S. cellulosum* (Table 2.4). The genomes of *M. fulvus* and *C. coralloides* in these analyses follow this trend and orthologous genes are found for 80.7% and 69.5%, respectively of the *M. xanthus* genes in these two genomes (Table 2.4).

Overall, the myxobacteria possess some of the largest chromosomes found so far in the Bacteria. The genetic content of the myxobacteria has apparently shifted significantly across the evolutionary time-span of the order, where the number of genes conserved between distantly related species approaches 30%, and nearly all of those genes are also found in non-myxobacteria genomes as well (Huntley *et al.*, 2011). Similarly, the genomic context of the order has also been the subject of significant change across the order. Symmetrical inter-replichore inversions and segmental genome duplications combined with gene acquisitions and losses have clearly contributed to the structure of the extant myxobacterial genomes. Thus, the myxobacterial genomes display a high degree of plasticity.

Myxobacteria whole-genome comparisons: signal transduction

All organisms require signal transduction systems to regulate diverse cellular processes. The evolutionary success of bacteria is arguably dependent on their ability to rapidly sense and respond to internal and external changes in their environment. Given that the developmental cycle of fruiting myxobacteria is triggered by environmental cues, it is hardly surprising that genes for signal transduction systems are enriched in their genomes (Table 2.3).

Since the dawn of the genomics era, signal transduction systems have been a target of comparative genomics analyses. The three classic paradigms of bacterial signal transduction systems are one-component systems (OCSs), two-component systems (TCSs), and chemotaxis (Che) systems (Stock *et al.*, 2000; Ulrich *et al.*, 2005; Wuichet and Zhulin, 2010). Serine-threonine protein kinases (STPKs) that typify eukaryotic signal transduction systems have also been shown to regulate processes such as development, differentiation, and stress responses in bacteria (Chaba *et al.*, 2002; Inouye and Nariya, 2008; Zorina *et al.*, 2011). Recently, extracytoplasmic function sigma factors (σ^{ECF}) have been called the 'third pillar' of bacterial signal transduction that primarily facilitates transmembrane signal transduction (Staron *et al.*, 2009). In this section, we focus on the analysis of myxobacteria signal transduction profiles and how they compare to universal, taxonomic, and functional relationships. The details of certain aspects of these systems are described further in Chapter 7.

Comparative genomic studies have found that OCS and TCS are ubiquitous in the bacterial genome universe and that the number of OCS and TCS genes in a genome is proportional to the size of the genome (Ulrich *et al.*, 2005). In contrast, there are no established linear relationships between Che systems, STPKs, and σ^{ECF} and genome size (Perez *et al.*, 2008; Staron *et al.*, 2009; Wuichet and Zhulin, 2010). The number of OCSs, TCSs, Che systems, STPKs, and σ^{ECF} encoded in myxobacteria genomes compared with the average numbers in other genomes as defined in the MiST2 database (Ulrich and Zhulin, 2010) are provided in Table 2.5.

Previous analysis revealed that OCSs are underrepresented in the *M. xanthus* genome in comparison to 15 other diverse genomes (Goldman *et al.*, 2006). Using the 1538 completely sequenced genomes available in July 2012 from the MiST2 database (Ulrich and Zhulin, 2010), we expanded the analysis of the relationship between OCS representation and genome size. OCS genes are underrepresented in all Deltaproteobacteria including the myxobacteria compared to universal trends; however, OCS abundance is proportional to genome size in all the Deltaproteobacteria (Fig. 2.3A and Table 2.5).

As for OCSs, TCS abundance is proportional to genome size (Ulrich *et al.*, 2005). TCS genes are overrepresented in NMD compared to the trend for all bacterial genomes (Fig. 2.3B). Interestingly, TCS gene numbers in myxobacteria follow the trend set by all bacterial genomes in relation to genome size and such genes are underrepresented in the large genomes of fruiting myxobacteria in comparison to NMD. Of note, the large genomes of fruiting myxobacteria are significantly enriched in STPKs whereas the NMD follow the trend for all bacterial genomes (Table 2.5 and Fig. 2.3C). This enrichment in STPKs may compensate for the lower numbers of TCS and could potentially also allow for greater signalling network complexity. Analysis of the myxobacteria kinome found that the number of STPKs increase exponentially relative to genome size in myxobacteria [(Perez *et al.*, 2008) and Chapter 7]. In agreement, we find that the relationship between genome size and STPK enrichment is non-linear (Fig. 2.3C). Although larger genomes are more likely to be enriched in STPKs, there is not a linear relationship between the number of STPK genes and genome size at the universal or taxonomic-specific level (Fig. 2.3C and Table 2.5) consistent with previous analyses (Perez *et al.*, 2008). Instead multicellular behaviour seems to be the main evolutionary driving force for their enrichment in select genomes (Perez *et al.*, 2008). STPKs are rare in bacteria, but fruiting myxobacteria encode more STPKs than any other available bacterial genomes, and many of these proteins have been linked to the developmental cycle (Chapter 7). Similar to the STPK trends, σ^{ECF} are enriched in all myxobacterial genomes whereas the NMD follow essentially follow the trend for all bacterial genomes (Fig. 2.3D and Table 2.5). σ^{ECF} genes are particularly enriched in the large genomes of fruiting myxobacteria (Fig. 2.3D). We find that σ^{ECF} genes are similarly enriched in

Table 2.5 Signal transduction protein profiles of myxobacteria

Genome(s)[a]	Number of genes[b] (% of total) Total	OCS	TCS	STPK	σ^{ecf}	Che[c] (% of total)
Myxococcus xanthus DK 1622	7316	315 (4.3%)	263 (3.6%)	98[d] (1.3%)	41 (0.6%)	8 (0.1%)
Myxococcus fulvus HW-1	7284	325 (4.5%)	268 (3.7%)	107 (1.5%)	38 (0.5%)	9 (0.1%)
Stigmatella aurantiaca DW4/3-1	8352	487 (5.8%)	322 (3.9%)	207 (2.5%)	33 (0.4%)	10 (0.1%)
Anaeromyxobacter dehalogenans 2CP-C	4346	159 (3.7%)	173 (4.0%)	19 (0.4%)	19 (0.4%)	7 (0.2%)
Haliangium ochraceum DSM 14365	6719	441 (6.6%)	174 (2.6%)	231 (3.4%)	39 (0.6%)	2 (0.03%)
Sorangium cellulosum So ce 56	9381	633 (6.7%)	270 (2.9%)	321 (3.4%)	87 (0.9%)	7 (0.1%)
NMD (33) [E]	3479	142 (4.1%)	139 (4.0%)	2 (0.1%)	2 (0.1%)	4 (0.1%)
Large non-myxobacteria genomes (36) [E]	7854	636 (8.1%)	193 (2.5%)	27 (0.3%)	38 (0.5%)	1 (<0.001%)
All (1538)	3258	177 (5.4%)	58 (1.8%)	4 (0.1%)	6 (0.2%)	1 (<0.001%)

[a]In July 2012, 1538 completely sequenced genomes were available from the MiST2 database (Ulrich and Zhulin, 2010). The *Corallococcus coralloides* genome was not part of the database at the time of information retrieval. Only one of the four closely related *Anaeromyxobacter* genomes is displayed although nearly identical information was retrieved from each. The average gene numbers for the 33 NMD genomes and for 36 large Non-myxobacteria genomes (>8Mb) are shown in addition to the average gene numbers for all 1538 genomes.
[b]OCS (one component systems), TCS (two component system), STPK (Ser/Thr protein kinase), and σ^{ECF} gene counts are provided for each genome and averaged for each genome group. OCS, TCS, and σ^{ECF} genes are identified by classifiers incorporated in the MiST database. STPK genes were identified in the MiST2 database by the Pkinase domain from Pfam (Punta *et al.*, 2012) in their corresponding amino acid sequences. Minor discrepancies between numbers provided here and those in Chapter 7 are presumably due to differences in the methods of identification, the latter of which often entailed additional manual curation.
[c]Che system counts are provided for each genome and averaged for each genome group. The number of Che systems is defined as the number of CheA genes (Wuichet and Zhulin, 2010) using classifiers that are incorporated in the MiST2 database.
[d]Although 99 STPK genes were identified in a previous analysis (Perez *et al.*, 2008), one is no longer annotated as a gene (MXAN_4338) and is absent from our analysis.
[e]Note that all numbers for the NMD and large non-myxobacteria genomes are average numbers.

other large genomes (Table 2.5), suggesting that their abundance may have some relationship to genome size and this relationship is non-linear in myxobacteria (Fig. 2.3D). However, even the small *A. dehalogenans* genome is enriched in σ^{ECF} genes when compared with the similarly sized genomes of NMD suggesting an additional link to taxonomy.

Che systems are abundant in most members of myxobacteria. The *S. aurantiaca* genome encodes 10 Che systems (Table 2.5), the highest number of any member of our 1538 genome set (data not shown). We do not find a relationship between genome size and Che system abundance (Table 2.5). Indeed, such a relationship would not be expected given that the primary role of Che systems is the regulation of motility. Motility is inextricably linked to development in myxobacteria, which may explain the abundance of Che systems in fruiting myxobacteria but not in *A. dehalogenans*. Furthermore, one of the Che systems in *M. xanthus* has been shown to be involved in the regulation of developmental gene expression (Kirby and Zusman, 2003; Willett and Kirby, 2011; Chapter 7). As described below, comparative genomics analyses support that there are many differences in the developmental pathways of Nannocystineae and Sorangiineae in comparison to Cystobacterineae (Huntley *et al.*, 2011; see also Fig. 2.4). The low abundance of Che systems in *H. ochraceum* may be another such example of their different developmental pathways. The ongoing efforts to understand the functions of the eight chemotaxis systems in *M. xanthus* will continue to shed light on their evolutionary relationships (Zusman *et al.*, 2007).

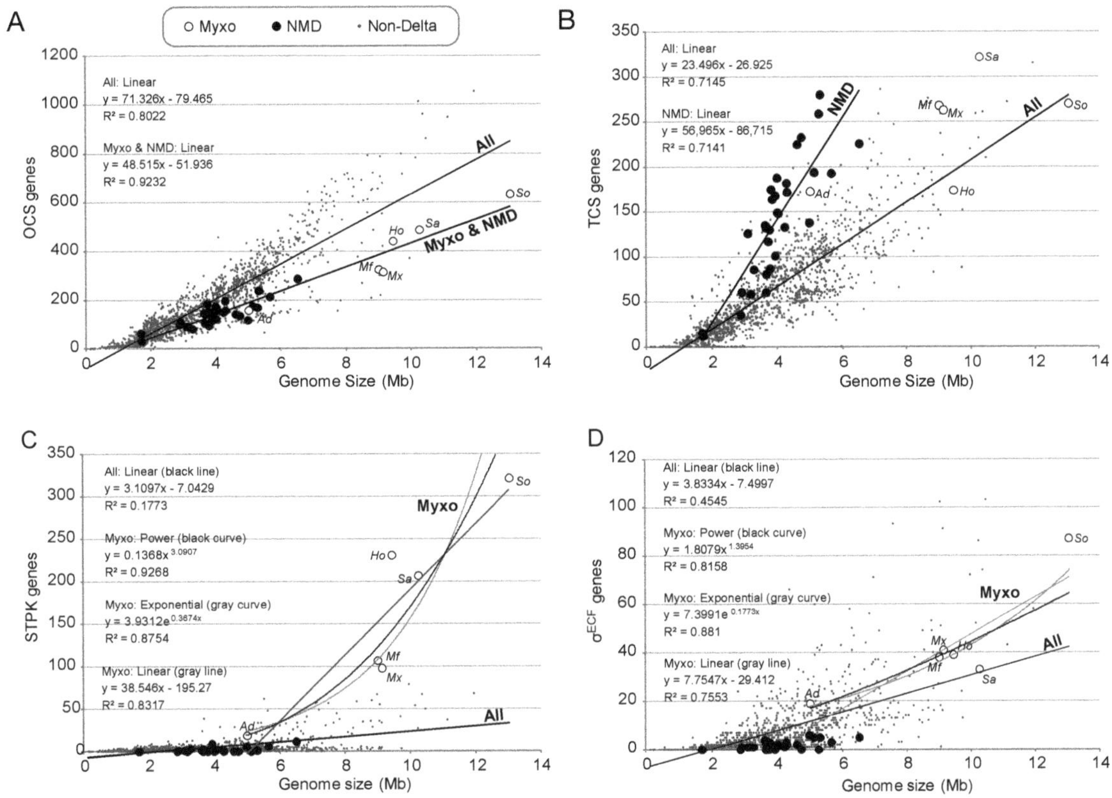

Figure 2.3 Signal transduction gene abundance in relation to genome size. (A), (B), (C), and (D) OCS, TCS, STPK, and σ^{ECF} protein encoding genes (Table 2.5), respectively, are plotted vs genome size for all genomes. The universal linear trendline (All) for gene abundance in relation to genome size is displayed. Linear trendlines of interest for TCS and OCS genes that are described in the text are shown, in addition to the three potential relationships between STPK and σ^{ECF} enrichment and myxobacteria genome size. Markers for all graphs are as depicted in the panel (A) key. Myxobacteria genomes in all graphs are labelled: *Ad*, *A. dehalogenans*; *Ho*, *H. ochraceum*; *Mf*, *M. fulvus*; *Mx*, *M. xanthus*; *Sa*, *S. aurantiaca*; and *So*, *S. cellulosum*. Data were extracted from the MiST2 database July 2012.

Myxobacteria whole-genome comparisons: the genetic programme for fruiting body formation

The myxobacteria owe most of their fame to their ability to undergo a multicellular developmental programme in response to starvation that culminates in the formation of spore-filled multicellular fruiting bodies. When cells detect a return to conditions favourable for vegetative growth, the myxospores germinate, resulting in the quick formation of a colony of cells with a population density optimal for predation, a lifestyle many members of the bacterial group are known to utilize. Indeed, most research on myxobacteria has been aimed at understanding the genetic programme resulting in fruiting body formation. Fruiting body morphology is species specific and, thus, genetically determined, and highly variable (Reichenbach, 1999; Dawid, 2000). The fruiting bodies of *M. xanthus* are relatively simple and resemble haystacks, while numerous myxobacterial species construct fruiting bodies that exhibit a high degree of morphological complexity, including stalks, sporangioles and appendages. All myxobacteria tested, with the exception of *A. dehalogenans*, initiate fruiting body formation in response to starvation, suggesting that the last common ancestor of the myxobacteria harboured a genetic programme for fruiting body formation. Moreover, in this evolutionary scenario, extant fruiting myxobacteria would be expected to share

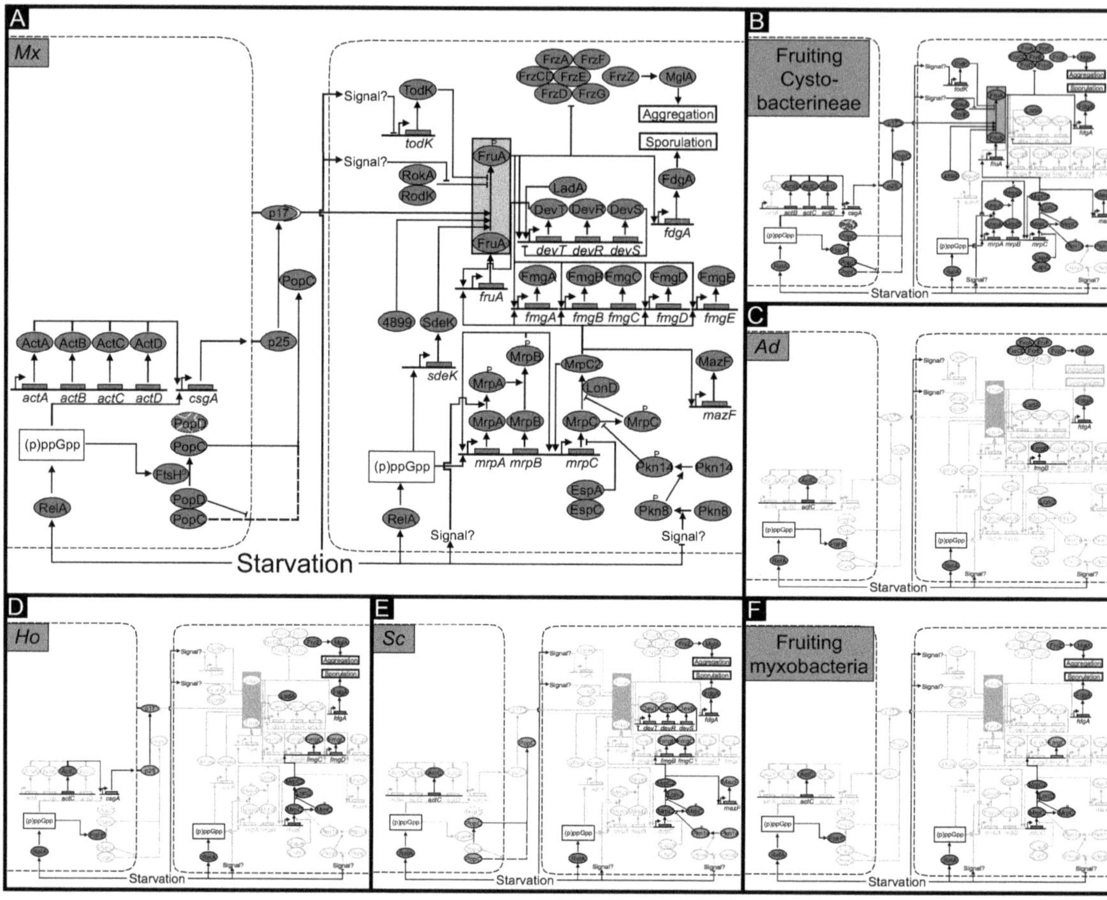

Figure 2.4 Conservation of the MrpC/FruA/C-signal transduction pathway of *M. xanthus* in other myxobacteria. (A) The MrpC/FruA/C-signal transduction pathway in *M. xanthus*. The conservation of the proteins shown in (A) in different species or groups of species is shown in (B–F). In these figures, the shading of a protein or gene is removed if the orthologue of the *M. xanthus* gene is absent from the given species (or from one member of the species group). Conservation is illustrated for fruiting Cystobacterineae (*M. fulvus*, *C. coralloides*, and *S. aurantiaca*, B), *A. dehalogenans* (*Ad*, C), *H. ochraceum* (*Ho*, D). *S. cellulosum* (*Sc*, E), and the fruiting myxobacteria (All species listed for (B) plus *H. ochraceum* and *S. cellulosum*, F).

in common a significant part of this programme. Clearly, given the long evolutionary history of myxobacteria and the different fruiting body morphologies, differences would also be expected.

As described in Huntley *et al.* (2011) many proteins have been identified as important for fruiting body formation in *M. xanthus*. Among these proteins, 95 have been characterized to function only in the developmental programme and not in vegetative growth (from here on this gene set is referred to as *M. xanthus* developer proteins). After identifying orthologues of these proteins, it was previously demonstrated that they are overrepresented in *S. aurantiaca* compared with the overall frequency of *M. xanthus* proteins having orthologues in *S. aurantiaca* (Huntley *et al.*, 2011). Similarly, it was found that the *M. xanthus* developer proteins are not more likely to be conserved than any other protein in *S. cellulosum*, *H. ochraceum* and *A. dehalogenans* (Huntley *et al.*, 2011) (Table 2.4). Including the genomes of *M. fulvus* and *C. coralloides* in this analysis, the same trend is observed (Table 2.4). If the myxobacteria would utilize a common set of proteins in fruiting body formation, the *M. xanthus* developer proteins should be, for the most part, conserved. The finding that the *M. xanthus* developer proteins are not highly conserved in *A. dehalogenans*, which does not form fruiting bodies, is probably to be expected, but for *S. cellulosum* and *H. ochraceum*, these observations may suggest that their developmental programme is significantly different from that of the Cystobacterineae.

A similar pattern is observed when the conservation of 408 *M. xanthus* genes transcriptionally up-regulated in response to starvation was examined in the other myxobacterial genomes (Huntley *et al.*, 2011) (Table 2.4). Adding *M. fulvus* and *C. coralloides* to this analysis, it is found that these 408 up-regulated genes are overrepresented in the genomes of *M. fulvus* and *C. coralloides.* On the other hand, 423 *M. xanthus* genes transcriptionally down-regulated in response to starvation are overirepresented in all the myxobacterial genomes as well as in the NMD. Most of these down-regulated genes are deeply conserved beyond the Deltaproteobacteria, and likely represent the starvation-induced stringent response (Huntley *et al.*, 2011). Combining the information on conservation of *M. xanthus* developer proteins and developmentally up-regulated genes, it appears that the Cystobacterineae may utilize a developmental programme that is significantly different from that in *H. ochraceum* or *S. cellulosum* (Huntley *et al.*, 2011).

The lack of conservation of the *M. xanthus* genetic programme for fruiting body formation in *H. ochraceum* or *S. cellulosum* is supported by the lack of conservation of the proteins in one of the best-studied developmental pathways in *M. xanthus*, the MrpC/FruA/C-signal transduction pathway (see also Chapter 6). As previously described (Huntley *et al.*, 2011), most of the components in this pathway are conserved in *S. aurantiaca* but not in *H. ochraceum* and *S. cellulosum* (Fig. 2.4). This pathway appears to be highly conserved amongst the fruiting Cystobacterineae (Fig. 2.4B) supporting the notion that the Cystobacterineae follow the same general programme during development. The lack of conservation of some of the corresponding genes may reflect lineage-specific additions and losses related to how each species interacts with and measures conditions in its preferred environment. For example, the halotolerant *M. fulvus* may have different detection methods for determining when to commit to development. There is a clear lack of the components of the MrpC/FruA/C-signal transduction pathway in *A. dehalogenans* (Fig. 2.4C). Both hypotheses for the reduced genome size of *A. dehalogenans* (all other species genomes grew, or *A. dehalogenans* underwent a reduction in genome size) could explain the conservation seen. If *A. dehalogenans* lost the need to develop, these genes would likely have been lost due to lack of selective pressure. If, however, the lineage leading to *A. dehalogenans* never had the ability to develop, these genes would never have been present in its ancestry. Importantly, in *H. ochraceum* and *S. cellulosum* only a few components of the MrpC/FruA/C-signal transduction pathway are conserved (Fig. 2.4D and E). From the information currently available, the only components in the MrpC/FruA/C-signal transduction shared by all fruiting myxobacteria are RelA, which is involved in the stringent response, the protease LonD, the transcriptional regulator MrpC, which is a homologue of the cyclic-AMP receptor (CRP) protein, ActC, which has homology to acetyltransferases, FmgC, which predicted to be an oxidase of a cytochrome P450 system, FdgA, which is involved in polysaccharide export, the response regulator FrzZ, and the small GTPase MglA (Fig. 2.4F). With the exception of FrzZ and MglA, all of these proteins are also widely distributed in non-myxobacteria and may be ancient components that are important in starvation and other stress response mechanisms. This more function-based approach to determine the conservation of the *M. xanthus* genetic programme for fruiting body formation in other fruiting myxobacteria suggests that the programme is conserved in the Cystobacterineae but not in the two other suborders. As discussed (Huntley *et al.*, 2011), this lack of conservation can be understood in two ways: The genetic programmes for fruiting body formation in myxobacteria evolved more than once within the order. In this scenario, the last common ancestor of the myxobacteria was not a fruiting bacterium, fruiting bodies are not homologous structures, and fruiting body formation in the myxobacteria is an example of convergent evolution. Alternatively, the last common ancestor of the myxobacteria was a fruiting bacterium. This programme would have been lost in *A. dehalogenans*, probably in parallel with a reduction in genome size. In the Cystobacterineae, the extant version of this programme would largely be shared because their divergence into separate species occurred relatively late. On the other hand, the *S. cellulosum* and the *H. ochraceum* lineages separated earlier from the Cystobacterineae lineage and during the course of evolution the initially shared programme may have diverged to such an extent that the overlap with the programmes in the Cystobacterineae is minor. According to this scheme, fruiting bodies are homologous structures generated from non-homologous proteins. To distinguish between these two scenarios, expression data over a developmental time-course from the other two Myxococcales suborders as well as more genetic and molecular data re required from species in the Nannocystineae and the Sorangiineae. Until such information is available, one may only guess the reasons the other suborders lack so much of the *M. xanthus* developmental programme.

Future of myxobacteria genomics

As of June, 2012, the NCBI lists four genome projects for which no data is currently available: two new *M. xanthus* genomes (strains DZ2, DZF1) by University of Iowa; and *Corallococcus (Myxococcus) macrosporus* and *Chondromyces apiculatus* by the Institute for Microbial Technology. In addition, MPI-Marburg will release the genome sequences of the two Cystobacterineae *Myxococcus stipitatus* and *M. macrosporus* by the end of 2012. Also, MPI-Marburg is currently planning genome projects for the two Cystobacterineae *Melittangium boletus* and *Cystobacter fulvus* as well as for the Nannocystineae *Nannocystis exedens* (Fig. 2.1B). These sequences should provide ever increasing refinement of our understanding of genome evolution and dynamics in the myxobacteria.

The genomes of *N. exedens* and *C. apiculatus* should provide valuable insight into the Nannocystineae and Sorangiineae. Gene conservation within those orders may give us clues into the differences discussed in the developmental process. By comparing the sequence regions of conserved genes, it may also be possible to improve the prediction for the 5′-ends of annotated genes, and may help in removing false positive and false negative (or pseudogene) gene calls from the annotations, especially as additional sequences from species in the less well represented Sorangiineae and Nannocystineae become available. This in turn will help in comparative analyses such as those described in this chapter.

The next several years should prove to be very exciting in regards to comparative genomics, both in the case of the Myxococcales and the broader world of (micro)biology. The continuous development of new, faster, cheaper, and more robust sequencing technologies is already leading to research possibilities, which were barely dreamt of a decade ago, such as re-sequencing of mutant genomes to identify mutations, or sequencing of complete cDNA pools (RNA-Seq) which can produce a high-quality, quantitative snapshot of gene expression in an organism. This increase in information should provide significant fuel for increased insight into how evolution within the Myxobacteria has affected individual components, complex processes, and even whole species and suborders.

References

Achaz, G., Coissac, E., Netter, P., and Rocha, E.P. (2003). Associations between inverted repeats and the structural evolution of bacterial genomes. Genetics *164*, 1279–1289.

Blattner, F.R., Plunkett, G., 3rd, Bloch, C.A., Perna, N.T., Burland, V., Riley, M., Collado-Vides, J., Glasner, J.D., Rode, C.K., Mayhew, G.F., *et al.* (1997). The complete genome sequence of *Escherichia coli* K-12. Science *277*, 1453–1462.

Chaba, R., Raje, M., and Chakraborti, P.K. (2002). Evidence that a eukaryotic-type serine/threonine protein kinase from *Mycobacterium tuberculosis* regulates morphological changes associated with cell division. Eur. J. Biochem. *269*, 1078–1085.

Cole, J.R., Wang, Q., Cardenas, E., Fish, J., Chai, B., Farris, R.J., Kulam-Syed-Mohideen, A.S., McGarrell, D.M., Marsh, T., Garrity, G.M., *et al.* (2009). The Ribosomal Database Project: improved alignments and new tools for rRNA analysis. Nucleic Acids Res. *37*, D141–145.

Curtis, P.D., Atwood III, J., Orlando, R., and Shimkets, L.J. (2007). Proteins associated with the *Myxococcus xanthus* extracellular matrix. J. Bacteriol. *189*, 7634–7642.

Dawid, W. (2000). Biology and global distribution of myxobacteria in soils. FEMS Microbiol. Rev. *24*, 403–427.

Dethlefsen, L., and Schmidt, T.M. (2007). Performance of the translational apparatus varies with the ecological strategies of bacteria. J. Bacteriol. *189*, 3237–3245.

Eisen, J.A., Heidelberg, J.F., White, O., and Salzberg, S.L. (2000). Evidence for symmetric chromosomal inversions around the replication origin in bacteria. Genome Biol. *1*, RESEARCH0011.

Fleischmann, R.D., Adams, M.D., White, O., Clayton, R.A., Kirkness, E.F., Kerlavage, A.R., Bult, C.J., Tomb, J.F., Dougherty, B.A., Merrick, J.M., *et al.* (1995). Whole-genome random sequencing and assembly of *Haemophilus influenzae* Rd. Science *269*, 496–512.

Garcia, R., Gerth, K., Stadler, M., Dogma, I.J., Jr., and Muller, R. (2010). Expanded phylogeny of myxobacteria and evidence for cultivation of the 'unculturables'. Mol. Phylogenet. Evol. *57*, 878–887.

Goldman, B.S., Nierman, W.C., Kaiser, D., Slater, S.C., Durkin, A.S., Eisen, J.A., Ronning, C.M., Barbazuk, W.B., Blanchard, M., Field, C., *et al.* (2006). Evolution of sensory complexity recorded in a myxobacterial genome. Proc. Natl. Acad. Sci. U.S.A. *103*, 15200–15205.

Goldman, B., Bhat, S., and Shimkets, L.J. (2007). Genome evolution and the emergence of fruiting body development in *Myxococcus xanthus*. PLoS One *2*, e1329.

Hedges, S.B., Dudley, J., and Kumar, S. (2006). TimeTree: a public knowledge-base of divergence times among organisms. Bioinformatics *22*, 2971–2972.

Huntley, S., Hamann, N., Wegener-Feldbrügge, S., Treuner-Lange, A., Kube, M., Reinhardt, R., Klages, S., Müller, R., Ronning, C.M., Nierman, W.C., *et al.* (2011). Comparative genomic analysis of fruiting body formation in Myxococcales. Mol. Biol. Evol. *28*, 1083–1097.

Huntley, S., Zhang, Y., Treuner-Lange, A., Kneip, S., Sensen, C.W., and Sogaard-Andersen, L. (2012). Complete genome sequence of the fruiting myxobacterium *Corallococcus coralloides* DSM 2259. J. Bacteriol. *194*, 3012–3013.

Inouye, S., and Nariya, H. (2008). Dual regulation with Ser/Thr kinase cascade and a His/Asp TCS in *Myxococcus xanthus*. Adv. Exp. Med. Biol. *631*, 111–121.

Jaillon, O., Aury, J.M., and Wincker, P. (2009). 'Changing by doubling', the impact of Whole Genome Duplications in the evolution of eukaryotes. C. R. Biol. *332*, 241–253.

Kahnt, J., Aguiluz, K., Koch, J., Treuner-Lange, A., Konovalova, A., Huntley, S., Hoppert, M., Søgaard-Andersen, L., and Hedderich, R. (2010). Profiling the outer membrane proteome during growth and development of the social bacterium *Myxococcus xanthus* by selective biotinylation and analyses of outer membrane vesicles. J. Proteome Res. *9*, 5197–5208.

Kellis, M., Birren, B.W., and Lander, E.S. (2004). Proof and evolutionary analysis of ancient genome duplication in the yeast *Saccharomyces cerevisiae*. Nature *428*, 617–624.

Kirby, J.R., and Zusman, D.R. (2003). Chemosensory regulation of developmental gene expression in *Myxococcus xanthus*. Proc. Natl. Acad. Sci. U.S.A. *100*, 2008–2013.

Klappenbach, J.A., Dunbar, J.M., and Schmidt, T.M. (2000). rRNA operon copy number reflects ecological strategies of bacteria. Appl Environ. Microbiol. *66*, 1328–1333.

Koonin, E.V., and Wolf, Y.I. (2008). Genomics of bacteria and archaea: the emerging dynamic view of the prokaryotic world. Nucleic Acids Res. *36*, 6688–6719.

Kuo, C.H., Moran, N.A., and Ochman, H. (2009). The consequences of genetic drift for bacterial genome complexity. Genome Res. *19*, 1450–1454.

Li, Z.-F., Li, X., Liu, H., Liu, X., Han, K., Wu, Z.-H., Hu, W., Li, F.-f., and Li, Y.-Z. (2011). Genome sequence of the halotolerant marine bacterium *Myxococcus fulvus* HW-1. J. Bacteriol. *193*, 5015–5016.

Loughney, K., Lund, E., and Dahlberg, J.E. (1983). Deletion of an rRNA gene set in *Bacillus subtilis*. J. Bacteriol. *154*, 529–532.

Luciano, J., Agrebi, R., Le Gall, A.V., Wartel, M., Fiegna, F., Ducret, A., Brochier-Armanet, C., and Mignot, T. (2011). Emergence and modular evolution of a novel motility machinery in bacteria. PLoS Genet. *7*, e1002268.

Mira, A., Ochman, H., and Moran, N.A. (2001). Deletional bias and the evolution of bacterial genomes. Trends Genet. *17*, 589–596.

Mrazek, J., and Karlin, S. (1998). Strand compositional asymmetry in bacterial and large viral genomes. Proc. Natl. Acad. Sci. U.S.A. *95*, 3720–3725.

Müller, F.-D., Treuner-Lange, A., Heider, J., Huntley, S., and Higgs, P. (2010). Global transcriptome analysis of spore formation in *Myxococcus xanthus* reveals a locus necessary for cell differentiation. BMC Genomics *11*, 264.

Natalia, I., Chris, D., Elke, L., Birte, A., Marcus, K., Elizabeth, S., Alla, L., Susan, L., Tijana Glavina Del, R., Matt, N., *et al.* (2010). Complete genome sequence of *Haliangium ochraceum* type strain (SMP-2T). Stand Genomic Sci. *2*, 96–106.

Ochman, H., and Davalos, L.M. (2006). The nature and dynamics of bacterial genomes. Science *311*, 1730–1733.

Ohno, S. (1970). Evolution by gene duplication (Berlin: Springer-Verlag).

Perez, J., Castaneda-Garcia, A., Jenke-Kodama, H., Muller, R., and Munoz-Dorado, J. (2008). Eukaryotic-like protein kinases in the prokaryotes and the myxobacterial kinome. Proc. Natl. Acad. Sci. U.S.A. *105*, 15950–15955.

Punta, M., Coggill, P.C., Eberhardt, R.Y., Mistry, J., Tate, J., Boursnell, C., Pang, N., Forslund, K., Ceric, G., Clements, J., *et al.* (2012). The Pfam protein families database. Nucleic Acids Res. *40*, D290–301.

Rainey, F.A., Ward-Rainey, N.L., Janssen, P.H., Hippe, H., and Stackebrandt, E. (1996). *Clostridium paradoxum* DSM 7308T contains multiple 16S rRNA genes with heterogeneous intervening sequences. Microbiology *142 (Pt 8)*, 2087–2095.

Reichenbach, H. (1999). The ecology of the myxobacteria. Environ. Microbiol. *1*, 15–21.

Reichenbach, H. (2005). Order VIII. Myxococcales Tchan, Pochon and Prevot 1948, 398. In Bergey's Manual of Systematic Bacteriology, 2nd Edition, D.R. Boone, R.W. Castenholz, G.M. Garrity, D.J. Brenner, N.R. Krieg, and S.J.T., eds. (New York: Springer), pp. 1059–1144.

Rocha, E.P. (2003). An appraisal of the potential for illegitimate recombination in bacterial genomes and its consequences: from duplications to genome reduction. Genome Res. *13*, 1123–1132.

Rocha, E.P. (2004). Order and disorder in bacterial genomes. Curr. Opin. Microbiol. *7*, 519–527.

Schneiker, S., Perlova, O., Kaiser, O., Gerth, K., Alici, A., Altmeyer, M.O., Bartels, D., Bekel, T., Beyer, S., Bode, E., *et al.* (2007). Complete genome sequence of the myxobacterium *Sorangium cellulosum*. Nat. Biotechnol. *25*, 1281–1289.

Semon, M., and Wolfe, K.H. (2007). Consequences of genome duplication. Curr. Opin. Genet. Dev. *17*, 505–512.

Shi, X., Wegener-Feldbrugge, S., Huntley, S., Hamann, N., Hedderich, R., and Sogaard-Andersen, L. (2008). Bioinformatics and experimental analysis of proteins of two-component systems in *Myxococcus xanthus*. J. Bacteriol. *190*, 613–624.

Shimkets, L.J., Dworkin, M., and Reichenbach, H. (2006). The Myxobacteria. In The prokaryotes: a Handbook on the Biology of Bacteria, M. Dworkin, S. Falkow, E. Rosenberg, K.-H. Schleifer, and E. Stackenbrandt, eds. (New York; London: Springer).

Staron, A., Sofia, H.J., Dietrich, S., Ulrich, L.E., Liesegang, H., and Mascher, T. (2009). The third pillar of bacterial signal transduction: classification of the extracytoplasmic function (ECF) sigma factor protein family. Mol. Microbiol. *74*, 557–581.

Stock, A.M., Robinson, V.L., and Goudreau, P.N. (2000). Two-component signal transduction. Ann. Rev. Biochem. *69*, 183–215.

Tatusov, R.L., Galperin, M.Y., Natale, D.A., and Koonin, E.V. (2000). The COG database: a tool for genome-scale analysis of protein functions and evolution. Nucleic Acids Res. *28*, 33–36.

Thomas, S.H., Wagner, R.D., Arakaki, A.K., Skolnick, J., Kirby, J.R., Shimkets, L.J., Sanford, R.A., and Loffler, F.E. (2008). The mosaic genome of *Anaeromyxobacter dehalogenans* strain 2CP-C suggests an aerobic common ancestor to the delta-proteobacteria. PLoS One 3, e2103.

Tillier, E.R., and Collins, R.A. (2000). Genome rearrangement by replication-directed translocation. Nat. Genet. *26*, 195–197.

Ulrich, L.E., and Zhulin, I.B. (2010). The MiST2 database: a comprehensive genomics resource on microbial signal transduction. Nucleic Acids Res. 38, D401–407.

Ulrich, L.E., Koonin, E.V., and Zhulin, I.B. (2005). One-component systems dominate signal transduction in prokaryotes. Trends in Microbiol. *13*, 52–56.

Willett, J.W., and Kirby, J.R. (2011). CrdS and CrdA comprise a two-component system that is cooperatively regulated by the Che3 chemosensory system in *Myxococcus xanthus*. MBio *2*.

Wuichet, K., and Zhulin, I.B. (2010). Origins and diversification of a complex signal transduction system in prokaryotes. Sci. Signal 3, ra50.

Zorina, A., Stepanchenko, N., Novikova, G.V., Sinetova, M., Panichkin, V.B., Moshkov, I.E., Zinchenko, V.V., Shestakov, S.V., Suzuki, I., Murata, N., *et al.* (2011). Eukaryotic-like Ser/Thr protein kinases SpkC/F/K are involved in phosphorylation of GroES in the Cyanobacterium *synechocystis*. DNA Res. *18*, 137–151.

Zusman, D.R., Scott, A.E., Yang, Z., and Kirby, J.R. (2007). Chemosensory pathways, motility and development in *Myxococcus xanthus*. Nat. Rev. Microbiol. *5*, 862–872.

Myxococcus xanthus Vegetative and Developmental Cell Heterogeneity

3

Penelope I. Higgs, Patricia L. Hartzell, Carina Holkenbrink and Egbert Hoiczyk

Abstract

The myxobacteria have long been considered model organisms for complex multicellular behaviour and Gram-negative differentiation. To effectively compete for nutrients in soil or aquatic environments, these bacteria have evolved multicellular hunting strategies and phenotypic heterogeneity within the population, such as phase variation and cell clustering, which likely contribute to the success of these organisms. To survive periods of nutrient limitation, most myxobacteria enter a complex developmental programme culminating in the formation of multicellular fruiting bodies filled with environmentally resistant spores. The developmental programme of *Myxococcus xanthus*, the model myxobacterium, involves segregation of the starving population into at least three distinct fates: aggregation into multicellular mounds followed by differentiation into spores, differentiation into a persister-like state termed peripheral rods, and developmentally induced cell lysis. This chapter describes what is currently known about the role of phenotypic heterogeneity and developmental cell fate differentiation in the *M. xanthus* life cycle. We describe the physiological characteristics as well as regulatory mechanisms involved in generating phenotypic heterogeneity and cell fate differentiation during both vegetative and developmental stages of these remarkably adaptable bacteria.

Introduction

Phenotypic heterogeneity (i.e. subpopulations with distinct features within a genetically identical population) is now recognized as a widespread phenomenon thought to provide an evolutionary advantage in survival of a species as a whole (Lopez *et al.*, 2009). In some cases, certain cells within a clonal population are primed with characteristics which enable survival of the species in the event of sudden environmental assaults. Many bacterial populations, for example, contain a subpopulation of dormant persister cells which can survive in the presence of certain antibiotics (Lewis, 2007). In other cases, the population benefits by 'division of labour' in which certain cells specialize to perform a specific function for the benefit of the entire population. In filamentous chains of *Anabaena* sp., some cells differentiate into heterocysts (specialized micro-aerobic nitrogen-fixing cells) which provide nitrogen to the remaining vegetative cells. In *B. subtilis,* some cells become 'minors' which specialize to secrete proteases (an energetically expensive process) that digest extracellular proteins to provide a nutrient source for the population (Lopez *et al.*, 2009). Under conditions where multicellular behaviour is favoured (e.g. biofilms), phenotypic heterogeneity may arise simply because spatial positioning within the community causes cells to respond to differences in the surrounding microenvironment. In more complex cases, bacteria have evolved dedicated regulatory programmes leading to the segregation of cells into distinct differentiated fates (Lopez *et al.*, 2010; Vlamakis *et al.*, 2008). The myxobacteria, an order within the deltaproteobacteria, display an amazing array of both phenotypic heterogeneity and programmed differentiation during their complex life cycles.

A hallmark feature of the myxobacteria is the prominent role of group behaviour during both vegetative and developmental phases of their complex life cycle. Most of the myxobacteria are so-called 'wolf-pack' feeders in which groups of cells swarm in search of prey microorganisms (or decaying organic material); these swarms paralyse and digest their prey by the collective secretion of a spectrum of antibiotics and digestive enzymes. As ubiquitous soil or aquatic organisms, the myxobacteria are not only subject to competition for resources, but must also be able to survive rapid changes in the local environment. Recent insights into both phase variation and cell cluster formation in *Myxococcus xanthus* suggest that heterogeneity within the population of vegetative cells may increase overall survival by division of labour and primed resistance (i.e. exhibiting characteristics which could enhance survival in the face of rapid environmental change). Under conditions of nutrient depletion, most of the myxobacteria also enter an elaborate developmental programme which culminates in the formation of multicellular fruiting bodies filled with environmentally resistant spores. This pathway has been best characterized in *M. xanthus* (see Chapter 6 for a detailed description of regulatory mechanisms of the developmental programme) (Fig. 3.1). During this programme, cells are segregated into one of at least three distinct cell fates: (1) cell lysis, (2) differentiation into environmentally resistant spores (which are produced within multicellular fruiting bodies), and (3) differentiation into peripheral rods (a persister-like state) (Fig. 3.1). Furthermore, phase-variation and cell clustering appear to play distinct roles during the developmental programme.

The aim of this chapter is to review what is currently understood about the role heterogeneity plays in the life cycle of the model myxobacterium, *M. xanthus*. We first address two known sources of heterogeneity within vegetative cells (phase variation and cell clusters) and review what is currently understood about their characteristics, their respective roles in the vegetative and developmental phases of the *M. xanthus* life cycle, and the regulatory mechanisms which govern production of each cell type. We next describe cell types specific to the developmental programme, including formation of fruiting bodies (and the spatial heterogeneity within), peripheral rods, and developmental cell lysis. Finally, we include a detailed review of the best characterized cell fate: the unique mechanism of *M. xanthus* spore differentiation. Open questions with respect to specific cell types are addressed at the end of each relevant section.

Phase variation

Phase variation is a manifestation of phenotypic plasticity that allows a single bacterial species to persist in alternate forms by expressing different sets of genes. Typically, each of the alternative genetic programmes results in production of different combinations of proteins, lipids, or carbohydrates on the cell surface that contribute to phase-specific changes in colony texture, colour or morphology (van der Woude and Baumler,

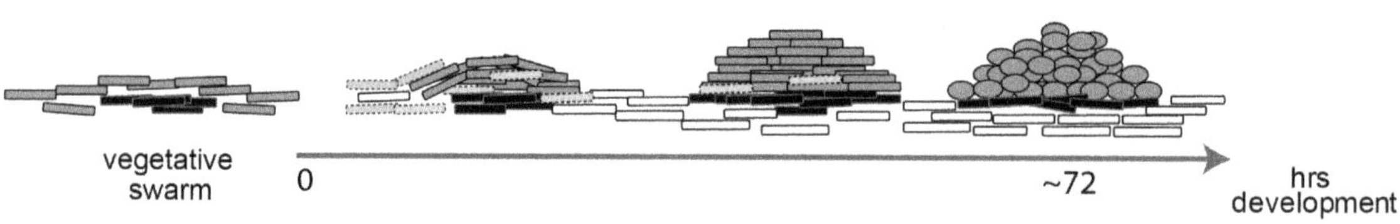

Figure 3.1 Schematic depicting cell heterogeneity and differentiation during the life cycle of *M. xanthus*. A vegetative swarm consists of rod shaped cells (grey rods), cell clusters (black rods), and tan phase variants (not shown). Upon sensing nutrient limitation (0h development), cells are induced to migrate into aggregation centres. Immediately prior to, or concomitant with aggregation, some cells lyse (grey dashed cells); lysis continues inside of the fruiting bodies. Cell clusters may form the base of the aggregation centre. Cells destined to become peripheral rods (white rods) do not aggregate and remain outside of the fruiting body. Inside the aggregation centres, cells differentiate into spherical environmentally resistant spores (grey spheres).

2004). Much is known about phase variation in pathogens because the ability to switch phases can contribute significantly to the virulence of an organism. Pathogenic strains that undergo phase variation produce cell types that typically evade recognition by the immune system, thus affording these pathogens enhanced opportunities to escape attack and survive in the host.

The role of phase variation in non-pathogens has been studied less thoroughly, yet evidence suggests that phase variation contributes to the environmental fitness of bacteria. Wild isolates and laboratory strains of *M. xanthus* undergo phase variation to produce colonies that are designated as yellow or tan. Although the underlying mechanism is unknown, the yellow variants can 'switch' to tan and vice versa; hence this reversible process is not due to a heritable mutation. Phase variation is complex in *M. xanthus* because it affects myriad properties including swarming, pigmentation, fruiting body formation, and sporulation (Wireman and Dworkin, 1975b). The yellow variants are the predominant form and are favoured in research for their robust phenotypes (described below). This section will summarize the current understanding of characteristics of yellow and tan variants, the relationship between the two cell types, and the environmental signals that stimulate phase switching.

Phase variation in *Myxococcus xanthus* generates two interactive cell types

While pathogenic bacteria use phase variation to evade host defences, phase variation in *M. xanthus* appears to have evolved as a mechanism to deploy two symbiotic cell types aimed at ensuring survival of the population. Yellow and tan colonies do not represent pure populations of either variant type, but rather are mixtures. This has been determined by inoculating individual yellow and tan colonies into liquid and directly plating for new colony formation from isolated cells. These data revealed that yellow colonies contain about 75–99% yellow and 1–25% tan cells. The percentages are reversed in the colonies of tan variants (Laue and Gill, 1995). The rate of switching from yellow to tan has been reported to be 10^{-2} to 10^{-3} per cell per generation; switching from tan to yellow is higher, so colonies of the yellow variant (the predominant cell type, variant or morphotype) outnumber those of the tan variant (Laue and Gill, 1994). The persistence of the yellow–tan ratio suggests that both cell types are needed for survival (Burchard and Dworkin, 1966).

Colonies are either yellow or tan because of the production of a yellow pigment called DKxanthene (DKX) (Meiser *et al.*, 2006). Pigmentation is only one of the visible features that distinguish phase variants. Two additional visual characteristics are swarm size and texture. [Surface translocation of a group of motile *M. xanthus* cells has historically been called swarming, although technically the term swarming refers to surface movement that involves flagella (Kearns, 2010). Another suggested name for group movement by a gliding organism is 'spreading', but until an alternative term is adopted, we will continue to use swarming in this chapter.] Swarming can be quantified as the rate of expansion of a group of cells over an agar surface. For example, an aliquot of $\approx 10^7$ cells (4 µl of 5×10^9 cells/ml) form a spot about 4 mm in diameter on an agar surface; if cells are swarm-proficient (Swr++), they can grow and swarm to about 30 mm in diameter in 4 days. A swarm-deficient strain might expand to only 15–20 mm in the same time period. WT-Y variants are proficient at swarming on agar surfaces [Swr++], and WT-T variants exhibit reduced swarming [Swr(+)]. One exception is the *dkxG* mutant, which is tan due to a mutation in the DKX biosynthesis pathway. This mutant produces both Swr++ and Swr(+) colony phenotypes (Furusawa *et al.*, 2011).

Phase variant colonies also have noticeable different surface properties; in general, colonies are predominantly either 'dull and textured' or 'shiny and smooth' (Fig. 3.2A and B). Colonies of wild-type yellow (WT-Y) variants look dull and have a richly textured surface, while colonies of wild-type tan (WT-T) variants generally look shiny and have a smooth surface. A few (<5%) dull WT-T colonies arise on occasion, but shiny WT-Y colonies are very rare. Some stable tan mutants may provide tools to understand the molecular differences between shiny and dull colony phenotypes. For example, the *asgB* mutant, first characterized as deficient in a

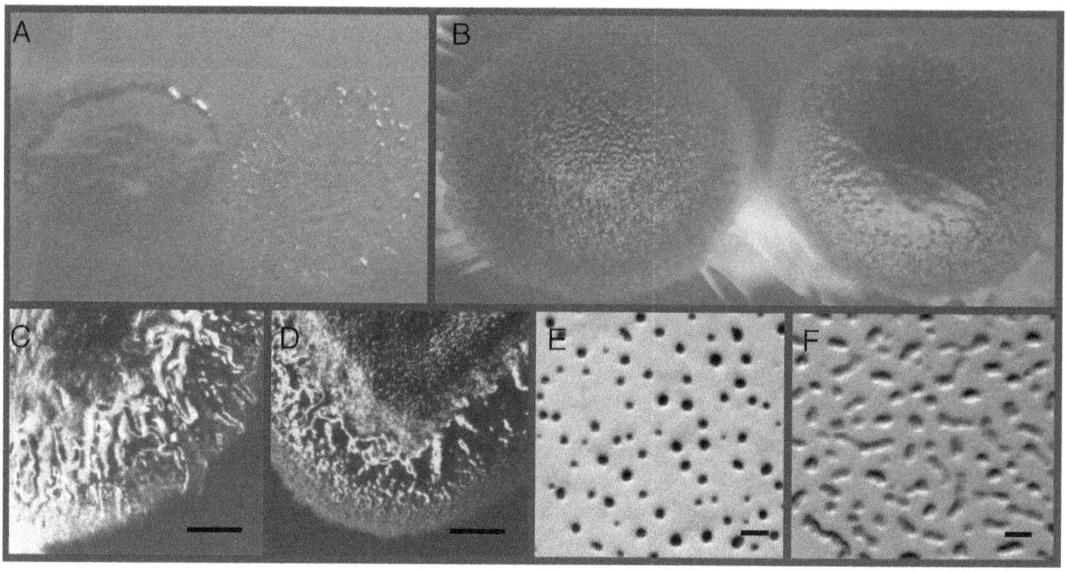

Figure 3.2 *Myxococcus* phase variation produces yellow and tan colonies. (A) Cells plated at high density show yellow (right) and tan (left) phenotypes. (B) Swarm spots (diameter 25mm) with dull matte (left) and shiny smooth (right) surface texture (Nikon coolpix image); Colony edge of a yellow (C) and tan (D) variant on CTPM agar medium (Nikon SMZ-U image toplit with a fiberoptic light; bar=2mm; Fruiting bodies after 24h at 32°C on TPM (starvation) agar for yellow variant (E) and tan variant (F); bar=0.5mm. A colour version of this figure is located in the plate section at the back of the book.

population density sensing during development, produces a stable tan phenotype with colonies that have a shiny surface (Mayo and Kaiser, 1989). In contrast, disruption of MXAN_0228 (encoding a HTH-Xre regulator; described below) produces a tan colony with a dull surface.

Rosenberg *et al.* (1977) reported that yellow and tan variants exhibit different density-dependent growth and swarming rates. At that time, it was not tested whether both single cell motility (A-motility; see Chapter 9) and group motility (S-motility: see Chapter 10), which are movements powered by distinct motility engines, are equally effected, because the density dependent swarm assays were performed only on a 1.5% agar surface, where A-motility predominates. Thus, the cell-density experiments have been recently re-examined. Swarming of WT-T is less than half that of WT-Y on 1.5% or 0.3% agar surfaces (Shi and Zusman, 1993) which favour A- and S-motility, respectively (Furusawa *et al.*, 2011). Time-lapse video microscopy, which provides a measure of gliding motor function and reversal frequency, did not reveal significant differences in the rate of A-motility gliding (10.5 μm/min WT-Y versus 8.9 μm/min WT-T) or reversal period (9.8 min/rev WT-Y and 11.2 min/rev WT-T) on 1.5% agar pads. Thus, the reduced ability of the tan variant to swarm at high cell density on 1.5% agar cannot be accounted for by changes in the A-motility motor (Furusawa *et al.*, 2011).

Furthermore, high cell densities (10^9 cells/ml) are critical for swarming for both WT-T and WT-Y on both agar surfaces. The effect is especially dramatic on 0.3% agar where S-motility (which requires cell–cell contact) predominates, because even a 10-fold decrease in initial cell density results in a dramatic (60%) decrease in swarming. Surprisingly, when the initial inoculum was at low cell density, swarms from both WT-T and WT-Y on both agar surfaces did not expand, even with prolonged incubation (5–7 days). After 5 days, the size of a swarm was smaller than the size of a typical isolated colony for each strain on a streak plate. Cells within the swarm were still viable, but did not move beyond the boundary of the swarm. This was an unexpected outcome of the swarm density-dependence experiments and suggests that there is a density threshold for swarming. It also suggests that gliding can be suppressed even under conditions that usually promote swarming.

Phase variation plays a role in development

Soil bacteria such as *M. xanthus* have evolved strategies to cope with the inevitable hardships of life in the wild. As described above, when faced with starvation conditions, *M. xanthus* cells undergo a complex developmental process to produce spores that are capable of surviving many unfavourable conditions (Harris *et al.*, 1998; Kaiser, 2003; Kim and Kaiser, 1990; Laue and Gill, 1995; Rolbetzki *et al.*, 2008; Shimkets, 1999; Sogaard-Andersen, 2004; Sogaard-Andersen *et al.*, 2003). When WT-Y are starved for nutrients, they aggregate and form fruiting bodies and spores. The response of WT-T to starvation is diminished in several ways. First, they fail to form dark mature fruiting bodies that are characteristic of the WT-Y. Second, they produce fewer spores. As shown in Fig. 3.2E, fruits of WT-Y are dense and dark, whereas the WT-T forms long ridged mounds that are lighter in colour. There is considerable variation in morphogenesis for reasons that are not yet apparent. WT-T generally forms aggregates (soft mounds); some form structures that resemble pale fruiting bodies. The pale colour is indicative of the lack of refractile spores. WT-Y cultures form 10–1000 times more heat resistant spores than the WT-T.

For many years, myxobiologists had noted that the number of heat-resistant spores recovered from a developing sample was similar to the number of tan cells that typically constituted the initial cell population. For example, the percentage of spores recovered from a starved WT culture was often about 10% of the input number, which correlated with the percentage of tan cells (5–10%) in that population. This led to the hypothesis that tan cells might be the progenitors of spores, but it was some time before this hypothesis could be tested. A study from the lab of Ron Gill provided the first genetic evidence that tan and yellow cells each play unique roles in the formation of heat-resistant spores (Laue and Gill, 1994, 1995). Using mixtures of antibiotic-tagged tan and yellow cells, Laue and Gill found that tan cells contribute disproportionately to the population of heat-resistant spores recovered from mixtures containing tan and yellow cells. The tan derivative used in the Gill study was a 'tan-locked' mutant (the molecular basis of which remains undetermined) that displayed a reduced frequency of phase variation. Since the tan cells produce only low numbers of spores when developed in the absence of yellow cells, this result suggested that yellow cells may provide some factor that the tan cells need in order to produce viable spores. Recent studies in the Dahl lab have confirmed the findings of Laue and Gill. These studies also show that tan cells are capable of forming spores through the glycerol induction pathway, which by-passes the fruiting body pathway (Dahl *et al.*, 2011). Interestingly, tan spores formed in this manner are more resistant to assault by sonication or lysozyme than yellow glycerol-induced spores.

Development in *M. xanthus* has often been compared with development of the eukaryotic slime mould *Dictyostelium*. The similarities of yellow and tan cells to stalk and spore cells that form during differentiation in *Dictyostelium* (Mohanty and Firtel, 1999; Weijer, 2004) reinforces this idea. One of the factors that yellow cells might 'feed' to tan variants to promote spore formation is DKX (Meiser *et al.*, 2006). *dkx* mutants are tan in colour and produce few heat-resistant spores (Meiser *et al.* 2006; Furusawa, 2011). Addition of purified DKX was able to partially rescue the sporulation defect of a *dkx* mutant (Meiser *et al.*, 2006). The mechanism underlying the ability of DKX to rescue the tan variants is a mystery at this time. The structure of DKX shows that it is an amphiphilic molecule; the lipophilic polyene side chain suggests that DKX could reside in the membrane, an idea supported by work in the Müller lab. As such, DKX could play a novel role in membrane stability or signal transduction.

Molecular methods reveal differences between tan and yellow variants

Similar to other microbes, phase switching in *M. xanthus* is regulated by factors in the environment. Exposure to UV and elevated temperatures (>35°C) appear to stimulate recovery of yellow cells in a population, suggesting that these environmental factors can regulate the tan-to-yellow switch (Rosenberg *et al.*, 1977). It was previously unclear whether these conditions selected for yellow variants by killing tan cells or whether they induced tan cells to convert to yellow, but more recent studies support the idea that UV exposure induces a tan-to-yellow switch (Furusawa *et al.*, 2011). Recently, DNA microarrays revealed that iron availability also regulates phase variation. mRNAs were prepared from WT-Y and WT-T variant populations grown under standard conditions while preventing exposure to UV or temperature shifts. Statistically significant data were collected for genes whose expression was increased in yellow cultures relative to tan cultures and genes whose expression was increased in tan cultures relative to yellow cultures (Furusawa *et al.*, 2011). As expected, yellow phase variants showed dramatic changes in expression of genes in operons required for synthesis of DKX. Disruption of the MXAN_4299 (*dkxG*) gene in this pathway blocks synthesis of the yellow pigment, but the mutant is still able to switch between a Swr++ and Swr(+) state, indicating that pigment production can be uncoupled from other phase-associated phenotypes such as swarming (Furusawa *et al.*, 2011).

Also increased in WT-Y variants were nine genes in the biosynthesis pathway of antibiotic TA (myxovirescin), a macrolactam lactone that targets the type II signal peptidase that processes pro-lipoproteins (Xiao *et al.*, 2012) and plays a critical role in the ability of *M. xanthus* cells to prey on other microbes (Xiao *et al.*, 2011). Moreover, a number of serine-threonine kinases or 'pseudo' kinases were also differentially regulated in yellow variants, raising the possibility that a kinase cascade may be part of the phase switch.

A gene encoding a helix–turn–helix xenobiotic response element (HTH-Xre, MXAN_0228) DNA binding protein is increased specifically in yellow variants (Furusawa *et al.*, 2011). Deletion of HTH-Xre results in a mutant that is tan (DKX$^-$) similar to a WT-T, yet swarming and fruiting phenotypes of the HTH-Xre mutant more closely resemble the WT-Y. The expression profile for the HTH-Xre mutant confirms this observation. RT-PCR analysis shows that expression of *dkx* genes is decreased in the HTH-Xre mutant, but expression of myxovirescin genes is similar to the WT-Y. HTH-Xre may be subject to autoregulation, since its expression is increased in the ΔMXAN_0228 (HTH-Xre) mutant.

The unique set of genes expressed in WT-T provides insight into the biology of phase variation. Expression of genes involved in the acquisition of iron was increased in tan variants (Hartzell, unpublished). These include genes involved in production of the siderophore myxochelin, a number of TonB homologues that likely energize outer membrane receptors, a FepA homologue for transport of iron-containing myxochelin across the outer membrane, haemin-binding and transport proteins, and efflux proteins. The data suggest that iron availability plays a role in the regulation of phase variation. The distribution of labour, with yellow cells specializing in producing DKX and myxovirescin and tan cells specializing in iron uptake, suggests that production of components such as the siderophore myxochelin may be incompatible with that of the secondary metabolites DKX and myxovirescin. If this is the case, phase variation may have evolved as a strategy to accommodate critical specialized functions in individual cells that promote colonial growth.

Cell clusters

An additional source of heterogeneity in vegetative *M. xanthus* cells is the observation that cells growing on surfaces can be separated on the basis of differential centrifugation into those that are found in tight aggregates *versus* those that are not (Lee *et al.*, 2011, 2012). The population of aggregated cells, which represents (under the conditions tested) 25% of the total population, has been termed cell clusters to distinguish them from cells which move into aggregation centres during the developmental programme (Fig. 3.1). These cell clusters display a number of characteristics which are different from the cells which do not sediment. First, the cell clusters appear to be encased by an extracellular polymeric substance (EPS) layer such that the average per cell EPS is approximately 3-fold higher than that of the non-sedimented cells (Berleman *et al.*, 2011; Lee *et al.*, 2012). Consistently, the predominant EPS-associated protein – the FibA metalloprotease (Kearns *et al.*, 2002) – is found exclusively in cell clusters (Lee *et al.*, 2012). It was also observed that the FrzCD methyl accepting chemosensory protein (MCP), which plays an important role in regulating *M. xanthus* motility (Blackhart and Zusman, 1985; McBride and Zusman, 1993), is more highly methylated in cell clusters. A possible link between these observations may be FrzS, a pseudo-response regulator which is thought to be a component of the Frz chemosensory pathway (Ward *et al.*, 2000). FrzS has been shown to be necessary for appropriate EPS production and *frzS* mutants appear to make fewer cell clusters (Berleman *et al.*, 2011). Additionally, Dif, a chemosensory pathway which somehow functions in a signalling pathway with FibA (Bonner *et al.*, 2006), and influences the methylation state of FrzCD (Xu *et al.*, 2008), regulates the production of EPS (Yang *et al.*, 2000).

After induction of starvation, at least until the onset of aggregation, cell clusters can still be isolated by differential centrifugation. Analysis of several developmental regulators and marker proteins suggests that cell clusters likely follow a different developmental programme (Lee *et al.*, 2012). For instance, in cell clusters, the major developmental regulators, MrpC and FruA, do not accumulate as rapidly nor to the same levels as in cells which do not sediment. In contrast, the accumulation of C-signal (CsgA p17), an important cell contact-dependent signalling protein (Kruse *et al.*, 2001), is similar in the two populations. The precise implication of the differential accumulation of these developmental regulators is currently not known, and awaits cell specific markers to track the fate of cell clusters after the onset of aggregation.

The observation that cell clusters represent a significant source of molecular heterogeneity is only recent, and the role, if any, that these cells play in the *M. xanthus* life cycle is not understood. During vegetative conditions, cell clusters may simply better weather changes in the environment in much the same way that epiphytic aggregates of *Pseudomonas syringae* withstand desiccation more readily than single cells (Monier and Lindow, 2003, 2004). During developmental conditions, several functions for cell clusters can be envisaged including, serving as a platform upon which aggregating cells build fruiting bodies (Curtis *et al.*, 2007), acting as a spacing determinant to control appropriate spacing between fruiting bodies (Xie *et al.*, 2011), or simply functioning as the *M. xanthus* version of the more elaborate stalks observed in related species, such as *M. stipitatis* (Reichenbach, 1993). One possibility which should be investigated is that cell clusters are related to phase variation (discussed above). These investigations await the identification of cell cluster-specific markers which would allow in situ single cell analysis throughout the developmental programme. Importantly, recognition that cell clusters are enriched in the accumulation or activation of several proteins involved in motility (FrzCD, FrzS), chemotaxis (FibA), and EPS production suggests the function of these proteins, most of which have been investigated either in the whole population or in isolated cells, should be reinvestigated in the context of cell clusters.

Fruiting bodies

To date, all known myxobacteria (with the possible exception of the *Anaeromyxobacter* species) produce multicellular spore-filled fruiting bodies under conditions of nutrient depletion. It is assumed that fruiting bodies enable dispersal of groups of spores to nutrient rich environments and/or increase resistance to environmental challenges (see Chapter 1 for further discussion on this topic). Germination of groups of spores would facilitate both cooperative feeding (Rosenberg *et al.*, 1977) and social (group) motility (Dworkin and Kaiser, 1985) of the emerging colony. Production of spore-filled fruiting bodies is directed by the starvation-induced developmental programme which induces certain cells in the population to aggregate into mounds of approximately 10^5–10^6 cells, and, exclusively within these mounds, differentiate into environmentally resistant spores (Fig. 3.1). Fruiting body production requires a solid surface, a threshold population density, motility, and a complex series of inter- and intracellular signals enabling the cells to be coordinated in space and time (reviewed in Shimkets, 1999). A detailed description of the genetic programme directing fruiting body formation is the subject of Chapter 6; the following section will focus only on the architecture and heterogeneity within the developing fruiting bodies.

Structure and overall architecture of *Myxococcus xanthus* fruiting bodies

Formation of multicellular fruiting bodies has been best studied in *M. xanthus*. Morphologically, within 24 h of the developmental programme (depending on both the strain and the degree of starvation), small foci of aggregated cells can be observed (Kuner and Kaiser, 1982; O'Connor and Zusman, 1989). Over the course of the next several hours, these initial foci may disintegrate, merge or increase in size to become larger aggregates. These aggregates eventually develop into mound-shaped mature spore-filled fruiting bodies, 0.1–0.2 mm wide and about 0.04–0.1 mm high (Fig. 3.3A, B

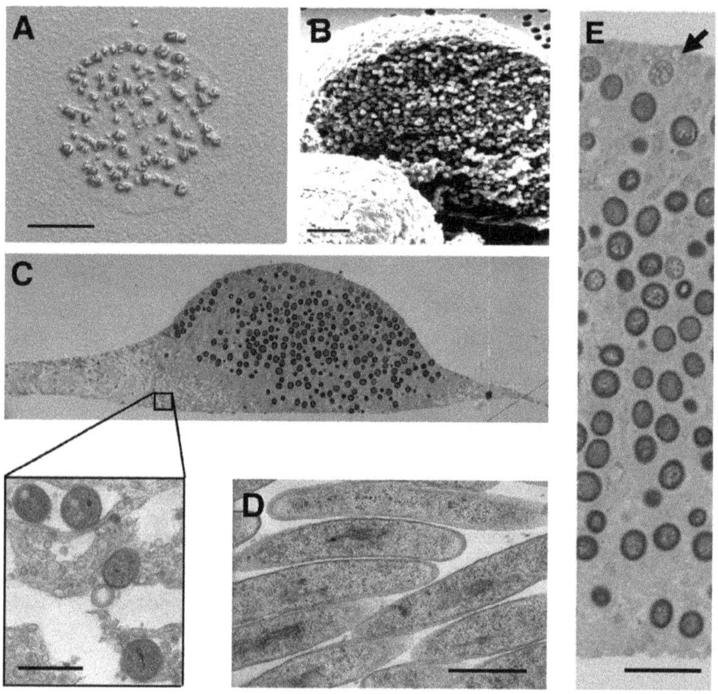

Figure 3.3 Light, scanning, and transmission electron micrographs of *M. xanthus* fruiting bodies. (A) Stereomicroscope image of *M. xanthus* fruiting bodies after 48h on nutrient poor agar. Mature mound-shaped fruiting bodies are approximately 0.1–0.2mm wide and 0.04–0.1mm high. (B) Scanning electron microscopic picture of a mature fruiting body that has been sliced open to reveal the densely packed spores in the interior of the structure. Note that due to the drying and shrinkage of the structure during preparation, the spores appear to be more densely packed than in the corresponding picture of a resin-embedded fruiting body. (C) Cross-section of a high pressure-frozen and resin-embedded 48h old fruiting body. The interior of the fruiting body is mostly filled with developing spores that contain lipid bodies signalling their commitment to the developmental pathway that eventually will result in the formation of mature fruiting body spores. In contrast, the cells at the base and perimeter of the fruiting body usually lack lipid bodies and signs of developmental differentiation (see also cells at higher magnification in the boxed picture on the lower left, the column on the right, and in picture D) showing the vegetative cell-like architecture characteristic of peripheral rods and/or cell clusters. A complex matrix formed of slime, cellular debris, and proteins fills the space between the cells in the fruiting body. As this matrix is not dried during the preparation of the structure, the spores appear further apart than in the scanning electron micrograph of picture (B). It is important to note that this arrangement is probably closer to life than the one visualized in the scanning electron micrograph. (D) Isolated and cryo-prepared peripheral rods lack lipid bodies and show a vegetative cell-like architecture. (E) A particular subpopulation of cells is visible at the surface of the fruiting body just beneath the pellicula-like layer that envelopes the entire fruit (black arrow). These cells develop lipid bodies and start to round up, but do not progress and form the characteristic thick fibrillar spore coat. It is tempting to speculate that the lack of the spore coat is the consequence of the formation of the carbohydrate pellicula by these cells. Bars in (A) 1mm, in (B) 15μm, in (D) 1μm and in (E) 5μm. (B) reproduced with permission from O'Connor and Zusman (1988) J. Bacteriol. *170*, 4103–4112. (C), D) and E) reproduced with permission from Hoiczyk *et al.* (2009). Mol. Microbiol. *74*, 497–517.

and C). A number of excellent reports (Shimkets, 1990; Ward and Zusman, 1999; White, 1993) and Chapter 8 in this book describe the overall macroscopic events of this process in great detail. This section will focus on the complex processes that unfold in the interior of the fruiting body during the maturation process. Because of the large size of these structures, most morphological studies have relied on scanning electron microscopy (SEM), a technique that allows studying surfaces at high resolution but is severely limited in its ability to reveal interior details. This limitation, together with the fact that SEM samples are dehydrated and, therefore, shrink, has often resulted in the impression that fruiting bodies are simply 'piles' of spores that are covered by a thin membrane-like layer (Fig. 3.3B). Although these pictures are aesthetically pleasing, they do not accurately reflect the complex architecture of these structures, particularly during the initial formation and maturation period (24–72 h). This architecture is only revealed when fruiting bodies are flash-frozen, thin-sectioned and examined by transmission electron microscopy (Hoiczyk *et al.*, 2009). Using this technique, a much more nuanced picture emerges, in which various groups of cells within the fruiting body show different morphologies that reflect their developmental trajectory and role during the formation of this structure (Fig. 3.3C, D and E).

The first group of such specialized cells form a layer at the bottom of the fruiting body and, even after many days, do not show any obvious signs of differentiation (e.g. change of their shape or the development of intracellular lipid bodies; see below). While these cells are often termed peripheral rods (O'Connor and Zusman, 1991c), they may instead be comprised of cell clusters, (see 'Cell cluster', above) (Lee *et al.*, 2012). In fact, these cells may represent the *M. xanthus* 'stalk'; a structure that is much more obvious in closely related *M. stipitatis* or *M. fulvis* species (Reichenbach, 1993). It should also be noted that a small number of undeveloped cells are also found throughout the interior of the fruiting body. From morphological studies, it is not clear whether these undifferentiated cells have entered the fruiting body at a later time point and have not yet started to convert into spores or whether they comprise a final differentiation state. The second, and by far largest group, are cells that make up the bulk of the cell mass of the fruiting body, and which eventually differentiate into spores. Morphologically, these cells are relatively easy to recognize because, shortly after entering the aggregation centre (6–12 h), they show a coordinated sequence of three highly characteristic morphological changes including (i) formation of large intracellular lipid bodies, (ii) cellular morphogenesis into spherical prespore cells, and (iii) formation of a thick, multilayered spore envelope. Spores and the mechanistic details of differentiation will be described in detail below.

The third subpopulation of cells found in thin sections of developing fruiting bodies form the surface layer of the fruiting body. These cells initially start developmental differentiation similarly to those differentiating into spores in the interior of the fruiting body (Fig. 3.3C and E). Surface layer cells form lipid bodies in their cytoplasm and begin converting to spherical prespore cells. At this stage, however, differentiation appears to come to a halt such that they lack the thick spore coat that is characteristic of mature fruiting body spores. Simultaneously, a thin pellicula-like layer forms which encases the entire mound-shaped structure (Fig. 3.3B, C and E). This presumably exopolysacharide-rich layer (Lux *et al.*, 2004) likely serves to keep the fruit together. The observation that certain mutant strains (i.e. Δ*pilA*; MXAN_5783) do not form this layer because of a reduced production of exopolysaccharides suggest that this structure is most likely dispensable for spore and fruiting body formation (Curtis *et al.*, 2007). Although the precise mechanism for formation of the pellicula has not been studied, it is tempting to speculate that the surface layer cells secrete the carbohydrates that form the pellicula (Lux *et al.*, 2004) and that this expenditure of polysaccharide material prevents them from forming their usual spore coat.

Finally, an additional structural feature observed in fruiting bodies is an amorphous matrix which fills the space between the individual cells. This matrix has traditionally been termed 'slime' (Zusman *et al.*, 2007). Although it is clear that this matrix contains the slime secreted during cell movements as a component of, it may also contain pili, proteins, and cellular debris of lysed cells (E. Hoiczyk, unpublished observations). As the fruiting body matures, the number of non-spore cells in its interior gradually declines, probably because some of the remaining cells have either left the fruiting body as peripheral rods or undergone lysis. However, even after prolonged time (120 h and longer), some non-differentiated surface layer and interior cells persist, indicating that a certain small number of cells never execute the developmental differentiation programme.

Myxococcus xanthus spores – methods of induction

Fruiting body-dependent sporulation occurs exclusively within the mounds and represents a minor proportion (1–15%) of the initial developing population. Mature spores can be detected between 24 and 72 h after onset of starvation, with maximal sporulation levels after approximately 5 days. Thus, the core sporulation mechanism has been difficult to characterize. For instance, mutants in genes involved in monitoring population density, directing or enabling cells to move in fruiting bodies (i.e. *asgA*, *csgA*, *fruA*) are unable to produce fruiting body spores, although these gene products are not involved in spore differentiation per se. Fortunately, sporulation in *M. xanthus* can also be initiated via chemical induction of cells in rich broth (Dworkin and Gibson, 1964). The ability to generate chemical-induced spores is of great advantage in analysis of *M. xanthus* spore differentiation mechanism not only because genes which are necessary for the core sporulation process can be distinguished from those which are exclusively necessary for fruiting body formation (Müller *et al.*, 2010), but also because the nearly complete and synchronized sporulation process enables metabolic and '-omic' analyses over the course of spore differentiation without contamination from cells following alternate developmental cell fates.

Chemical-induction was first described by Dworkin and Gibson (1964). These authors demonstrated that spore-like entities can be induced by the addition of 0.5 M glycerol to cells growing exponentially in nutrient rich broth. Under these conditions, virtually all cells synchronously produce spores within 4–8 h. These 'chemical-induced' cells rearrange into spherical, phase-bright (refractile) spores which are resistant to heat and sonic disruption (although to a lesser extent than starvation-induced spores) (Sudo and Dworkin, 1969), and can efficiently germinate once the inducer has been removed. For many years, it was unclear how and why sporulation could be chemically induced. Although 0.5 M glycerol is the most efficient, seemingly unrelated groups of chemicals (such as dimethyl sulphoxide, ethylene glycol, butanol, phenethyl alcohol, and glucosamine) (Bacon and Rosenberg, 1967; Dworkin and Sadler, 1966; Mueller and Dworkin, 1991) can induce sporulation usually at concentrations unlikely to exist in the natural environment. Early experiments suggested that sporulation induction was not due to an osmotic effect, nor to appreciable accumulation or metabolism of these chemical inducers inside the cell (Sadler and Dworkin, 1966). A satisfying explanation came with the discovery that addition of D-amino acids or cell wall disrupting antibiotics could likewise induce spore formation (O'Connor and Zusman, 1997). Furthermore, all methods of chemical induction induce beta-lactamase activity which has been associated with cell wall damage in other organisms. Thus, it was suggested that addition of these chemicals to exponentially growing cells either induces or mimics a state which is perceived as indicative of, cell wall damage. It was proposed that fruiting body-independent sporulation is a natural part of the life cycle and evolved to protect cells subject to cell wall damage.

A caveat here is that not all sporulation features are accurately represented by chemical-induced spores. For instance, the genomic content of starvation-induced spores is 2N, whereas that of chemical-induced spores is variable. Glycerol spores do not contain the fatty acid 1-*O*-13-methyltetradecylglycerol which is specific to fruiting body spores (Ring *et al.*, 2006). Certain proteins are either lacking in chemical-induced spores [e.g. the polyphosphate storage Protein W (Otani *et al.*, 1998)], or are not proteolytically cleared [e.g. the MreB cytoskeletal protein (Müller *et al.*, 2012)]. These features may be linked to starvation rather than spore differentiation *per se*. Differences can also be noted in the spore coat (described in detail below), but these differences do not preclude the use of chemical-induced spores as an excellent model system for characterization of the unique *M. xanthus* spore differentiation mechanism. Indeed, it is our view that spore differentiation is a distinct programme whose main features are represented by chemical induction.

Spore features

M. xanthus spores are produced to withstand periods of nutrient limitation and environmental stress, and their physical properties reflect this purpose. Individual spores are metabolically quiescent, and resistant to desiccation, heat (up to 60°C), detergents, UV irradiation, and enzymatic digestion (Dworkin and Sadler, 1966; Sudo and Dworkin, 1969). In the laboratory setting, the observation that spores are resistant to sonic disruption is used (typically in combination with heat resistance) to isolate spores from other developmental or vegetative cells (Higgs and Merlie, 2008). Spores can remain dormant for many years; germination and outgrowth are triggered by polypeptides and require magnesium and calcium ions (Otani *et al.*, 1995).

Analysis of spore structure has come from both chemical- and starvation-induced spores and a description of both spore types will be presented here. Chemical- and starvation-induced spores appear similar by light microscopy, but electron microscopy analysis reveals that they are not identical in certain features. Both spores contain a cell envelope which has been dramatically altered from that of typical Gram-negative cells. The vegetative cell envelope consists of a 6- to 8-nm-thick cytoplasmic membrane, an equally thick peptidoglycan layer, and an 8- to 10-nm-wide outer membrane (White, 1984). Most morphological descriptions of *M. xanthus* spore envelopes have either used this scheme or that of the more extensively studied *B. subtilis* spore envelope (reviewed in Henriques and Moran, 2007) to identify and assign the different layers within the envelope. Although a valid approach, it should be kept in mind that the resulting assignments are not without pitfalls, because the *M. xanthus* spore envelopes have not been isolated, separated into different individual sublayers, and biochemically analysed. For instance, the surprising finding that chemical-induced spores do not contain peptidoglycan (Bui *et al.*, 2009) although they possess a layer in the envelope which corresponds to the position of the peptidoglycan layer in the vegetative cell wall (Fig. 3.4e), reinforces the necessity for rigorous studies to biochemically characterize these different layers. In the following, we will therefore first morphologically describe the individual layers of the chemical- and starvation-induced spore envelopes before critically discussing their potential origin, composition, and function with respect to the underlying remodelling processes and the different properties of the two spore types.

Mature chemical-induced and fruiting body spores possess envelopes that consist of at least three different morphologically identifiable layers (Fig. 3.4). Starting from the cytoplasmic side outwards, the first is the homogenously stained (non-fibrous) inner layer (IL) of 20–30 nm (Fig. 3.4). The second is a thin (8–10 nm) membrane-like double layer referred to as outer membrane (OM) (Fig. 3.4a–d). The third layer or the coat (Ct) differs between chemical- and starvation-induced spores. In chemical-induced spores, it is ~10 nm thick (Fig. 3.4e), while in fruiting body spores it is a massive fibrous structure ranging from 70 nm to 130 nm (Ct in Fig. 3.4). Because of the differences in the coat layer, the total thickness of the spore envelope in chemical induced spores is ~50 nm, while in fruiting body spores it is approimately three times as thick (up to 160 nm). Of these three structural layers, the second is most likely the genuine outer membrane based on both its membrane-like appearance as well as its relative position within the spore envelope. While the inner layer is reminiscent of peptidoglycan, it was recently demonstrated that peptidoglycan could not be isolated or identified in chemical-induced spores (Bui *et al.*, 2009); therefore, this layer is either formed by a dramatically modified version of peptidoglycan which escaped detection, or by an unknown *de novo* synthesized polymer. In light of these results, assignment of this layer as peptidoglycan in fruiting body spore envelopes (O'Connor and Zusman, 1999) should be interpreted with caution. Clearly, more analyses are needed to examine whether peptidoglycan can be identified in fruiting body spores and to establish the chemical composition of this important spore envelope layer.

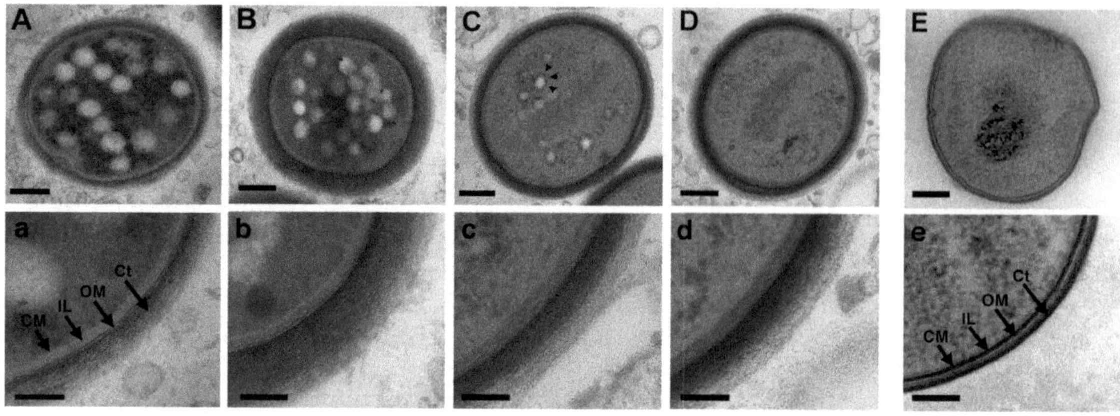

Figure 3.4 Electron micrographs of thin sections of fruiting body and glycerol spores of *M. xanthus*. (A–D) Representative pictures of developing spores after 24 (A), 48 (B), 72 (C), and 96 (D) hours of nutrient limitation. During the spore maturation process, the size of the intracellular lipid bodies decreases and lipid bodies are absent after ~72h (small black arrowheads). Concurrently, fruiting body spores form a massive outer exopolysaccharide layer. (E) Electron micrograph of a glycerol spore from *M. xanthus* strain DZ2 after growth in CYE medium and addition of 0.5M glycerol for 24h. Note that glycerol spores lack the intracellular lipid bodies and contain a thinner extracellular exopolysaccharide layer. While the exopolysaccharide layer is always thinner, some pictures in the literature show glycerol spores which contain lipid bodies (O'Connor and Zusman, 1999). The most likely explanation for this discrepancy is that these cells contained lipid bodies prior to induction of sporulation. Cells devoid of lipid bodies induced with glycerol do not appear to form lipid bodies *de novo* during the short conversion time. All bars, 0.25 µm. (a–e) Enlargements of representative cell envelope regions of the individual spores. For details see text. CM, cytoplasmic membrane; IL, inner layer; OM, outer membrane; Ct, coat. All bars, 100nm. (A–D) and (a–d) reproduced with permission from Hoiczyk *et al.* (2009). Mol. Microbiol. *74*, 497–517. (E) and (e) reproduced with permission from Müller *et al.* (2012). Mol. Microbiol. *83*, 486–505.

The cytoplasmic membrane should be located adjacent to the inner layer. Unfortunately, in most thin sections of spores, no clear cytoplasmic membrane can be seen, an observation that also holds true for the cell envelope of vegetative *M. xanthus* cells and the envelopes of other Gram-negative bacteria, such as *E. coli*. This lack of visualization has generally been attributed to preservation of the cell turgor during cryo-preparation which forces the cytoplasmic membrane tightly against the peptidoglycan layer preventing visualization as a separate layer in thin sections (Beveridge and Graham, 1991).

The coat layer, located outside of the outer membrane, shows the greatest difference between chemical- and fruiting body spores. In chemical-induced spores, this layer is thin and has, in conventional Epon-embedded specimens, a somewhat punctate appearance (e.g. Inouye *et al.*, 1979; O'Connor and Zusman, 1999). In cryo-embedded and -sectioned cells, however, this layer is smooth and homogenous (Fig. 3.4e). This difference in appearance is likely the result of the different preparation techniques. Interestingly, in chemical (glycerol)-induced spores, this layer is only as much as 10 nm thick which may be, in part, responsible for the lower resistance of these spores against chemical and physical assault (Sudo and Dworkin, 1969). In contrast, this layer is much thicker in fruiting body spores and accounts for the bulk of the spore envelope. The appearance of this layer varies between fruiting body spores. In most fruiting body spores, this coat layer appears homogeneously fibrillar in thin sections, but in up to 50% of spores, the inner most part of the layer is somewhat more darkly stained (compare Fig. 3.4a with 3.4b, c and d). This more darkly stained layer has sometimes been interpreted as evidence for the presence of yet another layer termed the cortex (Tengra *et al.*, 2006). However, terming this layer cortex is somewhat confusing, because in endospores the cortex refers to the thickened peptidoglycan layer. Moreover, without additional evidence, it is currently impossible to decide whether this layer is indeed a separate structure or, more likely, a zone in which the carbohydrate fibrils of the spore coat are more densely packed and therefore more intensely stained. Finally, biochemical evidence suggests that the fruiting body spores contain a final outer layer ('cuticula') which is comprised of at least Protein S (*tps*, MXAN_5432) and Protein C (MXAN_6106) (Inouye *et al.*, 1979; McCleary *et al.*, 1991). It has been suggested that this cuticula forms a layer of about 30 nm. However, a critical reexamination of the fruiting body spore coats after both Epon- or cryo-embedding does not reveal any discernible additional layer on top of the fibrous spore coat that would fit this description (Fig. 3.4). The most likely explanation for this lack of visualization of the 'cuticula' is that the layer formed by the two proteins is too thin to be seen in thin sections. This explanation is in line with the fact that both proteins S and C are small (< 20 kDa) and therefore should only give rise to a layer of ~2–3 nm thickness if they arrange in a 2D lattice similar to bacterial S-layers (Sara and Sleytr, 2000).

Mechanism of spore differentiation

M. xanthus sporulation involves a process fundamentally different from that of the well-characterized endosporulation model. Endospores are produced by an asymmetric cell division which produces a compartment (forespore) that is subsequently engulfed by the mother cell. The complex peptidoglycan- and protein-rich endospore spore coat is built onto the forespore surface within the protective environment of the mother cell; once the spore is mature, it is released by mother cell lysis (Driks, 1999) (Fig. 3.5A). In contrast, *M. xanthus* spore differentiation is septation-independent and involves rearrangement of the entire ~7 × 0.5 μm rod-shaped cell into an ~1.5 μm spherical spore. The peptidoglycan may be absent (Bui *et al.*, 2009), and the rigid spore coat is primarily carbohydrate rich (Kottel *et al.*, 1975). Deposition of the spore coat occurs without the benefit of a mother cell; it must be directed from within the differentiating *M. xanthus* cell and must circumvent the Gram-negative double membrane (Fig. 3.5B). Thus, *M. xanthus* spore differentiation represents an excellent model system for analysis of core physiological processes, such as metabolic rewiring, rod-to-sphere morphogenesis, production and assembly of the spore coat on the spore surface. Our current knowledge of these processes is reviewed below.

Sporulation-specific metabolic changes

The chemical-induced sporulation programme has been useful to characterize specific mechanistic details of the core spore differentiation process. Sporulation is associated with extensive metabolic rearrangements. Under vegetative conditions, *M. xanthus* obtains carbon and energy from amino acids and fatty acids (obtained by macromolecule digestion of prey organisms and decaying organic matter) which are imported into the cell and likely enter directly into the TCA cycle (Bretscher and Kaiser, 1978; Curtis and Shimkets, 2008; Watson and Dworkin, 1968). In contrast, during sporulation, both enzyme activity (Filer *et al.*, 1977a,b; Orlowski *et al.*, 1972) and global transcriptome analyses (Müller *et al.*, 2010) suggest that sporulating cells decrease expression of many genes encoding enzymes of the TCA cycle and increase expression of genes encoding enzymes of both the glyoxylate shunt and gluconeogenesis pathways presumably to produce carbohydrate precursors for storage compounds [e.g. trehalose (McBride and Zusman, 1989) and glycogen (Nariya and Inouye, 2003)] and spore coat polysaccharides, respectively (Kottel *et al.*, 1975). Microarray analysis also suggested that expression of genes assigned to the category of protein degradation was significantly increased (Müller *et al.*, 2010) suggesting turnover of vegetative proteins may fuel this metabolic rewiring. Consistent with formation of a quiescent spore, transcriptome analysis also suggested that genes encoding vegetative cellular processes (e.g. motility, and biosynthesis of natural products) as well as those involved in growth including protein synthesis and secretion/trafficking were overrepresented in the down-regulated genes. Biochemical analyses suggest that phospholipid biosynthesis is terminated upon glycerol-induced sporulation (Bacon and Rosenberg, 1967). Continued protein synthesis could be observed at least until the spherical prespore stage when refractivity is acquired, and

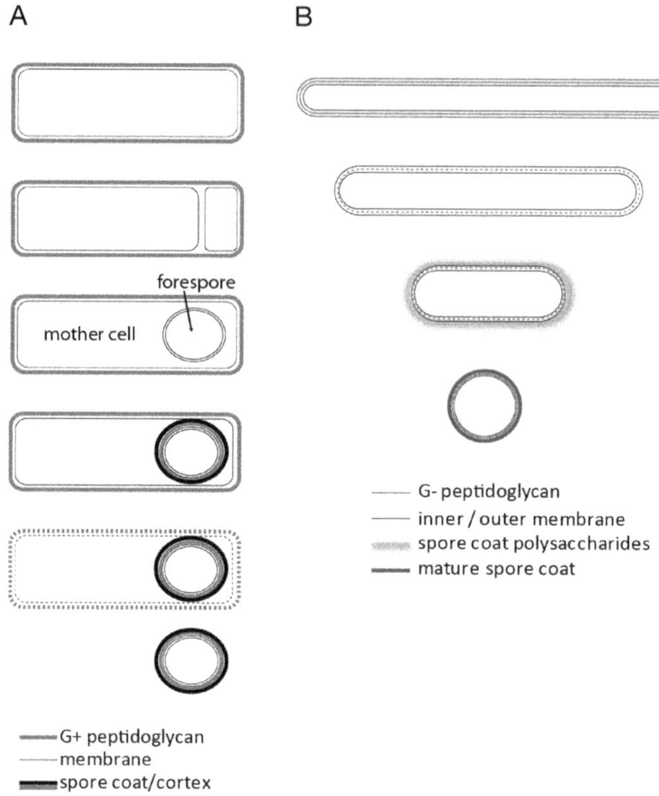

Figure 3.5 Schematic comparing the spore differentiation mechanisms of endospores and myxospores. (A) As typified by the Gram-positive *Bacillus subtilis*, endospores arise from an asymmetric septation event generating a prespore compartment which becomes engulfed by the mother cell 9reviewed in Higgins and Dworkin, 2012). The complex protein- and peptidoglycan-containing spore envelope layers are assembled within the protective environment of the mother cell. The mature endospore is released by lysis of the mother cell. (B) The spores of the Gram-negative *M. xanthus* are produced by septation-independent rearrangement of the entire rod-shaped cell into a spherical spore. The carbohydrate-rich spore coat is assembled outside the outer membrane in a process which is directed by cell envelope machinery of the prespore. At least in chemical-induced myxospores (see text), the spore coat replaces the function of the peptidoglycan (cell wall), which is degraded at some point during the sporulation process.

chloramphenicol, a protein synthesis inhibitor, blocked the sporulation process (Sadler and Dworkin, 1966).

Rod-to-sphere morphogenesis

The first obvious sign of sporulation is the shortening of the rod-shaped cells as they begin to rearrange into a sphere. At least during chemical induction of sporulation, the cells simultaneously widen as they shorten and the surface area of the cell continuously decreases (A. Herrmann, C. Holkenbrink and P.I. Higgs, unpublished data). Thus, rod-to-sphere morphogenesis is not simply a matter of degradation of the rod-shape determining peptidoglycan sacculus. The cell cytoskeletal protein, MreB, is involved in this process because addition of the MreB-polymerization inhibitor compound, A22, prevents continued cell shortening (Müller *et al.*, 2012). Because the initial stages of rod-to-sphere morphogenesis would require peptidoglycan rearrangement, we presume MreB is necessary to guide proteins (either penicillin binding proteins or lytic enzymes) involved in this process. It is not clear if the peptidoglycan is somehow contracted with concomitant thickening, or whether sections of the glycan backbone are removed in a process that is somehow the reverse of peptidoglycan synthesis during elongation. The lack of peptidoglycan in spores could be the result of consistent loss during rod-to-sphere morphogenesis, or due to a final degradation step after the rigid spore coat is assembled. Consistent with this hypothesis, mutants which fail to assemble a rigid spore coat are capable of morphogenesis to spherical prespores, but subsequently revert to rod-shape (Müller *et al.*, 2012). During the reversion process, the cells undergo a transient period of severe branched or spiral morphologies which are reminiscent of certain mutants defective in peptidoglycan assembly in *E. coli* and *C. crescentus* (Nilsen *et al.*, 2004; Takacs *et al.*, 2010). It is likely that the partially rearranged/degraded peptidoglycan of the mutant spherical prespores may lead to 'inert' patches of peptidoglycan such that new peptidoglycan is not equivalently inserted and shape deformations are produced. The observation that meso-diaminopimelic acid subunits are both actively incorporated and released during chemical-induced sporulation suggests extensive replacement of the existing wall with new material (Dawson and Jones, 1979), although the new material may not be peptidoglycan. In this context, it is interesting to note that muramic acid was identified as a component of fruiting body spore coats (Adye and Powelson, 1961). As for chemical (glycerol)-induced spores, however, this peptidoglycan signature molecule may not necessarily mean that the coat indeed contains peptidoglycan. Various antibiotics which target peptidoglycan synthesis prevent sporulation. For instance, cephalexin, which prevents septation in vegetative cells, causes sporulating cells to form spheroplasts and lyse shortly after initiation of shape change (Jones *et al.*, 1981). Addition of penicillin G blocked sporulation only very early during the induction (i.e. when cells were shortening), whereas addition of D-cycloserine blocked spore formation for a longer period (up to the point of the ovoid prespore) (White *et al.*, 1968). In contrast, mecillinam, whose target is penicillin binding protein 2 (PBP2), has no effect on production of resistant spores. Together, these studies support an active rearrangement of peptidoglycan during rod-to-spore morphogenesis. The enzymes involved in this process have not been identified, but candidates include certain PBPs whose expression is increased in chemical-induced sporulation transcriptome analysis (Müller *et al.*, 2010). Preliminary analyses suggest at least one of these candidates plays a relatively minor role during chemical-induced sporulation, with at least two necessary for spore germination (A. Herrmann, C. Holkenbrink and P.I. Higgs, unpublished data).

Spore coat production

During the rod-to-sphere morphogenesis, spore coat material begins to accumulate on the surface of the prespores (Müller *et al.*, 2012). The spore coat consists of 75% carbohydrate (one third glucose and two-thirds N-acetyl-galactosamine), 14% protein, 8% glycine, and under 1% organic phosphorous (Kottel *et al.*, 1975). Two genetic loci, *exo* and *nfs*, have been described which appear to be necessary for the secretion and assembly of the carbohydrate portion of the spore coat (Müller *et al.*, 2012). Interestingly, at least under chemical-induced sporulation, a properly assembled spore coat is an essential step in spore maturation, because both *exo* and *nfs* mutants cannot maintain the spherical morphology and instead revert to rod-shaped cells even in the presence of inducer (Müller *et al.*, 2012). Since the peptidoglycan is probably absent in chemical-induced spores (Bui *et al.*, 2009), it is assumed that assembly of the rigid spore coat must replace the role of the peptidoglycan in preserving cell integrity in the face of the internal cellular pressure. Under starvation-induction of sporulation, these mutants do not revert to rod-shaped cells, but produce reduced levels of spore-like entities which are not viable (Licking *et al.*, 2000; Müller *et al.*, 2010, 2012). The spore coat is lacking in *exo* mutants (Müller *et al.*, 2012). The *exo* cluster encodes nine co-transcribed genes (Müller *et al.*, 2012) whose expression is increased under both starvation and chemical induction of sporulation (Licking *et al.*, 2000; Müller *et al.*, 2010; Ueki and Inouye, 2005). ExoA (FdgA) and ExoCD share homology to the OPX and PCP-2a like components of 'Wzy-dependent' polysaccharide export machinery (Cuthbertson *et al.*, 2009), and it has been demonstrated that these genes are necessary for sporulation under both starvation and chemical induction (Kimura *et al.*, 2011; Licking *et al.*, 2000; Müller *et al.*, 2012; Ueki and Inouye, 2005). Thus, it is likely that the Exo proteins are necessary to secrete glucose and/or N-acetylgalactosamine containing polysaccharides to the sporulating cell surface; these carbohydrates likely form some, or all, of the spore envelope layer 3 (Fig. 3.4).

A model for spore coat polysaccharide secretion which is based on the mechanism of Wzy-dependent polysaccharide export systems (Whitfield and Paiment, 2003; Whitfield, 2006) is as follows. ExoE, a polyisoprenyl phosphatase hexose-1-phosphate (PHPT) transferase homologue, generates undecaprenyl-activated sugars to which additional repeat units may be added by glycosyltransferases such as ExoH, a sugar nucleotide aminotransferase homologue. These repeat units are flipped across the inner membrane by a flippase (Wzx), and the repeat units are then polymerized on the periplasmic side of the inner membrane by Wzy. Wzx and Wzy homologues are not encoded in the *exo* locus, but MXAN_3026 and MXAN_1035, respectively, are good candidates. Expression of MXAN_3026 appears to be increased during chemical-induced sporulation (Müller *et al.*, 2010) and the mutant fails to export spore coat material to the cell surface (Müller *et al.*, 2012). The polymerized repeat units are then exported through the PCP-2a/OPX channel consisting of ExoCD and ExoA (FdgA). ExoC and ExoD together constitute a split PCP homologue consisting of the inner membrane copolymerizing protein whose chain length determination activity is regulated by a tyrosine kinase (ExoD, aka BtkA). BtkA has been shown to phosphorylate ExoC *in vitro* (Kimura *et al.*, 2011). Although this model must be demonstrated biochemically, it is consistent with previous experiments indicating that glycine and *N*-acetylgalactosamine accumulation into the spore coat is sensitive to bacitracin, an inhibitor of undecaprenyl transfer (Filer *et al.*, 1977a).

Once spore coat polysaccharides have been exported to the surface, protein machinery encoded by the *nfs* locus is necessary to arrange the polysaccharides into a rigid, stress-bearing layer (Müller *et al.*, 2012). The *nfs* locus consists of eight genes (*nfsA–H*) which are likely co-transcribed and whose expression is increased during chemical- and starvation-induced sporulation (Müller *et al.*, 2010). Most of the genes are annotated as proteins of unknown function which may share ancestry with proteins involved in A-motility in *M. xanthus* (Luciano *et al.*, 2011; Müller *et al.*, 2010, 2012). Deletion of the entire *nfs* locus prevents viable spore formation under both starvation- and chemical-induction (Müller *et al.*, 2010). The observation that *nfs exo* double mutants behave identically to each single mutant with respect to reversion to rod shape after formation of spherical prespores, suggested that the Nfs proteins likely function in spore coat production (Müller *et al.*, 2012). In contrast to *exo* mutants, however, both EM and anti-spore coat immunofluorescence analyses indicated that in the *nfs* mutant, spore coat material is secreted to the spore surface, but is detected as loosely associated amorphous material rather than the wild type tightly associated rigid layer (Müller *et al.*, 2012). Thus, it is hypothesized that the Nfs proteins are necessary to arrange the flexible capsular-like polysaccharides secreted by the Exo proteins into a stress-bearing rigid layer. Analysis of individual *nfs* gene deletions indicates that at least seven of the eight Nfs proteins are necessary for sporulation, and it is likely that these proteins form a complex that resides in the cell envelope (Müller *et al.*, 2012). Recent data suggest that Nfs function requires productive interaction with a TolQ homologue which may serve to couple proton-motive force to conformational energization of the Nfs complex (M. Wartel and T. Mignot, personal communication). It is likely the Nfs complex is involved in modifying spore coat polysaccharides because carbohydrate subunit linkages in the spore coat are significantly different from the wild type (C. Holkenbrink, E. Hoiczyk and P. I. Higgs, unpublished data), but it is unclear if the Nfs proteins mediate spore coat modification directly, or are necessary to secrete or organize proteins or carbohydrates involved in this process.

Based on the appearance of perturbed spore coat production in mutant strains, several additional proteins are thought to play a role in spore coat production, but it is yet unclear how they function. Protein S-1 (*ops*) is produced during the sporulation phase of starvation-induced development (Teintze *et al.*, 1985). S-1 is highly similar, and partially functionally redundant to Protein S (see below), but, unlike Protein S, it may remain in the cytoplasm (Furuichi *et al.*, 1985). CbgA was identified as a homologue of *B. subtilis* SpoVR which is involved in endospore cortex formation (Tengra *et al.*, 2006). While this is noteworthy because very few homologues to proteins involved in endosporulation can be identified in *M. xanthus*, SpoVR homologues can be identified in a wide-range of non-sporulating bacteria. Fruiting body spores produced by the *cbgA* mutant either lacked, or had a reduced levels of, the darkly stained region of spore coat layer 3, which was attributed to alteration of the cortex (Tengra *et al.*, 2006). However, the term 'cortex', which refers to the peptidoglycan layer in endospores, should be used with caution, since it is not clear whether fruiting body spores, like chemical-induced spores, lack peptidoglycan. It is possible that *cbgA* mutant spores fail to mature properly since they contain large lipid body structures which are already absent in the wild type, and this mutant also displayed severe defects in fruiting body shape. The MspA, B, and C proteins were identified in a proteomic study of starvation-induced spores (Dahl *et al.*, 2007). These mutants also display an altered spore coat in which the electron-dense layer adjacent to the outer membrane is absent or reduced. Spores produced by *cbgA*, *mspA*, *mspB*, or *mspC* mutants display decreased resistance to detergents and heat. Interestingly, under chemical induction of sporulation, MspB and MspC seem to be additionally important for UV resistance (Dahl and Fordice, 2011; Dahl *et al.*, 2007). In exactly what functional mechanism these proteins participate is not yet understood.

Three spore coat proteins (termed S, C, and U) have been identified on the basis of release by boiling spores in the presence of SDS (internal spore proteins are not released by this method) (Inouye *et al.*, 1979; McCleary *et al.*, 1991). Two of these proteins, Protein S and C, cannot be detected in chemical-induced spores, and form the spore cuticula – a proteinaceous outer layer. Consistently, Protein S and C can be quantitatively released from mature spores in the presence of 0.1 N NaOH, and Protein S can be additionally released in the presence of Ca^{2+} chelators (Inouye *et al.*, 1979; McCleary *et al.*, 1991). Both proteins can individually spontaneously self-assemble on spore coat surfaces in the presence of Ca^{2+} and Mg^{2+}, respectively. Protein S shows striking homology to members of the family of γ-crystallins found in the lens and cornea of vertebrate eyes (Bagby *et al.*, 1994; Wistow *et al.*, 1985). Protein C was recently identified as a ~31 kDa C-terminal fragment of the extracellular metalloprotease, FibA (Lee *et al.*, 2011). This fragment consists of two PPC repeats which are regions of poorly defined function that are often associated with extracellular proteases (Yan *et al.*, 2009). Interestingly, PPC domains share putative structural and functional features with PKD (polycystic kidney disease) domains (Yeats *et al.*, 2003) and the PKD protein fold is found in the surface layer protein (SLP) of the archaeon *Methanosarcina mazei* where it is thought to mediate cell–cell attachment (Jing *et al.*, 2002). Deletion of the genes encoding Protein S or Protein C does not have any significant effect on the production, resistance, or viability of starvation induced spores, and it has been proposed that the cuticula is important for spore-to-spore adhesion inside of fruiting bodies, rather than in spore formation or resistance (Komano *et al.*, 1984; Lee *et al.*, 2011). Both proteins are produced during the aggregation stage of the developmental programme (Inouye *et al.*, 1979; McCleary *et al.*, 1991). Interestingly, Protein S contains no signal sequence suggesting the necessity of a non-canonical secretion system for export to the cell surface. It has been shown that Protein S first accumulates as a soluble cytoplasmic protein; later during development, it is apparently rapidly transported across the cytoplasmic membrane and then across the OM (Nelson and Zusman, 1983).

Protein U can also be released from spore coats by boiling in SDS (Gollop *et al.*, 1991). Unlike Proteins S and C, however, U is found in both chemical- and starvation-induced spores. Recently, MXAN_3885 from *M. xanthus* strain DK1622 was characterized as a homologue of Protein U in *M. xanthus* strain DZF1, because it shared 82% identity, but had an earlier developmental protein accumulation profile (Leng *et al.*, 2011). It is likely that this gene encodes Protein U, since there are no paralogues of MXAN_3885 in the sequenced DK1622 genome, and the difference in timing of accumulation can be explained by the fact that DK1622 develops more quickly that DZF1 (Lee *et al.*, 2012). We suggest the sequence differences are either strain specific or due to manual sequencing errors in the original publication (Gollop *et al.*, 1991). Interestingly, MXAN_3885 is likely secreted by a chaperone-usher pathway (Leng *et al.*, 2011), the first description of such a secretion system used for a spore coat protein. Disruption of MXAN_3885 produces fruiting bodies in which the spores appear to be less well connected, and spores which display disruptions in the darkly staining region of the coat layer which was attributed to an alteration of the 'cortex' (Leng *et al.*, 2011).

Spore resistance and storage

Like the spores or cysts from other organisms, *M. xanthus* spores have been shown to accumulate storage compounds which can function in resistance or as a fuel source for sporulation or germination. The *M. xanthus* Protein W, encoded by *prw* (MXAN_2491), was identified as an abundant spore-specific protein which forms a single, large (200 to 400 nm) electron-dense particle inside the spores (Otani *et al.*, 1998). X-ray microanalysis of purified protein W particles suggested they contained high levels of polyphosphate. It was therefore proposed that Protein W is a polyphosphate storage protein, and it was presumed that this high energy storage compound may be consumed to fuel sporulation or germination. However, *prw* mutants, although delayed in aggregation and sporulation, produced wild type levels of resistant and viable spores. Myxospores were also shown to accumulate trehalose during sporulation which, as in the resting stages from other organisms, was proposed to contribute to heat and desiccation resistance (McBride and Zusman, 1989). It was further shown that the trehalose content rapidly decreases during germination. It is probable that trehalose is synthesized from glycogen (Pan *et al.*, 2008) which has been shown to accumulate prior to sporulation and then to decrease during sporulation (Nariya and Inouye, 2003). Myxospores also accumulate glycine betaine, which is, uncommonly for non-halophilic bacteria, likely synthesized by S-adenosylmethionine dependent methylation of glycine (Kimura *et al.*, 2010). Glycine betaine accumulation in spores is proposed to facilitate germination under conditions of osmotic stress.

Like Gram-positive endospores, myxospores are resistant against DNA-damage by UV light (Sudo and Dworkin, 1969). In endospores, UV resistance is mediated by small acid-soluble proteins (SASPs), which cover and protect the DNA (Setlow, 2006). Although, the genome of *M. xanthus* lacks orthologues of the *B. subtilis* SASPs, two small acid-soluble proteins, SapA and SapB, were identified in *M. xanthus* spore lysates (Dahl and Fordice, 2011). The disruption of at least *sapA* led to reduced UV resistance of starvation-induced spores. Interestingly, in chemical-induced spores, SapA could not be identified in the acid soluble fraction whereas MspB and MspC could. Like endospore SASPs, at least SapA has predicted disordered regions; however, unlike endospores SASPs, which are soluble proteins, SapA, MrpB, and MrpC have predicted membrane-spanning segments. These data suggest that UV resistance may be an aspect of the spore cell envelope rather than by direct protection of the DNA.

Perhaps the most important goal in further characterization of spore differentiation and resistance characteristics is a careful biochemical examination of the spore cell envelope layers. Are there multiple layers composed of different material, or is this a staining artefact? Do fruiting body spores, as well as chemical-induced spores, lack the peptidoglycan layer? Since the production and assembly of the spore coat are important for maintaining morphogenesis and likely replace the function of the peptidoglycan, it will be interesting to investigate this process thoroughly. For instance, how are the spore coat polysaccharides assembled into a compact and rigid layer on the spore surface? How is the peptidoglycan reorganized during morphogenesis from rod shaped cell to spherical spores. If spores lack peptidoglycan, how is the peptidoglycan synthesized during spore outgrowth?

Peripheral rods

In 1991, O'Connor and Zusman first documented and characterized peripheral rods as a distinct differentiated state within the developmental population (O'Connor and Zusman, 1991c). Starting from the observation that cells can be identified between and surrounding mature fruiting bodies by both light and electron microscopy, O'Connor and Zusman demonstrated that single rod-shaped cells can be separated from the mature multicellular fruiting bodies by differential centrifugation. Specifically, while the multicellular fruiting bodies pellet after short, low speed centrifugation, peripheral rods remain in the supernatant as single cells. This approach was used to document that peripheral rods have been identified in all strains for which they have been examined (Lee *et al.*, 2012; O'Connor and Zusman, 1991c) including several laboratory strains [DZF1 aka DK101 and FB (Dworkin, 1962), DK1622 (Wall *et al.*, 1999), DZ2 (Campos and Zusman, 1975)] and environmental isolates (DSW2, DSW4, DSW5; O'Connor and Zusman, 1991c). Interestingly, O'Connor and Zusman (1991c) demonstrated that the proportion of peripheral rods produced is reproducibly constant when assayed on nutrient limited clone fruiting plates, but varies between strains in the range of 4–30% (7-day developmental cells). At least in strain DZF1, the proportion of peripheral rods increases with increasing levels of nutrients in the developmental media; for example, addition of 0.5% casitone to starvation media increases the proportion of peripheral rods from ~20% to 57% (O'Connor and Zusman, 1991b). Based on these observations, O'Connor and Zusman proposed that peripheral rods have evolved to take advantage of low levels of nutrients which are insufficient to either promote growth or to induce germination of the spores inside fruiting bodies. In this respect, they may be similar to a persister state (Balaban, 2011). In situ, peripheral rods do not appear to be motile or to undergo cell division (O'Connor and Zusman, 1991b). However, both motility and division can be restored by addition of nutrients suggesting these cells are not permanently differentiated. It was suggested that peripheral rods retain the capacity to aggregate if they are harvested and plated on developmental media at a sufficiently high density (O'Connor and Zusman, 1991b). In contrast to these results, however, Hoiczyk *et al.* (2009) observed that peripheral rods were not capable of producing fruiting bodies when plated on fresh developmental media. The reason may be the experimental conditions and/or the strains used in different studies.

Little has been investigated regarding the structure of peripheral rods, but based on the few available ultrastructural data, it appears that peripheral rods do not noticeably change from vegetative cells (Fig. 3.3C and D) including in the cell envelope. In line with the importance of intracellular lipid bodies as markers for the initiation of differentiation, these cells appear to not form these structures (Hoiczyk *et al.*, 2009). It has been observed by electron microscopy analyses that peripheral rods appear to become hyperpiliated (E. Hoiczyk, unpublished results), and by dye binding assays that the peripheral rod population does not significantly accumulate extracellular polymeric substances (EPS), while the fruiting body population produces at least 12-fold more EPS than vegetative cells (Lee *et al.*, 2012). Analysis of the total protein resolved by 2D gel electrophoresis demonstrated that the peripheral rod proteome is significantly different from that of both vegetative and fruiting body populations (O'Connor and Zusman, 1991a). Flow cytometry analysis of DNA-stained cells suggested that peripheral rods contain a 1N genomic content, whereas developmental spores are 2N (Tzeng and Singer, 2005) (see Chapter 4 for details).

Several specific proteins have been examined in peripheral rods including proteins that are produced exclusively under developmental conditions, such as myxobacterial haemagglutinin (MBHA), spore coat Protein S (Tps), as well as Mg-independent neutral and alkaline phosphatase activities (O'Connor and Zusman, 1991a,c). Spore coat Protein C was also detected in peripheral rods (O'Connor and Zusman, 1991b), although this protein was recently identified as the 30 kDa fragment of the metalloprotease FibA, and it was found to be heterogeneously produced in vegetative cells (Lee *et al.*, 2011, 2012). Although these developmental proteins can be detected in peripheral rods, a careful comparison of equivalent numbers of peripheral rods vs. fruiting body cells suggests that these markers are produced in drastically different levels in the two cells types (Lee *et al.*, 2012). For example, the 30 kDa FibA fragment, Protein S, the developmental transcriptional regulators MrpC/C2 and FruA, and the cell-contact sensor/regulator C-signal were produced at drastically reduced levels in peripheral rods (defined as non-aggregated cells after 36h of development in submerged culture) relative to fruiting body cells (Julien *et al.*, 2000; Lee *et al.*, 2012). Additionally, these studies suggested differences in the relative levels of components of the Type-4 pilus social motility machinery in these two developmental populations. In peripheral rods, the pilin component, PilA, is produced at similar levels as in vegetative cells whereas the inner membrane PilC component is reduced by approximately 2.5-fold. In contrast, in fruiting body cells, PilA is 5-fold more abundant than in vegetative cells, but PilC is undetectable under the conditions measured (Lee *et al.*, 2011).

Although production of peripheral rods is linked to the developmental programme, it is not yet clear which molecular regulatory mechanisms control their production. Several models have been proposed which are not necessarily mutually exclusive. One model suggests that peripheral rods may be produced by cells which do not make sufficient end-to-end contacts to efficiently exchange C-signal. Therefore, C-signal dependent processes such as stimulation of aggregation, positive feedback regulation of the C-signal precursor, CsgA, and sporulation are not triggered (Julien *et al.*, 2000). In support of this model, Julien *et al.* observed that end-to-end contacts amongst peripheral rods are approximately 35% of that in aggregation centres and that CsgA levels are reduced in peripheral rods. An additional suggestion for control of peripheral rods comes from the observation that early during the developmental programme, every cell replicates its genome to become 2N; it was proposed that the low level of C-signalling observed in peripheral rods could be insufficient to prevent a septation event leading to 1N peripheral rods (Tzeng *et al.*, 2006). It has also been proposed that production of lipid bodies, which are formed very early after nutrient depletion, is a determinant for fruiting body production because they are necessary as energy reserves to fuel aggregation and sporulation (Hoiczyk *et al.*, 2009); however, the proteins necessary for production of lipid bodies have not yet been identified. More recently, based on the examination of several additional regulatory proteins, it was proposed that peripheral rods may arise from cells that fail to accumulate sufficient levels of the development transcriptional regulators, MrpC and its target FruA (Lee *et al.*, 2012), which are essential to induce aggregation into fruiting bodies and sporulation (see Chapter 6 for details). It remains unclear at this point exactly when MrpC and FruA levels begin to differ in peripheral rods, and whether this is due to proteolytic turnover of MrpC (and perhaps FruA) specifically in cells which are destined to become peripheral rods, or whether *mrpC* is not expressed in the first place and therefore does not induce expression of *fruA*. In order to distinguish between these possible mechanisms of regulation, it is imperative to understand exactly when during the developmental programme, cells are committed (destined) to differentiate into this cell type. Identification of markers suitable for single cell analyses within the developing population would enable these analyses. Good marker candidates include MrpC and the proteins involved in lipid body formation.

Programmed cell death/ developmental autolysis

For over 30 years, whether, when and to what extent *M. xanthus* developmental cell lysis occurs has been a matter of some debate. While a decrease in cell number during development has been observed by several groups, definitively assigning this phenomenon to directed cell lysis has been complicated both by the facts that cells in fruiting bodies are difficult to separate and enumerate, and that cells undergoing spore morphogenesis (prespores) are extremely fragile. Additionally, while several factors have been proposed to mediate developmental cell death, to date, no definitive mechanism of action has been elucidated. The current understanding of developmental lysis is described below.

Wireman and Dworkin first proposed developmental lysis after counts of developing cells revealed a reproducible pattern in which, after induction of starvation on solid medium, the cell number first increases by approximately one doubling, and then subsequently decreases such that 65% of the maximum number of cells were absent at completion of development (Wireman and Dworkin, 1975a, 1977). To date, this overall pattern has been observed under different developmental conditions and with several *M. xanthus* background strains, although both the relative timing and degree of the decrease in cell number can vary (Janssen and Dworkin, 1985; Lee *et al.*, 2012; Nariya and Inouye, 2008; O'Connor and Zusman, 1988; Wireman and Dworkin, 1975a, 1977). To circumvent cell counting inaccuracies, these results were supported by measuring [methyl-^{3}H]thymidine-labelled DNA retained in cells (Wireman and Dworkin, 1975a, 1977). Together, however, the interpretation of these results as a programmed lysis event was called into question by O'Connor and Zusman with the argument that manipulation of fragile developmental cells resulted in underestimation of developmental cell numbers, and that the [methyl-^{3}H] signal was subject to quenching particularly in spores (O'Connor and Zusman, 1988) (and, in retrospect, because it was not then recognized that spores are in fact 2N; Tzeng and Singer, 2005). The observation that little [methyl-^{3}H]DNA label was released into the supernatant of developing cultures further supported the argument that lysis was not a significant developmental cell fate. However, this observation should also be interpreted with caution since released DNA could be trapped by the extracellular matrix of neighbouring developing cells (Wang *et al.*, 2011); extracellular DNA is, in fact, a component of the matrix in many biofilms (Montanaro *et al.*, 2011). As a counter argument to lysis as an investigator-mediated artefact of cell manipulation, it was subsequently observed that lysis could still be detected when cells were allowed to develop while encased in agarose microbeads, and that this phenomenon was likely dependent on both diffusible and contact dependent developmental signals (Rosenbluh *et al.*, 1989). Consistently, several mutants have been identified which do not lyse (discussed below), and lysis may be complemented by addition of other strains suggesting it to be a staged event (Janssen and Dworkin, 1985; Lee *et al.*, 2012; Nariya and Inouye, 2008; Rosenbluh and Rosenberg, 1989). Thus, the best interpretation of these combined data is that lysis is a developmental cell fate, but that the degree of cell lysis may be magnified by cell manipulation.

Two different, but not necessarily unrelated, mechanisms have been proposed to induce developmental cell lysis: production of autocides and a novel toxin–antitoxin system. An initial mechanism suggested the involvement of autocides (AMI and AMV) which were characterized as a mixture of fatty acids and phosphatidyl ethanolamine (PE), respectively (Gelvan *et al.*, 1987; Varon *et al.*, 1984, 1986). It was demonstrated that PE could be converted into fatty acids by the action of phospholipase activity and suggested that these fatty acids acted to permeabilize cells which ultimately led to lysis. It remains unclear whether all cells are induced to activate phospholipases, and/or are resistant to lysis, but this phospholipase activity could be induced by addition of glucosamine or mannosamine, two components that could restore lysis in lysis-deficient strains (Janssen and Dworkin, 1985; Mueller and Dworkin, 1991). Nearly 20 years later, Nariya and Inouye (2008) proposed that developmental cell lysis was a programmed event controlled by a novel toxin–antitoxin (TA) system, consisting of MazF and MrpC. In classical TA systems, a MazF endoribonuclease is inactivated by binding to a MazE antitoxin, and MazEF autoregulate expression of the *mazEF* operon (Magnuson, 2007). Stress conditions promote the turnover of the proteolytically labile MazE proteins such that MazF is released to exert toxic activity. These systems have been proposed to mediate either altruistic cell lysis, or alternatively, a persister-like state which promotes population survival under adverse conditions (Gerdes *et al.*, 2005; Magnuson, 2007). Interestingly, the *M. xanthus mazF* is genetically an orphan and no MazE homologue could be identified in the genomic sequence. Instead, it appears that MrpC, an important developmental transcriptional regulator, acts as an antitoxin to MazF (Nariya and Inouye, 2008). Consistently, Δ*mazF* mutants failed to lyse during development, whereas Δ*mrpC* mutants constitutively producing MazF, produce higher levels of dead cells. Thus, it was proposed that cell death is mediated by MazF's endoribonuclease activity after specific turnover of MrpC in certain cells. A major caveat to this mechanism is the more recent discovery that MazF appears to play a significant role in cell lysis only in the DZF1 (DK101) strain background which bears a lesion in the *pilQ* locus leading to defects in social motility and cell adhesiveness (Lee *et al.*, 2012). The observation that deletion of *mazF* in two other wild type *M. xanthus* strains produces a very minor, delayed developmental phenotype relative to the respective wild type suggests that MazF's function is redundant in wild type cells. However, based on the more recent proposal that MazF proteins play a role in mediating growth arrest rather than cell lysis (Maisonneuve *et al.*, 2011), it is also worth investigating whether MrpC-MazF instead functions in this capacity. Consistently, under developmental conditions, *mrpC* mutants do not show the initial increase in cell number which precedes the onset of lysis in the wild type (Lee *et al.*, 2012). An intriguing possibility is that the MrpC-MazF system is necessary for formation of peripheral rods. It is unknown whether a defect in production of peripheral rods would have a discernible developmental phenotype under the conditions examined (Lee *et al.*, 2012). Consistent with this hypothesis, MrpC levels appear to be low in peripheral rods (Lee *et al.*, 2012), and it would be interesting to examine MazF levels within peripheral rods.

Cell lysis was originally proposed to occur after completion of aggregation, concomitant with or immediately preceding the onset of spore differentiation (Wireman and Dworkin, 1977). Examination of the data suggests that under the conditions used in this study (strain FB on strict starvation plates), the cell number began to decrease between 24–30 h after induction of starvation and continued to drop linearly over at least 96 h; aggregates were apparently observed starting from 24 h (Wireman and Dworkin, 1977). A recent reexamination of the timing of cell lysis and aggregation indicated that under the conditions tested (strain DZ2 developed under submerged culture), a similar pattern of cell number decrease to the original study was observed. However, simultaneous characterization of aggregation indicated that the onset of cell lysis correlated with a sudden shift of cells into aggregation centres (Lee *et al.*, 2012). Specifically, between 24–30 h of development, there was both a 19% decrease in total cell number and a 17% increase in sedimented (i.e. aggregated) cells, correlating with the production of visible aggregation centres at 30 h of development. Thus, we argue that the onset of lysis immediately proceeds, or is concomitant with, the onset of aggregation, but likely continues during maturation of the fruiting bodies. This timing raises the possibility that lysis may be necessary to stimulate aggregation, instead of, or in addition to, it's long-touted role in stimulating spore differentiation.

A possible connection between CsgA, the precursor to the contact dependent C-signal which is necessary for developmental aggregation and sporulation, and cell death remains to be fully explored. *csgA* mutants do not exhibit the developmental specific decrease in cell number associated with cell lysis (Janssen and Dworkin, 1985; Lee *et al.*, 2012). Interestingly, the disruption of *fruA*, which encodes a protein proposed to function downstream in the C-signal cascade, did not significantly affect the cell lysis event (C. Holkenbrink and P.I. Higgs, unpublished data) suggesting that CsgA (or C-signal) may have a direct role in cell lysis. Cell lysis (and development) can be rescued in the *csgA* mutant by co-culture with wild-type cell-free lysates, or by addition of glucosamine, suggesting a possible link to stimulation of the production of autocides (Janssen and Dworkin, 1985; Mueller and Dworkin, 1991).

Despite the recent description of the developmentally induced MazF toxin which seemed to control developmental cell lysis, the observation that MazF is not absolutely necessary in wild type strains indicates that the mechanism of cell lysis must be reinvestigated, and it is imperative to identify the regulatory and functional mechanism(s) which control this developmental cell fate. There are several questions to pursue: what is the role of MrpC and CsgA in regulating developmental cell lysis? Is the MazF protein redundant and, if so, what are the additional players? What is the link to autocide production? Insights to these questions will clarify the larger picture whether developmental lysis really is an altruistic form of programmed cell death which releases nutrients/signals essential for the remaining cells to complete development.

Concluding remarks

Given the preference for cooperative multicellular behaviour, is it no surprise that population heterogeneity is an integral part of myxobacterial behaviour. In general, population heterogeneity is an area of research which, in the last several years, has sparked renewed interest as molecular tools have been developed which can be used to examine single-cell expression of cell fate and population heterogeneity determination factors. In *M. xanthus*, the model myxobacterium, important advances, both in the identification of cell fate-specific characteristics and in the regulatory pathways governing cell fate determination and population heterogeneity, have been identified, and the next few years will see a focus on identification of single-cell reporters which can be used to track the real time onset and localization of heterogeneity. A largely unexplored, but highly exciting, frontier would be the development of other myxobacteria as model systems for heterogeneity. *S. aurantiaca*, for instance, produces much more elaborate fruiting bodies containing additional features such as a stalk, branches, and individual spore-containing sporangioles (Reichenbach, 1993). This complex fruiting body is certainly the result of an ever more elaborate differentiation process that involves more signals, more receptors, and processes that are likely absent in *M. xanthus*.

Acknowledgements

The authors would like to thank members of their laboratories for critical reading of the manuscript and discussions with the *Myxococcus* research community that helped shape the concept of myxobacterial developmental differentiation discussed here. E.H. was supported by the Public Health Service Grant GM085024 from the National Institutes of Health. P.L.H. is support by the National Science Foundation (MCB1052525); P.I.H. and C.H. are supported by the Max Planck Society, and the Deutsche Forschungsgemeinschaft (HI 1593/2-1), and C.H. was supported by a fellowship from the International Max Planck Research School for Environmental, Cellular, and Molecular Microbiology.

References

Adye, J.C., and Powelson, D.M. (1961). Microcyst of *Myxococcus xanthus*. Chemical composition of the wall. J. Bacteriol. *81*, 780–785.

Bacon, K., and Rosenberg, E. (1967). Ribonucleic acid synthesis during morphogenesis in *Myxococcus xanthus*. J. Bacteriol. *94*, 1883–1889.

Bagby, S., Harvey, T.S., Eagle, S.G., Inouye, S., and Ikura, M. (1994). Structural similarity of a developmentally regulated bacterial spore coat protein to beta gamma-crystallins of the vertebrate eye lens. Proc. Natl. Acad. Sci. U.S.A. *91*, 4308–4312.

Balaban, N.Q. (2011). Persistence: mechanisms for triggering and enhancing phenotypic variability. Curr. Opin. Genet. Dev. *21*, 768–775.

Berleman, J.E., Vicente, J.J., Davis, A.E., Jiang, S.Y., Seo, Y.E., and Zusman, D.R. (2011). FrzS regulates social motility in *Myxococcus xanthus* by controlling exopolysaccharide production. PLoS ONE *6*, e23920.

Beveridge, T.J., and Graham, L.L. (1991). Surface layers of bacteria. Microbiol. Rev. *55*, 684–705.

Blackhart, B.D., and Zusman, D.R. (1985). 'Frizzy' genes of *Myxococcus xanthus* are involved in control of frequency of reversal of gliding motility. Proc Natl Acad Sci U S A *82*, 8767–8770.

Bonner, P.J., Black, W.P., Yang, Z., and Shimkets, L.J. (2006). FibA and PilA act cooperatively during fruiting body formation of *Myxococcus xanthus*. Mol. Microbiol. *61*, 1283–1293.

Bretscher, A.P., and Kaiser, D. (1978). Nutrition of *Myxococcus xanthus*, a fruiting myxobacterium. J. Bacteriol. *133*, 763–768.

Bui, N.K., Gray, J., Schwarz, H., Schumann, P., Blanot, D., and Vollmer, W. (2009). The peptidoglycan sacculus of *Myxococcus xanthus* has unusual structural features and is degraded during glycerol-induced myxospore development. J. Bacteriol. *191*, 494–505.

Burchard, R.P., and Dworkin, M. (1966). A bacteriophage for *Myxococcus xanthus*: isolation, characterization and relation of infectivity to host morphogenesis. J. Bacteriol. *91*, 1305–1313.

Campos, J.M., and Zusman, D.R. (1975). Regulation of development in *Myxococcus xanthus*: effect of 3′:5′-cyclic AMP, ADP, and nutrition. Proc. Natl. Acad. Sci. U.S.A. *72*, 518–522.

Curtis, P.D., and Shimkets, L.J. (2008). Metabolic pathways relevant to predation, signaling, and development. In Myxobacteria:Multicellularity and Differentiation, D.E. Whitworth, ed. (Washington, DC: ASM Press), pp. 241–258.

Curtis, P.D., Taylor, R.G., Welch, R.D., and Shimkets, L.J. (2007). Spatial organization of *Myxococcus xanthus* during fruiting body formation. J. Bacteriol. *189*, 9126–9130.

Cuthbertson, L., Mainprize, I.L., Naismith, J.H., and Whitfield, C. (2009). Pivotal roles of the outer membrane polysaccharide export and polysaccharide copolymerase protein families in export of extracellular polysaccharides in Gram-negative bacteria. Microbiol. Mol. Biol. Rev. *73*, 155–177.

Dahl, J.L., and Fordice, D. (2011). Small acid-soluble proteins with intrinsic disorder are required for UV resistance in *Myxococcus xanthus* spores. J. Bacteriol. *193*, 3042–3048.

Dahl, J.L., Tengra, F.K., Dutton, D., Yan, J., Andacht, T.M., Coyne, L., Windell, V., and Garza, A.G. (2007). Identification of major sporulation proteins of *Myxococcus xanthus* using a proteomic approach. J. Bacteriol. *189*, 3187–3197.

Dahl, J.L., Ulrich, C.H., and Kroft, T.L. (2011). Role of phase variation in the resistance of *Myxococcus xanthus* fruiting bodies to Caenorhabditis elegans predation. J. Bacteriol. *193*, 5081–5089.

Dawson, H., and Jones, M.V. (1979). Cell wall turnover during myxospore formation in *Myxococcus xanthus*. J. General Microbiol. *112*, 143–148.

Driks, A. (1999). *Bacillus subtilis* spore coat. Microbiol. Mol. Biol. Rev. *63*, 1–20.

Dworkin, M. (1962). Nutritional requirements for vegetative growth of *Myxococcus xanthus*. J. Bacteriol. *84*, 250–257.

Dworkin, M., and Gibson, S.M. (1964). A system for studying microbial morphogenesis: rapid formation of microcysts in *Myxococcus xanthus*. Science *146*, 243–244.

Dworkin, M., and Kaiser, D. (1985). Cell interactions in myxobacterial growth and development. Science *230*, 18–24.

Dworkin, M., and Sadler, W. (1966). Induction of cellular morphogenesis in *Myxococcus xanthus*. I. General description. J. Bacteriol. *91*, 1516–1519.

Filer, D., Kindler, S.H., and Rosenberg, E. (1977a). Myxospore coat synthesis in *Myxococcus xanthus*: enzymes associated with uridine 5′-diphosphate-N-acetylgalactosamine formation during myxospore development. J. Bacteriol. *131*, 745–750.

Filer, D., White, D., Kindler, S.H., and Rosenberg, E. (1977b). Myxospore coat synthesis in *Myxococcus xanthus: in vivo* incorporation of acetate and glycine. J. Bacteriol. *131*, 751–758.

Furuichi, T., Komano, T., Inouye, M., and Inouye, S. (1985). Functional complementation between the two homologous genes, *ops* and *tps*, during differentiation of *Myxococcus xanthus*. Mol. Gen. Genet. *199*, 434–439.

Furusawa, G., Dziewanowska, K., Stone, H., Settles, M., and Hartzell, P. (2011). Global analysis of phase variation in *Myxococcus xanthus*. Mol. Microbiol. *81*, 784–804.

Gelvan, I., Varon, M., and Rosenberg, E. (1987). Cell-density-dependent killing of *Myxococcus xanthus* by autocide AMV. J. Bacteriol. *169*, 844–848.

Gerdes, K., Christensen, S.K., and Lobner-Olesen, A. (2005). Prokaryotic toxin–antitoxin stress response loci. Nat. Rev. Microbiol. *3*, 371–382.

Gollop, R., Inouye, M., and Inouye, S. (1991). Protein U, a late-developmental spore coat protein of *Myxococcus xanthus*, is a secretory protein. J. Bacteriol. *173*, 3597–3600.

Harris, B.Z., Kaiser, D., and Singer, M. (1998). The guanosine nucleotide (p)ppGpp initiates development and A-factor production in *Myxococcus xanthus*. Genes. Dev. *12*, 1022–1035.

Henriques, A.O., and Moran, C.P., Jr. (2007). Structure, assembly, and function of the spore surface layers. Annu. Rev. Microbiol. *61*, 555–588.

Higgins, D., and Dworkin, J. (2012). Recent progress in *Bacillus subtilis* sporulation. FEMS Microbiol. Rev. *36*, 131–148.

Higgs, P.I., and Merlie, J.P., Jr. (2008). *Myxococcus xanthus*: cultivation, motility, and development. In Myxobacteria: Multicellularity and Differentiation, D.E. Whitworth, ed. (Washington, DC: ASM Press), pp. 465–478.

Hoiczyk, E., Ring, M.W., McHugh, C.A., Schwar, G., Bode, E., Krug, D., Altmeyer, M.O., Lu, J.Z., and Bode, H.B. (2009). Lipid body formation plays a central role in cell fate determination during developmental differentiation of *Myxococcus xanthus*. Mol. Microbiol. *74*, 497–517.

Inouye, M., Inouye, S., and Zusman, D.R. (1979). Biosynthesis and self-assembly of protein S, a development-specific protein of *Myxococcus xanthus*. Proc. Natl. Acad. Sci. U.S.A. *76*, 209–213.

Janssen, G.R., and Dworkin, M. (1985). Cell–cell interactions in developmental lysis of *Myxococcus xanthus*. Dev. Biol. *112*, 194–202.

Jing, H., Takagi, J., Liu, J.H., Lindgren, S., Zhang, R.G., Joachimiak, A., Wang, J.H., and Springer, T.A. (2002). Archaeal surface layer proteins contain beta propeller, PKD, and beta helix domains and are related to metazoan cell surface proteins. Structure *10*, 1453–1464.

Jones, M.V., Dawson, H., Wells, V.E., and Perkins, H.R. (1981). Growth and cellular differentiation of *Myxococcus xanthus* n the presence of beta-lactam antibiotics. J. General Microbiol. *124*, 281–290.

Julien, B., Kaiser, A.D., and Garza, A. (2000). Spatial control of cell differentiation in *Myxococcus xanthus*. Proc. Natl. Acad. Sci. U.S.A. *97*, 9098–9103.

Kaiser, D. (2003). Coupling cell movement to multicellular development in myxobacteria. Nat. Rev. Microbiol. *1*, 45–54.

Kearns, D.B. (2010). A field guide to bacterial swarming motility. Nat. Rev. Microbiol. *8*, 634–644.

Kearns, D.B., Bonner, P.J., Smith, D.R., and Shimkets, L.J. (2002). An extracellular matrix-associated zinc metalloprotease is required for dilauroyl phosphatidylethanolamine chemotactic excitation in *Myxococcus xanthus*. J. Bacteriol. *184*, 1678–1684.

Kim, S.K., and Kaiser, D. (1990). C-factor: a cell–cell signaling protein required for fruiting body morphogenesis of *M. xanthus*. Cell *61*, 19–26.

Kimura, Y., Kawasaki, S., Yoshimoto, H., and Takegawa, K. (2010). Glycine betaine biosynthesized from glycine provides an osmolyte for cell growth and spore germination during osmotic stress in *Myxococcus xanthus*. J. Bacteriol. *192*, 1467–1470.

Kimura, Y., Yamashita, S., Mori, Y., Kitajima, Y., and Takegawa, K. (2011). A *Myxococcus xanthus* bacterial tyrosine kinase, BtkA, is required for the formation of mature spores. J. Bacteriol. *193*, 5853–5857.

Komano, T., Furuichi, T., Teintze, M., Inouye, M., and Inouye, S. (1984). Effects of deletion of the gene for the development-specific protein S on differentiation in *Myxococcus xanthus*. J. Bacteriol. *158*, 1195–1197.

Kottel, R.H., Bacon, K., Clutter, D., and White, D. (1975). Coats from *Myxococcus xanthus*: characterization and synthesis during myxospore differentiation. J. Bacteriol. *124*, 550–557.

Kruse, T., Lobedanz, S., Berthelsen, N.M., and Sogaard-Andersen, L. (2001). C-signal: a cell surface-associated morphogen that induces and co-ordinates multicellular fruiting body morphogenesis and sporulation in *Myxococcus xanthus*. Mol. Microbiol. *40*, 156–168.

Kuner, J.M., and Kaiser, D. (1982). Fruiting body morphogenesis in submerged cultures of *Myxococcus xanthus*. J. Bacteriol. *151*, 458–461.

Laue, B.E., and Gill, R.E. (1994). Use of a phase variation-specific promoter of *Myxococcus xanthus* in a strategy for isolating a phase-locked mutant. J. Bacteriol. *176*, 5341–5349.

Laue, B.E., and Gill, R.E. (1995). Using a phase-locked mutant of *Myxococcus xanthus* to study the role of phase variation in development. J. Bacteriol. *177*, 4089–4096.

Lee, B., Mann, P., Grover, V., Treuner-Lange, A., Kahnt, J., and Higgs, P.I. (2011). The *Myxococcus xanthus* spore cuticula protein C is a fragment of FibA, an extracellular metalloprotease produced exclusively in aggregated cells. PLoS ONE *6*, e28968.

Lee, B., Holkenbrink, C., Treuner-Lange, A., and Higgs, P.I. (2012). *Myxococcus xanthus* developmental cell fate production: heterogeneous accumulation of developmental regulatory proteins and reexamination of the role of MazF in developmental lysis. J. Bacteriol. *194*, 3058–3068.

Leng, X., Zhu, W., Jin, J., and Mao, X. (2011). Evidence that a chaperone-usher-like pathway of *Myxococcus xanthus* functions in spore coat formation. Microbiology *157*, 1886–1896.

Lewis, K. (2007). Persister cells, dormancy and infectious disease. Nat. Rev. Microbiol. *5*, 48–56.

Licking, E., Gorski, L., and Kaiser, D. (2000). A common step for changing cell shape in fruiting body and starvation-independent sporulation of *Myxococcus xanthus*. J. Bacteriol. *182*, 3553–3558.

Lopez, D., Vlamakis, H., and Kolter, R. (2009). Generation of multiple cell types in *Bacillus subtilis*. FEMS Microbiol. Rev. *33*, 152–163.

Lopez, D., Vlamakis, H., and Kolter, R. (2010). Biofilms. Cold Spring Harb. Perspect. Biol. *2*, a000398.

Luciano, J., Agrebi, R., Le Gall, A.V., Wartel, M., Fiegna, F., Ducret, A., Brochier-Armanet, C., and Mignot, T. (2011). Emergence and modular evolution of a novel motility machinery in bacteria. PLoS Genet. *7*, e1002268.

Lux, R., Li, Y., Lu, A., and Shi, W. (2004). Detailed three-dimensional analysis of structural features of *Myxococcus xanthus* fruiting bodies using confocal laser scanning microscopy. Biofilms *1*, 293–303.

Magnuson, R.D. (2007). Hypothetical functions of toxin-antitoxin systems. J Bacteriol *189*, 6089–6092.

Maisonneuve, E., Shakespeare, L.J., Jorgensen, M.G., and Gerdes, K. (2011). Bacterial persistence by RNA endonucleases. Proc. Natl. Acad. Sci. U.S.A. *108*, 13206–13211.

Mayo, K.A., and Kaiser, D. (1989). *asgB*, a gene required early for developmental signalling, aggregation, and sporulation of *Myxococcus xanthus*. Mol. Gen. Genet. *218*, 409–418.

McBride, M.J., and Zusman, D.R. (1989). Trehalose accumulation in vegetative cells and spores of *Myxococcus xanthus*. J. Bacteriol. *171*, 6383–6386.

McBride, M.J., and Zusman, D.R. (1993). FrzCD, a methyl-accepting taxis protein from *Myxococcus xanthus*, shows modulated methylation during fruiting body formation. J. Bacteriol. *175*, 4936–4940.

McCleary, W.R., Esmon, B., and Zusman, D.R. (1991). *Myxococcus xanthus* protein C is a major spore surface protein. J. Bacteriol. *173*, 2141–2145.

Meiser, P., Bode, H.B., and Muller, R. (2006). The unique DKxanthene secondary metabolite family from the myxobacterium *Myxococcus xanthus* is required for developmental sporulation. Proc. Natl. Acad. Sci. U.S.A. *103*, 19128–19133.

Mohanty, S., and Firtel, R.A. (1999). Control of spatial patterning and cell-type proportioning in *Dictyostelium*. Semin. Cell Dev. Biol. *10*, 597–607.

Monier, J.M., and Lindow, S.E. (2003). Differential survival of solitary and aggregated bacterial cells promotes aggregate formation on leaf surfaces. Proc. Natl. Acad. Sci. U.S.A. *100*, 15977–15982.

Monier, J.M., and Lindow, S.E. (2004). Frequency, size, and localization of bacterial aggregates on bean leaf surfaces. Appl. Environ. Microbiol. *70*, 346–355.

Montanaro, L., Poggi, A., Visai, L., Ravaioli, S., Campoccia, D., Speziale, P., and Arciola, C.R. (2011). Extracellular DNA in biofilms. Int. J. Artificial Organs *34*, 824–831.

Mueller, C., and Dworkin, M. (1991). Effects of glucosamine on lysis, glycerol formation, and sporulation in *Myxococcus xanthus*. J. Bacteriol. *173*, 7164–7175.

Müller, F.D., Treuner-Lange, A., Heider, J., Huntley, S.M., and Higgs, P.I. (2010). Global transcriptome analysis of spore formation in *Myxococcus xanthus* reveals a locus necessary for cell differentiation. BMC Genomics *11*, 264.

Müller, F.D., Schink, C.W., Hoiczyk, E., Cserti, E., and Higgs, P.I. (2012). Spore formation in *Myxococcus xanthus* is tied to cytoskeleton functions and polysaccharide spore coat deposition. Mol. Microbiol. *83*, 486–505.

Nariya, H., and Inouye, S. (2003). An effective sporulation of *Myxococcus xanthus* requires glycogen consumption via Pkn4-activated 6-phosphofructokinase. Mol. Microbiol. *49*, 517–528.

Nariya, H., and Inouye, M. (2008). MazF, an mRNA interferase, mediates programmed cell death during multicellular *Myxococcus* development. Cell *132*, 55–66.

Nelson, D.R., and Zusman, D.R. (1983). Transport and localization of protein S, a spore coat protein, during fruiting body formation by *Myxococcus xanthus*. J. Bacteriol. *154*, 547–553.

Nilsen, T., Ghosh, A.S., Goldberg, M.B., and Young, K.D. (2004). Branching sites and morphological abnormalities behave as ectopic poles in shape-defective *Escherichia coli*. Mol. Microbiol. *52*, 1045–1054.

O'Connor, K.A., and Zusman, D.R. (1988). Reexamination of the role of autolysis in the development of *Myxococcus xanthus*. J. Bacteriol. *170*, 4103–4112.

O'Connor, K.A., and Zusman, D.R. (1989). Patterns of cellular interactions during fruiting-body formation in *Myxococcus xanthus*. J. Bacteriol. *171*, 6013–6024.

O'Connor, K.A., and Zusman, D.R. (1991a). Analysis of *Myxococcus xanthus* cell types by two-dimensional polyacrylamide gel electrophoresis. J. Bacteriol. *173*, 3334–3341.

O'Connor, K.A., and Zusman, D.R. (1991b). Behavior of peripheral rods and their role in the life cycle of *Myxococcus xanthus*. J. Bacteriol. *173*, 3342–3355.

O'Connor, K.A., and Zusman, D.R. (1991c). Development in *Myxococcus xanthus* involves differentiation into two cell types, peripheral rods and spores. J. Bacteriol. *173*, 3318–3333.

O'Connor, K.A., and Zusman, D.R. (1997). Starvation-independent sporulation in *Myxococcus xanthus* involves the pathway for beta-lactamase induction and provides a mechanism for competitive cell survival. Mol. Microbiol. *24*, 839–850.

O'Connor, K.A., and Zusman, D.R. (1999). Induction of beta-lactamase influences the course of development in *Myxococcus xanthus*. J. Bacteriol. *181*, 6319–6331.

Orlowski, M., Martin, P., White, D., and Wong, M.C. (1972). Changes in activity of glyoxylate cycle enzymes during myxospore development in *Myxococcus xanthus*. J. Bacteriol. *111*, 784–790.

Otani, M., Inouye, M., and Inouye, S. (1995). Germination of myxospores from the fruiting bodies of *Myxococcus xanthus*. J. Bacteriol. *177*, 4261–4265.

Otani, M., Kozuka, S., Xu, C., Umezawa, C., Sano, K., and Inouye, S. (1998). Protein W, a spore-specific protein in *Myxococcus xanthus*, formation of a large electron-dense particle in a spore. Mol. Microbiol. *30*, 57–66.

Pan, Y.T., Carroll, J.D., Asano, N., Pastuszak, I., Edavana, V.K., and Elbein, A.D. (2008). Trehalose synthase converts glycogen to trehalose. Febs. J. *275*, 3408–3420.

Reichenbach, H. (1993). Biology of the Myxobacteria: Ecology and Taxonomy. In Myxobacteria II, M. Dworkin, and A.D. Kaiser, eds. (Washington, DC: American Society for Microbilology), pp. 13–62.

Ring, M.W., Schwar, G., Thiel, V., Dickschat, J.S., Kroppenstedt, R.M., Schulz, S., and Bode, H.B. (2006). Novel iso-branched ether lipids as specific markers of developmental sporulation in the myxobacterium *Myxococcus xanthus*. J. Biol. Chem. *281*, 36691–36700.

Rolbetzki, A., Ammon, M., Jakovljevic, V., Konovalova, A., and Sogaard-Andersen, L. (2008). Regulated secretion of a protease activates intercellular signaling during fruiting body formation in *M. xanthus*. Dev. Cell *15*, 627–634.

Rosenberg, E., Keller, K.H., and Dworkin, M. (1977). Cell density-dependent growth of *Myxococcus xanthus* on casein. J. Bacteriol. *129*, 770–777.

Rosenbluh, A., and Rosenberg, E. (1989). Autocide AMI rescues development in *dsg* mutants of *Myxococcus xanthus*. J. Bacteriol. *171*, 1513–1518.

Rosenbluh, A., Nir, R., Sahar, E., and Rosenberg, E. (1989). Cell-density-dependent lysis and sporulation of *Myxococcus xanthus* in agarose microbeads. J. Bacteriol. *171*, 4923–4929.

Sadler, W., and Dworkin, M. (1966). Induction of cellular morphogenesis in *Myxococcus xanthus*. II. Macromolecular synthesis and mechanism of inducer action. J. Bacteriol. *91*, 1520–1525.

Sara, M., and Sleytr, U.B. (2000). S-Layer proteins. J. Bacteriol. *182*, 859–868.

Setlow, P. (2006). Spores of *Bacillus subtilis*: their resistance to and killing by radiation, heat and chemicals. J. Appl. Microbiol. *101*, 514–525.

Shi, W., and Zusman, D.R. (1993). The two motility systems of *Myxococcus xanthus* show different selective advantages on various surfaces. Proc. Natl. Acad. Sci. U.S.A. *90*, 3378–3382.

Shimkets, L.J. (1990). Social and developmental biology of the myxobacteria. Microbiol. Rev. *54*, 473–501.

Shimkets, L.J. (1999). Intercellular signaling during fruiting-body development of *Myxococcus xanthus*. Annu. Rev. Microbiol. *53*, 525–549.

Sogaard-Andersen, L. (2004). Cell polarity, intercellular signalling and morphogenetic cell movements in *Myxococcus xanthus*. Curr. Opin. Microbiol. *7*, 587–593.

Sogaard-Andersen, L., Overgaard, M., Lobedanz, S., Ellehauge, E., Jelsbak, L., and Rasmussen, A.A. (2003). Coupling gene expression and multicellular morphogenesis during fruiting body formation in *Myxococcus xanthus*. Mol. Microbiol. *48*, 1–8.

Sudo, S.Z., and Dworkin, M. (1969). Resistance of vegetative cells and microcysts of *Myxococcus xanthus*. J. Bacteriol. *98*, 883–887.

Takacs, C.N., Poggio, S., Charbon, G., Pucheault, M., Vollmer, W., and Jacobs-Wagner, C. (2010). MreB drives de novo rod morphogenesis in *Caulobacter crescentus* via remodeling of the cell wall. J. Bacteriol. *192*, 1671–1684.

Teintze, M., Thomas, R., Furuichi, T., Inouye, M., and Inouye, S. (1985). Two homologous genes coding for spore-specific proteins are expressed at different times during development of *Myxococcus xanthus*. J. Bacteriol. *163*, 121–125.

Tengra, F.K., Dahl, J.L., Dutton, D., Caberoy, N.B., Coyne, L., and Garza, A.G. (2006). CbgA, a protein involved in cortex formation and stress resistance in *Myxococcus xanthus* spores. J. Bacteriol. *188*, 8299–8302.

Tzeng, L., and Singer, M. (2005). DNA replication during sporulation in *Myxococcus xanthus* fruiting bodies. Proc. Natl. Acad. Sci. U.S.A. *102*, 14428–14433.

Tzeng, L., Ellis, T.N., and Singer, M. (2006). DNA replication during aggregation phase is essential for *Myxococcus xanthus* development. J. Bacteriol. *188*, 2774–2779.

Ueki, T., and Inouye, S. (2005). Identification of a gene involved in polysaccharide export as a transcription target of FruA, an essential factor for *Myxococcus xanthus* development. J. Biol. Chem. *280*, 32279–32284.

Varon, M., Cohen, S., and Rosenberg, E. (1984). Autocides produced by *Myxococcus xanthus*. J. Bacteriol. *160*, 1146–1150.

Varon, M., Tietz, A., and Rosenberg, E. (1986). *Myxococcus xanthus* autocide AMI. J. Bacteriol. *167*, 356–361.

Vlamakis, H., Aguilar, C., Losick, R., and Kolter, R. (2008). Control of cell fate by the formation of an architecturally complex bacterial community. Genes. Dev. *22*, 945–953.

Wall, D., Kolenbrander, P.E., and Kaiser, D. (1999). The *Myxococcus xanthus pilQ* (*sglA*) gene encodes a secretin homolog required for type IV pilus biogenesis, social motility, and development. J. Bacteriol. *181*, 24–33.

Wang, J., Hu, W., Lux, R., He, X., Li, Y., and Shi, W. (2011). Natural transformation of *Myxococcus xanthus*. J. Bacteriol. *193*, 2122–2132.

Ward, M.J., and Zusman, D.R. (1999). Motility in *Myxococcus xanthus* and its role in developmental aggregation. Curr. Opin. Microbiol. *2*, 624–629.

Ward, M.J., Lew, H., and Zusman, D.R. (2000). Social motility in *Myxococcus xanthus* requires FrzS, a protein with an extensive coiled-coil domain. Mol. Microbiol. *37*, 1357–1371.

Watson, B.F., and Dworkin, M. (1968). Comparative intermediary metabolism of vegetative cells and microcysts of *Myxococcus xanthus*. J. Bacteriol. *96*, 1465–1473.

Weijer, C.J. (2004). *Dictyostelium* morphogenesis. Curr. Opin. Genet. Dev. *14*, 392–398.

White, D. (1984). Structure and function of myxobacteria cells and fruiting bodies. In Myxobacteria: development and cell interactions, E. Rosenberg, ed. (New York: Springer Verlag), pp. 51–67.

White, D. (1993). Myxospore and Fruiting Body Morphogenesis. In Myxobacteria II, M. Dworkin, and D. Kaiser, eds. (Washington, DC: American Society for Microbiology), pp. 307–332.

White, D., Dworkin, M., and Tipper, D.J. (1968). Peptidoglycan of *Myxococcus xanthus*: structure and relation to morphogenesis. J. Bacteriol. *95*, 2186–2197.

Whitfield, C. (2006). Biosynthesis and assembly of capsular polysaccharides in *Escherichia coli*. Annu. Rev. Biochem. *75*, 39–68.

Whitfield, C., and Paiment, A. (2003). Biosynthesis and assembly of Group 1 capsular polysaccharides in *Escherichia coli* and related extracellular polysaccharides in other bacteria. Carbohydr. Res. *338*, 2491–2502.

Wireman, J.W., and Dworkin, M. (1975a). Morphogenesis and developmental interactions in myxobacteria. Science *189*, 516–523.

Wireman, J.W., and Dworkin, M. (1975b). Morphogenesis and developmental interactions in *Myxococcus xanthus*. Science *189*, 516–522.

Wireman, J.W., and Dworkin, M. (1977). Developmentally induced autolysis during fruiting body formation by *Myxococcus xanthus*. J. Bacteriol. *129*, 798–802.

Wistow, G., Summers, L., and Blundell, T. (1985). *Myxococcus xanthus* spore coat protein S may have a similar structure to vertebrate lens beta gamma-crystallins. Nature *315*, 771–773.

van der Woude, M.W., and Baumler, A.J. (2004). Phase and antigenic variation in bacteria. Clin. Microbiol. Rev. *17*, 581–611.

Xiao, Y., Wei, X., Ebright, R., and Wall, D. (2011). Antibiotic production by myxobacteria plays a role in predation. J. Bacteriol. *193*, 4626–4633.

Xiao, Y., Gerth, K., Muller, R., and Wall, D. (2012). Myxobacterium-produced antibiotic TA (myxovirescin) inhibits type II signal peptidase. Antimicrob. Agents Chemother. *56*, 2014–2021.

Xie, C., Zhang, H., Shimkets, L.J., and Igoshin, O.A. (2011). Statistical image analysis reveals features affecting fates of *Myxococcus xanthus* developmental aggregates. Proc. Natl. Acad. Sci. U.S.A. *108*, 5915–5920.

Xu, Q., Black, W.P., Cadieux, C.L., and Yang, Z. (2008). Independence and interdependence of Dif and Frz chemosensory pathways in *Myxococcus xanthus* chemotaxis. Mol. Microbiol. *69*, 714–723.

Yan, B.Q., Chen, X.L., Hou, X.Y., He, H., Zhou, B.C., and Zhang, Y.Z. (2009). Molecular analysis of the gene encoding a cold-adapted halophilic subtilase from deep-sea psychrotolerant bacterium *Pseudoalteromonas* sp. SM9913: cloning, expression, characterization and function analysis of the C-terminal PPC domains. Extremophiles *13*, 725–733.

Yang, Z., Ma, X., Tong, L., Kaplan, H.B., Shimkets, L.J., and Shi, W. (2000). *Myxococcus xanthus dif* genes are required for biogenesis of cell surface fibrils essential for social gliding motility. J. Bacteriol. *182*, 5793–5798.

Yeats, C., Bentley, S., and Bateman, A. (2003). New knowledge from old: in silico discovery of novel protein domains in *Streptomyces coelicolor*. BMC Microbiol. *3*, 3.

Zusman, D.R., Scott, A.E., Yang, Z., and Kirby, J.R. (2007). Chemosensory pathways, motility and development in *Myxococcus xanthus*. Nat. Rev. Microbiol. *5*, 862–872.

Cell Cycle Regulation in *Myxococcus xanthus* During Vegetative Growth and Development: Regulatory Links Between DNA Replication and Cell Division

4

Anke Treuner-Lange, Lotte Søgaard-Andersen and Mitchell Singer

Abstract

In response to starvation *Myxococcus xanthus* cells initiate a developmental programme that culminates in the formation of multicellular fruiting bodies inside which the rod-shaped cells differentiate into spherical myxospores. Whereas the rod-shaped vegetative cells contain one or two chromosomes, depending on the stage of the cell cycle, mature myxospores contain two chromosomes and peripheral rods, a differentiated cell type localized outside fruiting bodies, contain one chromosome. Moreover, during development, DNA replication occurs in the pre-aggregation stage and a new round of replication and termination of replication in the pre-aggregation stage are essential for fruiting body formation and myxospore differentiation. These observations suggest that the cell cycle is strictly regulated during fruiting body formation. Here, we describe the current understanding of the *M. xanthus* cell cycle with special emphasis on replication and chromosome organization as well as cell division. It is currently unknown how myxospores with two chromosomes and peripheral rods with one chromosome arise during development. Importantly, with the recent introduction of cell biology methods, the tools are now available to address how DNA replication, chromosome organization, cell division and cell fate determination are coupled during *M. xanthus* fruiting body formation.

Introduction

One of the best-studied examples of bacterial differentiation in response to environmental changes is spore formation. Bacterial species that are phylogenetically widely separated have evolved the ability to form spores in response to starvation (Shimkets and Brun, 2000). Spore formation has been most extensively studied in three very different model organisms: *Bacillus subtilis* (low G+C Gram-positive), *Streptomyces coelicolor* (high G+C Gram-positive) and *M. xanthus* (Deltaproteobacterium). Given the large phylogenetic distances between these three species, it is hardly surprising that the mechanisms underlying spore formation and the actual spores (endospore, exospore and myxospore, respectively) are morphologically very different in these three species.

Interestingly, cell cycle control during spore formation is also very different in these three species: In *B. subtilis* (Kroos, 2007), spore formation is initiated by an asymmetric cell division in a rod-shaped cell resulting in the formation of a large mother cell and a small forspore (Fig. 4.1). The forspore is then engulfed by the mother cell and ultimately the mother cell lyses resulting in the release of the haploid mature spore. In *S. coelicolor*, chains of haploid spores are formed by synchronous cell divisions in aerial hyphae (Flärdh and Buttner, 2009) (Fig. 4.1). In *M. xanthus*, spore formation is coupled to the formation of multicellular fruiting bodies and individual rod-shaped cells that have accumulated inside a

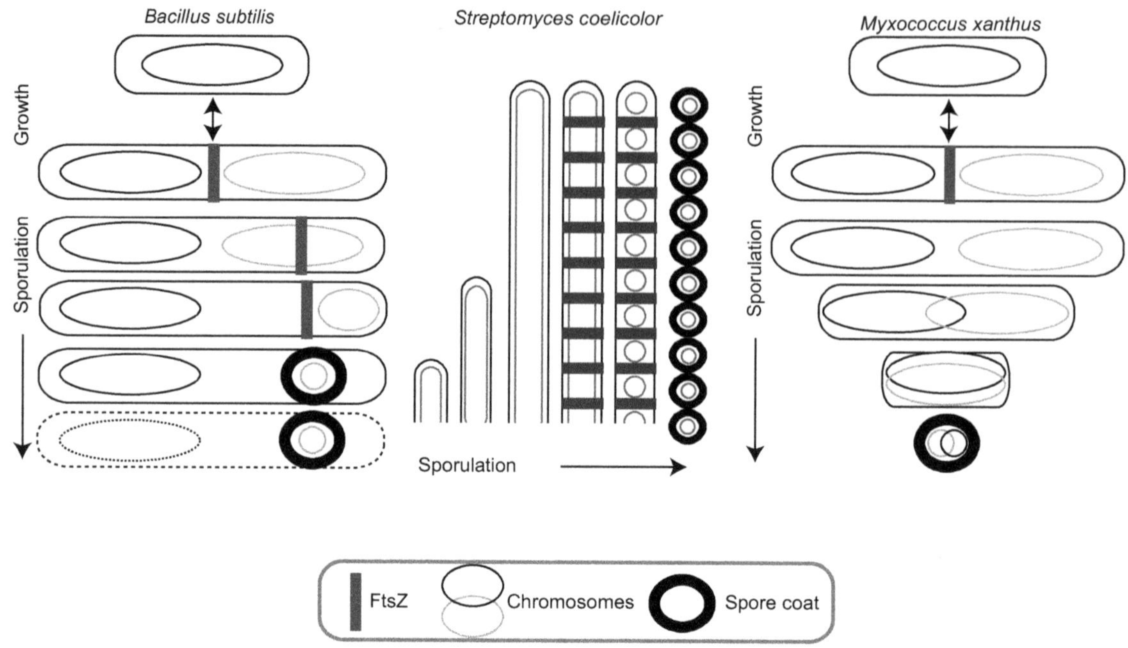

Figure 4.1 Sporulation in *M. xanthus*, *B. subtilis* and *S. coelicolor*. The *M. xanthus* cell is shown with two chromosomes at the initiation of sporulation. It is not known whether cells that differentiate to myxospores contain two chromosomes at the onset of starvation. The lysed *B. subtilis* mother cell and its chromosome are indicated in stippled lines. See text for details.

fruiting body differentiate to spherical diploid spores without a prior cell division (Kroos, 2007) (Fig. 4.1). These differences in the sporulation process suggest that in contrast to *B. subtilis* and *S. coelicolor*, no specific developmental cell division occurs in *M. xanthus* in order for the sporulation process to successfully occur. Rather an inhibition of cell division must occur in those *M. xanthus* cells destined to become spores. While much work has been done over the past 30 years to understand cell motility, intercellular communication and signal transduction mechanisms in *M. xanthus* and how these processes are coordinated during fruiting body formation, the basic cell cycle processes of replication and cell division as well as how these processes are modified during fruiting body formation with myxospore formation have only received limited attention. In this chapter, we describe recent data that are beginning to reveal how the *M. xanthus* cell cycle is regulated with special emphasis on replication, chromosome organization and cell division in vegetative and developing cells.

Organization of the *Myxococcus xanthus* chromosome

Bacteria typically contain a single circular chromosome. Replication initiates at a well-defined origin of replication (*ori*) (Kaguni, 2011), proceeds bidirectionally, and terminates in the terminus (*ter*) region located opposite to the origin (Hendrickson and Lawrence, 2007). Following replication termination, daughter chromosomes are physically separated by the action of topoisomerases to resolve catenates, and site specific recombinases acting at the *dif* site in the *ter* region to resolve chromosome dimers (Espeli and Marians, 2004; Sherratt *et al.*, 2010). Bioinformatic approaches, such as cumulative GC skew analysis, can help to identify *ori* and *ter* regions. GC skew typically divides a chromosome into two regions corresponding to the two replichores: one with an excess of G over C, and the other with an excess of C over G (Salzberg *et al.*, 1998). Moreover, the shift points of the GC skew correlate with *ori* and *ter* (Lobry, 1996).

The 9,139,763 bp *M. xanthus* genome is circular and with the *ori* predicted to be located by GC nucleotide skew at position 1 (Goldman *et al.*, 2006). Using a similar approach, the *ter* region is predicted to be at bp 4,547,166 (corresponding to 49.75% of the chromosome) (Carnoy and Roten, 2009), suggesting that the *ter* region of the *M. xanthus* genome is opposite to *ori* on the chromosome (Fig. 4.2A). In support of this location of the *ter* region, the *dif* site is at position 4,489,697 bp (49.12% of the chromosome) (Carnoy and Roten, 2009).

Bacterial chromosomes are spatially organized with different chromosomal loci reproducibly localizing to the same cellular regions (Viollier *et al.*, 2004). The spatial organization of bacterial chromosomes in living cells can be mapped using fluorescently tagged DNA-binding proteins that bind to their cognate binding sequences on the chromosome (Wang *et al.*, 2008). In FROS

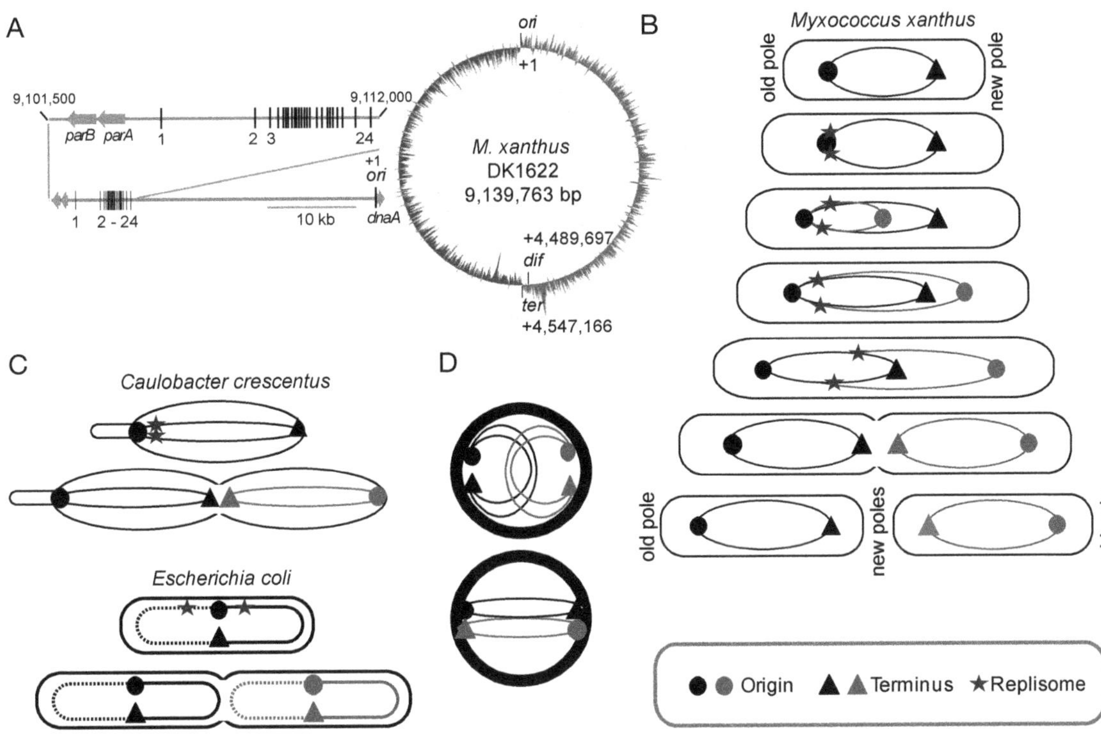

Figure 4.2 Chromosome organization and dynamics in *M. xanthus*. (A) Overall organization of the *M. xanthus* chromosome. GC skew is indicated on the circle representing the *M. xanthus* chromosome (right) together with the *ori* and *ter* regions, including *dif*. The GC skew is calculated as (C-G)/(C+G). On the circular chromosome, there is more C than G along the right half, i.e. replichore, of the chromosome (light grey) and less C than G on the left replichore (dark grey). In the diagram on the left, the location of the *dnaA* gene, the *parAB* genes and the 24 *parS* sites (black bars) are also indicated. (B) Chromosome organization and dynamics in *M. xanthus* during replication and segregation. The diagrams illustrate a cell through six stages of chromosome replication and segregation during a cell cycle. The two chromosomes with their associated *ori* and *ter* are shown in black and grey, respectively. (C) Chromosome organization in *C. crescentus* and *E. coli*. The diagrams illustrate (upper) a cell with one chromosome and two replisomes and (lower) a predivisional cell after completion of replication and segregation. The two chromosomes with their associated *ori* and *ter* are shown in black and grey, respectively. The stalk in *C. crescentus* is indicated. In *E. coli*, the two replichores are shown as stippled and non-stippled lines. (D) Chromosome organization in myxospores. Shown are schematics of two possible organizations of the two chromosomes in *M. xanthus* myxospores. Upper panel: if *ori* and *ter* on the same chromosome is localized to the same region, then the intermediate regions of the chromosome would loop into the interior of the myxospore; lower panel, if the *ori* of one chromosome and the *ter* of the second chromosome are localized to the same region, then the two chromosomes would run antiparallel to each other.

(fluorescent–repressor–operator system), different GFP variants are fused to the DNA binding proteins LacI and TetR and expressed in a cell. In parallel, the cognate operators are introduced as tandem repeats in several hundred copies on the chromosome. The binding of the fluorescently tagged repressors to their cognate operators allows the visualization of the genetic loci by fluorescence microscopy. In the case of the ParB/*parS* system, ParB is the DNA binding protein that is fluorescently tagged. ParB binds to *parS* sequences that are found as tandem repeats on the chromosome (see below) and fluorescently tagged ParB can be visualized as a fluorescent focus when it binds to the *parS* sequences. The FROS, as well as the ParB/*parS* system, have been used to map the spatial organization of the *M. xanthus* chromosome.

The *parAB* locus of *M. xanthus* localizes in the vicinity of *ori* (Koch *et al.*, 2011; Treuner-Lange *et al.*, 2013). Using a bioinformatic approach, 12 *parS* sequences were identified in the *M. xanthus* chromosome (Livny *et al.*, 2007). Recently, 25 *parS* sequences were identified using a less stringent approach, and 24 of these *parS* sequences cluster 30 kb away from *ori* and close to the *parAB* operon (Treuner-Lange *et al.*, 2013) (Fig. 4.2A). *In vitro*, ParB from *M. xanthus* binds specifically to a consensus *parS* sequence (A. Harms *et al.*, 2013). By following ParB localization using a ParB-YFP fusion protein it was found that in short cells with a single chromosome *ori* was located in a subpolar region 1.4 ± 0.3 μm from the nearest pole (corresponding to 27 ± 6% of the cell length) and close to the old cell pole (Fig. 4.2B). To visualize the *ter* region of the *M. xanthus* chromosome, the fluorescent TetR protein was followed in a *M. xanthus* strain containing an array of ~250 TetR-binding sites (*tetO*-array) at a position representing 53.3% of the *M. xanthus* genome (A. Harms *et al.*, 2013). In short cells with one chromosome, the *ter* region also localizes in a subpolar region but close to the new cell pole with a mean distance of 0.9 ± 0.3 μm to the nearest pole (corresponding to 17±5% of the cell length). This pattern of *ori* and *ter* localization suggest that the *M. xanthus* chromosome is organized around a longitudinal-axis with *ori* and *ter* in opposite subpolar regions (Fig. 4.2B).

Chromosome replication and segregation in *Myxococcus xanthus*

In bacteria, chromosome replication and segregation are temporally closely coupled, and segregation of daughter chromosomes occurs in parallel with replication. Moreover, the spatial organization of bacterial chromosomes is laid down during the segregation process (Viollier *et al.*, 2004). Work in several different bacterial species have demonstrated that chromosomally encoded orthologues of the plasmid partitioning proteins ParA and ParB function in chromosome segregation as well as in the spatial organization of bacterial chromosomes (Gerdes *et al.*, 2010). These two proteins function together with the *parS* sequences, which are often present in multiple copies in the *ori* region. In these systems, ParB binds the *parS* sequences. ParA proteins are ATPases and interact with the ParB/*parS* complex to mediate chromosome partitioning – most likely by pulling daughter chromosomes apart (Fogel and Waldor, 2006; Ptacin *et al.*, 2010).

Early studies by Zusman and Rosenberg demonstrated that vegetative *M. xanthus* cells contain on average 1.7 copies of the chromosome and replicate their DNA during 80% of the cell cycle (Zusman and Rosenberg, 1970). Replication and chromosome segregation in *M. xanthus* have been visualized by following fluorescently tagged ParB/*parS* complexes and fluorescently tagged replisomes by time-lapse fluorescence microscopy (Treuner-Lange *et al*, 2013; A. Harms *et al.*, 2013). As described, the *M. xanthus* genome encodes a ParA and ParB pair of proteins and 24 *parS* sequences close to *ori*. As discussed, in short cells, ParB-YFP localizes to a subpolar region close to the old cell pole. Shortly after cell division, the single ParB focus splits into two foci followed by segregation of one focus from the original subpolar region to the opposite subpolar region. Once the segregation is complete, the two ParB-YFP foci are essentially stationary and the two daughter chromosomes display mirror symmetry (Fig. 4.2B). Using the FROS system and the *tetO* array integrated close to the *ter* region to follow replication and segregation of this part of the chromosome, it has been shown that a single focus splits into two foci and then one focus segregates to the opposite cell half. As expected, replication of the *ter* region occurs later in the cell cycle than in the case of *ori*. In newborn cells, the *ter* region is in a subpolar region whereas the *ter* region is close to midcell when it is replicated. Overall, these findings suggest that during chromosome replication and segregation, one of the ParB/*parS* complex segregates from one subpolar region to the opposite subpolar region. Subsequently, the remainder of the segregating daughter chromosome follows and is 'laid down' between the segregated ParB/*parS* and the unreplicated mother chromosome. Presumably, this process also results in the *ter* region of the unreplicated chromosome being displaced towards midcell. In total, this segregation pattern results in two daughter chromosomes that are organized about a transverse axis at midcell and display mirror symmetry (Fig. 4.2B).

To visualize the replisome in *M. xanthus,* a fluorescent fusion protein of the single-strand binding protein (Ssb) was used (A. Harms *et al.,* 2013). Ssb has previously been used for localization of replisomes (Reyes-Lamothe *et al.,* 2008) because it binds tightly and cooperatively to single-stranded DNA during replication (Shereda *et al.,* 2008). Fluorescence microscopy revealed that even newborn cells contain Ssb-YFP foci suggesting that replication initiates during or shortly after cell division. As expected, the Ssb-YFP foci in newborn cells localize similarly (1.4 ± 0.3 µm to nearest pole, 26±5% of cell length) to that of the single ParB-YFP focus in newborn cells indicating that replication is indeed initiated at *ori* in the subpolar location close to the old pole. Time-lapse microscopy showed that the Ssb-YFP focus/foci move from the subpolar region towards midcell and reaches the midcell region when the *ter* region has relocated to the midcell region. In time-lapse recordings, Ssb-YFP is sometimes observed in one focus and sometimes in two foci, dynamically changing from one focus to two foci and back to one focus. These Ssb-YFP foci dynamics suggest that the two replisomes move independently of each other and that the fusion events may be the result of the two replisomes moving close together and appearing as a single focus before moving apart resulting in the single focus splitting in to two foci. Thus, these data are in agreement with previous analyses showing that chromosome replication in *M. xanthus* occurs through approximately 80% of the cell cycle (Zusman and Rosenberg, 1970).

In total, these observations lead to the following overall picture of *M. xanthus* chromosome replication and segregation (Fig. 4.2B): Replication initiates during or shortly after cell division at *ori,* which is in a subpolar region close to the old cell pole. During the chromosome segregation process, one daughter chromosome remains stationary whereas the second daughter chromosome segregates asymmetrically and unidirectionally towards the opposite cell pole. Specifically, the newly replicated and segregated *ori* (as determined following the ParB/*parS* complex) is positioned in the subpolar region close to the new pole. Subsequently, the remaining segregating daughter chromosome is laid down progressively and sequentially between this subpolar region and midcell. In parallel, and likely as a consequence of the segregating daughter chromosome being refolded into a compact chromosomal domain, the *ter* region on the mother chromosome is displaced towards midcell. Thus, after replication and segregation are complete the two chromosomes in the predivisional cell display mirror symmetry about a transverse axis at midcell. The two replisomes move from *ori* towards *ter,* which is in parallel displaced towards midcell from the opposite pole. The two replisomes meet when *ter* is at midcell and disassemble before cell division.

Chromosome organization and dynamics have been described in detail in two bacteria, the Alphaproteobacterium *Caulobacter crescentus* and the Gammaproteobacterium *Escherichia coli.* The Deltaproteobacterium *M. xanthus* chromosome organization is overall similar to that of *C. crescentus* but also has novel, unique features. Similar to the *C. crescentus* genome, the *M. xanthus* chromosome is organized about a longitudinal axis in newborn cells and predivisional cells contain two chromosomes displaying mirror symmetry about a transverse axis at midcell (Viollier *et al.,* 2004). However, in *C. crescentus* the ParB/*parS* complex is at the extreme pole and anchored to this pole by the PopZ protein (Bowman *et al.,* 2008; Ebersbach *et al.,* 2008) and the *ter* region is in close proximity to the opposite pole (Viollier *et al.,* 2004) (Fig. 4.2C). In comparison, the

ParB/*parS* complex and *ter* region in *M. xanthus* are localized in the subpolar regions and with a distance to the poles of approximately 1 μm in newborn cells. *M. xanthus* does not have a homologue of PopZ. The localization of the ParB/*parS* complex in *M. xanthus* suggests that factor(s) analogous to that of PopZ function to anchor the ParB/*parS* complex in the subpolar regions in *M. xanthus*. In sporulating *B. subtilis* cells, the origins are at the extreme poles and anchored to the poles by the RacA protein (Ben-Yehuda *et al.*, 2003); however, the *M. xanthus* genome does not encode a RacA homologue. Similarly, in the actinobacterium *Corynebacterium glutamicum* (high G+C Gram-positive) the ParB/*parS* complex is anchored to the cell poles by the DivIVA protein (Donovan *et al.*, 2012). *M. xanthus* contains a DivIVA homologue, however, an in-frame deletion of this gene does not cause a chromosome segregation defect nor a change in chromosome organization (Treuner-Lange *et al.*, 2013). In slow-growing *E. coli* cells containing a single chromosome, the chromosome is organized about a transverse axis with *ori* at midcell, *ter* broadly distributed around midcell, and the two replichores occupying separate cell halves (left–right asymmetry) in newborn cells (Wang *et al.*, 2006) (Fig. 4.2C). After replication and during segregation, the duplicated replichores segregate bidirectionally and symmetrically with the *ori* segregating from midcell to ¼ and ¾ positions ultimately generating a predivisional cell with the two segregated chromosomes displaying translational symmetry (left–right–left–right) (Wang *et al.*, 2005, 2006) (Fig. 4.2C). Upon cell division, this pattern of chromosome organization results in two daughter cells with the left–right asymmetry and the origin at midcell (Wang *et al.*, 2006).

The *M. xanthus* replisome dynamics share a unique blend of characteristics known from *C. crescentus*, *E. coli* and *B. subtilis* (Fig. 4.2C). Similar to the replisomes in *E. coli* (Reyes-Lamothe *et al.*, 2008) and *B. subtilis* (Migocki *et al.*, 2004; Berkmen and Grossman, 2006), the two replisomes in *M. xanthus* track independently along the two replichores whereas in *C. crescentus* the two replisomes form a single 'moving factory' (Jensen *et al.*, 2001). On the other hand, the overall direction of movement of the two replisomes in *M. xanthus* from a subpolar region towards midcell is similar to that in *C. crescentus* whereas in *E. coli* (Reyes-Lamothe *et al.*, 2008) and *B. subtilis* (Migocki *et al.*, 2004; Berkmen and Grossman, 2006), the two replisomes track along replichores localized in the two cell halves.

Cell division in *Myxococcus xanthus*

Cell division is a highly complex process in which the tubulin homologue FtsZ has a central function (Lutkenhaus, 2007). FtsZ is a cytoplasmic protein that polymerizes to form a ring-like structure at the division site, the Z-ring. Formation of the Z-ring at the incipient division site is the earliest detectable event during cell division and leads to the recruitment and assembly of the remaining division machinery. This machinery involves proteins, which drive the invagination of the cell membranes and synthesis of cell wall material. It is essential for the cells to ensure that Z-ring formation occurs at the right time, i.e. after the chromosome replication and segregation have been completed, and at the right position, between the segregated chromosomes. Only then will appropriately sized daughter cells with the correct chromosomes number be generated. Accordingly, bacteria contain sophisticated regulatory mechanisms to regulate formation of the Z-ring spatiotemporally.

Vegetative *M. xanthus* cells divide at midcell by binary fission to produce two equally sized daughter cells each with one chromosome. The *M. xanthus* genome encodes homologues of many cell division proteins including FtsZ. As in other bacteria, FtsZ in *M. xanthus* is essential (Garcia-Moreno *et al.*, 2009; Treuner-Lange *et al.*, 2013). FtsZ localizes in a cell cycle-dependent manner: In short cells with one chromosome or two incompletely segregated chromosomes FtsZ localizes in a diffused/speckled pattern and in long cells with two segregated chromosomes as a transverse band at midcell (Fig. 4.3) (Treuner-Lange *et al.*, 2013).

The correct positioning of the Z-ring in *M. xanthus* depends on the ParA-like protein PomZ, which is a so-called orphan ParA-like protein because it is not adjacent to a *parB* gene (Treuner-Lange *et al.*, 2013). A Δ*pomZ* mutation

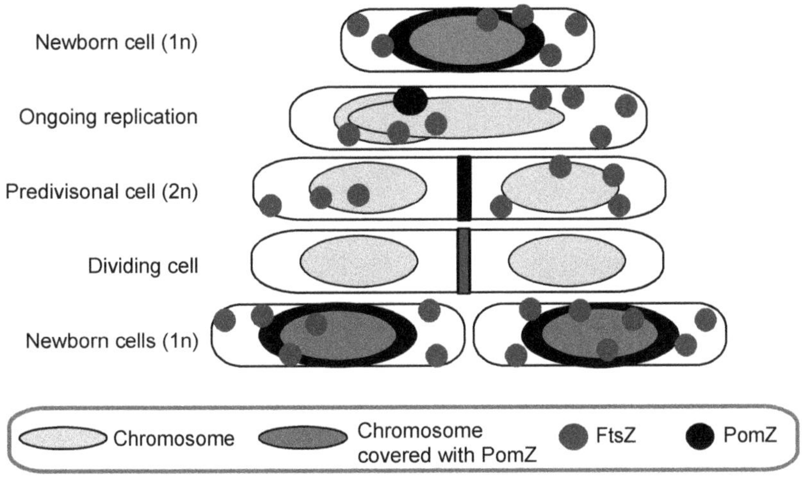

Figure 4.3 Cell division in *M. xanthus*. Diagrams illustrating the localization of FtsZ and the regulator of Z-ring formation, PomZ, during a cell cycle.

results in the formation of filamentous cells and chromosome-free minicells, a typical cell division phenotype. Indeed, lack of PomZ results in cell division defects and reduced formation and incorrect positioning of the few Z-rings formed (Treuner-Lange *et al.*, 2013). The localization of PomZ correlates with cell cycle progression. In short cells with one chromosome PomZ localizes in a patchy pattern over the chromosome (Fig. 4.3, upper cell); in longer cells with one or two incompletely segregated chromosomes PomZ localizes in a single cluster over the nucleoid (Fig. 4.3, second cell from top); and, in the longest cells with two completely segregated chromosomes it localizes at midcell in a cluster which spans the cell width (Treuner-Lange *et al.*, 2013) (Fig. 4.3, third cell from top). PomZ also localizes at midcell in constricting cells and seems to be an integral part of the divisome (Treuner-Lange *et al.*, 2013). Most importantly, PomZ localizes to the future division site before FtsZ as well as in the absence of FtsZ (Treuner-Lange *et al.*, 2013) (Fig. 4.3, third and fourth cells from top). The cell cycle-dependent localization of PomZ suggests a mechanism that couples cell cycle progression and cell division in *M. xanthus* (Fig. 4.3) (Treuner-Lange *et al.*, 2013). In total, these findings suggest that PomZ may recruit FtsZ to midcell and, thus, formation of the Z-ring at midcell is positively regulated in *M. xanthus* during growth.

The model for positive regulation of FtsZ recruitment to midcell in growing *M. xanthus* cells is in stark contrast to the models for the spatiotemporal regulation of Z-ring formation in other bacteria. In the model organisms *E. coli*, *B. subtilis* and *C. crescentus*, the spatiotemporal regulation of Z-ring formation at midcell is thought to rely on negative regulators (Lutkenhaus, 2007; Adams and Errington, 2009; Thanbichler, 2009) that inhibit Z-ring formation at the poles and over the nucleoid leaving only midcell free for the formation of the Z-ring. Specifically, in *E. coli* and *B. subtilis*, the Min system inhibits Z-ring formation at the cell poles and the nucleoid occlusion proteins, SlmA and Noc, inhibit Z-ring formation over the nucleoids. In *C. crescentus*, the MipZ protein inhibits Z-ring formation at the poles and over the nucleoid.

A crucial event in the *B. subtilis* sporulation process is the formation of an asymmetrically localized Z-ring over one of the chromosomes. The formation of the asymmetrically localized Z-ring is followed by an asymmetric division event and the chromosome destined for the spore is transported into the forspore (Errington, 2003) (Fig. 4.1). This shift in Z-ring formation depends on a Spo0A~P-dependent increase in transcription of the *ftsZ* gene and of the gene encoding the SpoIIE protein. SpoIIE and FtsZ localize to the asymmetric cell division site and localization of SpoIIE depends on FtsZ. How FtsZ recognizes the asymmetrical, incipient division site is not known. In *S. coelicolor* FtsZ is dispensable for growth, however, it is indispensable for the generation of the uninucleoid spores (Flärdh and Buttner, 2009; McCormick, 2009). In aerial hyphae, FtsZ first assembles into a helical filament and later into regularly positioned Z-rings (Fig. 4.1). During sporulation, FtsZ is recruited to the future division sites by the membrane-associated SsgB protein, which also stimulates the polymerization of FtsZ *in vitro* (Willemse *et al.*, 2011). Thus, in *S. coelicolor*, Z-ring formation is positively regulated.

The cell cycle and DNA synthesis during sporulation

In response to starvation, *M. xanthus* cells initiate a developmental programme that culminates in the formation of haystack-shaped fruiting bodies. Inside the fruiting bodies, the rod-shaped cells differentiate into myxospores. This developmental programme can be divided into three stages: The pre-aggregation stage from 0 to 6 h; the aggregation stage from 6 to 24 h during which cells move into aggregation centres; and, cell differentiation during which cells inside fruiting bodies differentiate into myxospores (note that the specific timing is dependent on the wild type strain used and the conditions under which cells are starved). Only a fraction of starving cells become myxospores, the majority of the cells either remains as peripheral rods outside of the fruiting bodies (O'Connor and Zusman, 1991) or undergo lysis (Rosenbluh *et al.*, 1989) (see Chapter 3 for details). *M. xanthus* can be artificially induced to form spore-like structures by addition of 0.5 M glycerol and in such a case nearly all cells form spores within 2 to 8 h (Dworkin and Gibson, 1964; Rosenberg *et al.*, 1967). Work on the DNA replication cycle of *M. xanthus* was pioneered in 1960s and 1970s by Zusman and Rosenberg studying the relationships between DNA content, vegetative growth, glycerol-induced spore formation and the subsequent germination of these chemically induced spores (Zusman and Rosenberg, 1968, 1970; Zusman *et al.*, 1978). After addition of glycerol to vegetative cells, an increase of DNA content of about ~ 20% was observed and led to a model that upon glycerol addition ongoing rounds of replication were completed but that no re-initiation of replication would occur, so that glycerol-induced spores would contain between 1–4 chromosomes (Rosenberg *et al.*, 1967; Zusman and Rosenberg, 1968). Recent studies using flow cytometry confirmed that glycerol-induced spores contain a wide range of DNA content, from 1 to at least 3 copies of the chromosome. Importantly, glycerol-induced spores differ from starvation-induced spores with respect to the spore coat and protein composition (Kottel *et al.*, 1975; Inouye *et al.*, 1979a,b; McCleary *et al.*, 1991) and the flow cytometry-based studies of chromosome content indicated differences in chromosome content between fruiting body derived myxospores and glycerol-induced spores (Tzeng and Singer, 2005). The differences between the two types of spores motivated several laboratories to reexamine the roles of DNA replication, cell division and the cell cycle in *M. xanthus* development.

Tzeng and Singer observed that after completion of fruiting body formation, cells either contained one or two chromosomes (Tzeng and Singer, 2005). Importantly, using flow cytometry purified myxospores were found to possess two copies of the chromosome whereas the peripheral rods contained a single copy of the chromosome, similar to stationary phase cells. Moreover, quantitative fluorescence microscopy and fluorescence in situ hybridization confirmed that myxospores were diploid. Interestingly, Tzeng and Singer also reported that the diploid myxospore not only contains two copies of the chromosomes, but that the *ori* and *ter* regions of the chromosome are associated with the peripheral regions of the myxospore and apparently in close proximity (Fig. 4.2D). These observations could suggest that the *ori* and *ter* regions on the same chromosome are bound to adjacent areas in the myxospore with the rest of the chromosome looping into the interior (Fig. 4.2D). Alternatively, *ori* and *ter* of the same chromosome are at opposite 'ends' of a myxospore with the rest of the chromosome in the interior and with the two chromosomes running antiparallel to each other across the 'length' of the myxospore (Fig. 4.2D).

DNA replication is required for fruiting body formation and myxospore formation

The observation that the two differentiated cell types that arise during fruiting body formation have different chromosome content raises many interesting questions with respect to cell cycle regulation and cell fate during fruiting body formation. During fruiting body formation, DNA replication occurs in the pre-aggregation phase and until 6 to 8 h of starvation (Tzeng *et al.*, 2006) and ceases before the aggregation stage begins. Tzeng *et al.* (2006) observed that addition of DNA replication inhibitors (hydroxyurea, nalidixic acid, and novobiocin) before 6 h of starvation blocked development at the early aggregation stage and also blocked sporulation. Further analyses demonstrated that addition of DNA replication inhibitors at later stages of fruiting body formation did not interfere with fruiting body formation and myxospore formation. Because the replication inhibitors used block ongoing rounds of replication, one interpretation of these data would be that completion of ongoing DNA replication is required for development, but new rounds of replication are not.

To address whether new rounds of replication are required for fruiting body formation, Rosario and Singer (Rosario and Singer, 2007) used a conditional lethal, temperature sensitive mutation of *dnaB* ($DnaB^{A116V}$). At the permissive temperature, 32°C, the mutant grows similarly to wild type. Moreover, the mutant forms wild type fruiting bodies with comparable sporulation frequencies at the permissive temperature. At the non-permissive temperature (37°C), cell growth stops within 6 h and cells arrest at the haploid 1n state. Moreover, DNA replication analysis of the mutant revealed a 'slow-stop' DNA synthesis phenotype. More importantly, after 6 to 12 h at the non-permissive temperature (and now being 1n), cells were blocked for fruiting body development and myxospore formation. Furthermore, this block in fruiting body formation was reversed by shifting the mutant back to the permissive temperature. Thus, *M. xanthus* cells require a new round of DNA replication in order to form fruiting bodies and differentiate to form myxospores.

DNA replication and checkpoint control of development

Current evidence strongly points to the pre-aggregation stage as being the singular most important stage in development in terms of replication. This is based on (i) the DNA replication inhibitor studies, which demonstrated that these inhibitors block development if added to pre-aggregating cells but not if added to aggregating cells (Tzeng *et al.*, 2006); (ii) the studies that demonstrated replication occurring in the pre-aggregation stage and prior to the aggregation stage (Tzeng *et al.*, 2006), and (iii) the *dnaB* mutant studies, which demonstrated that a new round of replication is needed to proceed into the aggregation stage of fruiting body formation (Rosario and Singer, 2007). These observations strongly imply the existence of a checkpoint mechanism at the early aggregation stage that halts development in response to errors in DNA replication.

An essential player in bacterial DNA replication is the universally conserved protein for initiation of DNA replication, DnaA (Messer, 2002; Mott and Berger, 2007). The *M. xanthus* DnaA protein shares strong sequence homology to those of many other proteobacteria (Goldman *et al.*, 2006). Rosario and Singer used qRT-PCR to analyse *M. xanthus dnaA* expression during development in wild type and in a variety of developmental mutants. In *E. coli*, *dnaA* expression is controlled by the RelA-dependent stringent response and expression decreases under nutrient limitation. In *M. xanthus*, like *E. coli*, *dnaA* expression initially decreases when cells are starved for nutrients, however, this decrease is not RelA-dependent (Rosario and Singer, 2010). In addition, 3 to 4 h post starvation, a burst in *dnaA* expression is detected by qRT-PCR (Rosario and Singer, 2010). This burst in expression of *dnaA* is consistent with initiation of DNA replication during the pre-aggregation stage of development. Moreover, it is during this stage of fruiting body formation that a variety of signal transduction pathways are activated and ultimately converge to stimulate aggregation, gene expression and sporulation (Whitworth, 2008). Interestingly, Rosario and Singer found that the developmental burst of *dnaA* expression is dependent on three of these early regulators, *relA*, *sigD* and *sdeK*. Taken together, these data suggest that DnaA could have a function in the checkpoint that couples the chromosomal status to progression of development.

How do myxospores with two chromosomes and peripheral rods with one chromosome arise during development? Clearly, the formation of myxospores with two chromosomes must depend on an inhibition of cell division. Based on DNA replication studies in vegetative cells, most cells are either replicating or have two complete chromosomes. Moreover, a new round of replication is required for development. Therefore, a complete inhibition of cell division during development is difficult to reconcile with the observed chromosome distribution because it would result in a population of cells with two or more chromosomes but not cells with one chromosome (as observed in peripheral rods).

Do all cells have to undergo a round of replication during development? If so, then it appears that cell division must be inhibited in some cells (those becoming spores) and allowed in others (those becoming peripheral rods). We currently prefer as a working model in which some cells initiate a new round of replication and in which cell division is blocked in those cells that eventually become spores. This working model brings up the important question of cell fate determination, i.e. are different cell cycle events regulated in a deterministic or stochastic manner during development.

An important question that remains to be addressed is why *M. xanthus* spores have two copies of the chromosomes, when other common soil-dwelling spore formers, such as *B. subtilis* and *S. coelicolor*, do not? Tzeng and Singer (2005) proposed that such an arrangement could enhance spore survival in response to DNA damage. Having two copies of the chromosome would provide a higher probability for recombination-mediated DNA repair and in this way increase the survival rate in response to DNA damage. There is precedent for this model: *B. megaterium*, *B. cereus* and *B. thuringiensis* spores have been reported to contain two copies of the chromosome and to be more resistant to DNA damage than its close relative *B. subtilis* (Woese, 1958; Hauser and Karamata, 1992).

Perspectives

While recent work on cell division, DNA replication and the cell cycle has provided us with a better understanding of these processes in *M. xanthus*, it is clear we still have a long way to go before these processes are understood at the molecular and cellular levels. What is clear, however, is that extrapolations from other bacteria with regard to controlling replication and cell division cannot be transferred one-to-one to *M. xanthus*. Importantly, with the recent introduction of cell biology tools for single cell analysis and cell fate determination we now have the tools to address how DNA replication, cell division and cell fate determination are coupled during *M. xanthus* fruiting body formation and myxospore differentiation.

Clearly, bacterial cell function can only be appreciated in the light of evolution and diversity. Extant bacteria are the result of billions of years of evolution. Over this amazing time span, bacteria have often evolved different solutions to the same problem, i.e. they have a variety of cell shapes, intermediary metabolisms and ways of responding to environmental changes. While cell shape, metabolism and signal transduction has been studied in many different bacteria; we are only scratching the surface with respect to the regulation of basic cellular processes such as cell cycle control, chromosome organization and dynamics, and cell division. However, the few studies done on these topics are clearly illuminating that also here different bacteria have evolved different solutions to the same basic problems of organizing the chromosome and finding the place to divide. These are exciting times in microbiology.

Acknowledgements

Work in the authors' laboratories is supported by the Max Planck Society, the German Research Council within the framework of the Graduate School 'Intra- and Intercellular Transport and Communication', and by the National Science Foundation grant MCB-1024989 and the California Agricultural Experiment Station.

References

Adams, D.W., and Errington, J. (2009). Bacterial cell division: assembly, maintenance and disassembly of the Z ring. Nat. Rev. Microbiol. 7, 642–653.

Ben-Yehuda, S., Rudner, D.Z., and Losick, R. (2003). RacA, a bacterial protein that anchors chromosomes to the cell poles. Science *299*, 532–536.

Berkmen, M.B., and Grossman, A.D. (2006). Spatial and temporal organization of the *Bacillus subtilis* replication cycle. Mol. Microbiol. *62*, 57–71.

Bowman, G.R., Comolli, L.R., Zhu, J., Eckart, M., Koenig, M., Downing, K.H., Moerner, W.E., Earnest, T., and Shapiro, L. (2008). A polymeric protein anchors the chromosomal origin/ParB complex at a bacterial cell pole. Cell *134*, 945–955.

Carnoy, C., and Roten, C.A. (2009). The *dif*/Xer recombination systems in proteobacteria. PLoS One *4*, e6531.

Donovan, C., Sieger, B., Krämer, R., and Bramkamp, M. (2012). A synthetic *Escherichia coli* system identifies a conserved origin tethering factor in Actinobacteria. Mol. Microbiol. *84*, 105–116.

Dworkin, M., and Gibson, S.M. (1964). A system for studying microbial morphogenesis: rapid formation of microcysts in *Myxococcus xanthus*. Science *146*, 243–244.

Ebersbach, G., Briegel, A., Jensen, G.J., and Jacobs-Wagner, C. (2008). A self-associating protein critical for chromosome attachment, division, and polar organization in *Caulobacter*. Cell *134*, 956–968.

Errington, J. (2003). Regulation of endospore formation in *Bacillus subtilis*. Nat. Rev. Micro. *1*, 117–126.

Espeli, O., and Marians, K.J. (2004). Untangling intracellular DNA topology. Mol. Microbiol. *52*, 925–931.

Flärdh, K., and Buttner, M.J. (2009). *Streptomyces* morphogenetics: dissecting differentiation in a filamentous bacterium. Nat. Rev. Micro. *7*, 36–49.

Fogel, M.A., and Waldor, M.K. (2006). A dynamic, mitotic-like mechanism for bacterial chromosome segregation. Genes Dev. *20*, 3269–3282.

Garcia-Moreno, D., Polanco, M.C., Navarro-Aviles, G., Murillo, F.J., Padmanabhan, S., and Elias-Arnanz, M. (2009). A vitamin B12-based system for conditional expression reveals *dksA* to be an essential gene in *Myxococcus xanthus*. J. Bacteriol. *191*, 3108–3119.

Gerdes, K., Howard, M., and Szardenings, F. (2010). Pushing and pulling in prokaryotic DNA segregation. Cell *141*, 927–942.

Goldman, B.S., Nierman, W.C., Kaiser, D., Slater, S.C., Durkin, A.S., Eisen, J.A., Ronning, C.M., Barbazuk, W.B., Blanchard, M., Field, C., *et al.* (2006). Evolution of sensory complexity recorded in a myxobacterial genome. Proc. Natl. Acad. Sci. U.S.A. *103*, 15200–15205.

Harms, A., Treuner-Lange, A., Schumacher, D., and Søgaard-Andersen, L. (2013). Tracking of chromosome and replisome dynamics reveals a novel chromosome arrangement. PLoS Genet. *in press*.

Hauser, P.M., and Karamata, D. (1992). A method for the determination of bacterial spore DNA content based on isotopic labelling, spore germination and diphenylamine assay; ploidy of spores of several *Bacillus* species. Biochimie *74*, 723–733.

Hendrickson, H., and Lawrence, J.G. (2007). Mutational bias suggests that replication termination occurs near the *dif* site, not at Ter sites. Mol. Microbiol. *64*, 42–56.

Inouye, M., Inouye, S., and Zusman, D.R. (1979a). Biosynthesis and self-assembly of protein S, a development-specific protein of *Myxococcus xanthus*. Proc. Natl. Acad. Sci. U.S.A. *76*, 209–213.

Inouye, M., Inouye, S., and Zusman, D.R. (1979b). Gene expression during development of *Myxococcus xanthus*: pattern of protein synthesis. Dev. Biol. *68*, 579–591.

Jensen, R.B., Wang, S.C., and Shapiro, L. (2001). A moving DNA replication factory in *Caulobacter crescentus*. EMBO J. *20*, 4952–4963.

Kaguni, J.M. (2011). Replication initiation at the *Escherichia coli* chromosomal origin. Curr. Opin. Chem. Biol. *15*, 606–613.

Koch, M.K., McHugh, C.A., and Hoiczyk, E. (2011). BacM, an N-terminally processed bactofilin of *Myxococcus xanthus*, is crucial for proper cell shape. Mol. Microbiol. *80*, 1031–1051.

Kottel, R.H., Bacon, K., Clutter, D., and White, D. (1975). Coats from *Myxococcus xanthus*: characterization and synthesis during myxospore differentiation. J. Bacteriol. *124*, 550–557.

Kroos, L. (2007). The *Bacillus* and *Myxococcus* developmental networks and their transcriptional regulators. Annu. Rev. Genet. *41*, 13–39.

Livny, J., Yamaichi, Y., and Waldor, M.K. (2007). Distribution of centromere-like *parS* sites in bacteria: insights from comparative genomics. J. Bacteriol. *189*, 8693–8703.

Lobry, J.R. (1996). Asymmetric substitution patterns in the two DNA strands of bacteria. Mol. Biol. Evol. *13*, 660–665.

Lutkenhaus, J. (2007). Assembly dynamics of the bacterial MinCDE system and spatial regulation of the Z ring. Annu. Rev. Biochem. *76*, 539–562.

McCleary, W.R., Esmon, B., and Zusman, D.R. (1991). *Myxococcus xanthus protein* C is a major spore surface protein. J. Bacteriol. *173*, 2141–2145.

McCormick, J.R. (2009). Cell division is dispensable but not irrelevant in *Streptomyces*. Curr. Opin. Microbiol. *12*, 689–698.

Messer, W. (2002). The bacterial replication initiator DnaA. DnaA and *oriC*, the bacterial mode to initiate DNA replication. FEMS Microbial. Rev. *26*, 355–374.

Migocki, M.D., Lewis, P.J., Wake, R.G., and Harry, E.J. (2004). The midcell replication factory in *Bacillus subtilis* is highly mobile: implications for coordinating chromosome replication with other cell cycle events. Mol. Microbiol. *54*, 452–463.

Mott, M.L., and Berger, J.M. (2007). DNA replication initiation: mechanisms and regulation in bacteria. Nat. Rev. Micro. *5*, 343–354.

O'Connor, K.A., and Zusman, D.R. (1991). Development in *Myxococcus xanthus* involves differentiation into two cell types, peripheral rods and spores. J. Bacteriol. *173*, 3318–3333.

Ptacin, J.L., Lee, S.F., Garner, E.C., Toro, E., Eckart, M., Comolli, L.R., Moerner, W.E., and Shapiro, L. (2010).

A spindle-like apparatus guides bacterial chromosome segregation. Nat. Cell Biol. *12*, 791–798.

Reyes-Lamothe, R., Possoz, C., Danilova, O., and Sherratt, D.J. (2008). Independent positioning and action of *Escherichia coli* replisomes in live cells. Cell *133*, 90–102.

Rosario, C.J., and Singer, M. (2007). The *Myxococcus xanthus* developmental program can be delayed by inhibition of DNA replication. J. Bacteriol. *189*, 8793–8800.

Rosario, C.J., and Singer, M. (2010). Developmental expression of *dnaA* is required for sporulation and timing of fruiting body formation in *Myxococcus xanthus*. Mol. Microbiol. *76*, 1322–1333.

Rosenberg, E., Katarski, M., and Gottlieb, P. (1967). Deoxyribonucleic acid synthesis during exponential growth and microcyst formation in *Myxococcus xanthus*. J. Bacteriol. *93*, 1402–1408.

Rosenbluh, A., Nir, R., Sahar, E., and Rosenberg, E. (1989). Cell-density-dependent lysis and sporulation of *Myxococcus xanthus* in agarose microbeads. J. Bacteriol. *171*, 4923–4929.

Salzberg, S.L., Salzberg, A.J., Kerlavage, A.R., and Tomb, J.F. (1998). Skewed oligomers and origins of replication. Gene *217*, 57–67.

Shereda, R.D., Kozlov, A.G., Lohman, T.M., Cox, M.M., and Keck, J.L. (2008). SSB as an organizer/mobilizer of genome maintenance complexes. Crit. Rev. Biochem. Mol. Biol. *43*, 289–318.

Sherratt, D.J., Arciszewska, L.K., Crozat, E., Graham, J.E., and Grainge, I. (2010). The *Escherichia coli* DNA translocase FtsK. Biochem. Soc. Trans. *38*, 395–398.

Shimkets, L.J., and Brun, Y.V. (2000). Prokaryotic development: Strategies to enhance survival. In Prokaryotic development, Y.V. Brun, and L.J. Shimkets, eds. (Washington, DC: American Society for Microbiology), pp. 1–7.

Thanbichler, M. (2009). Spatial regulation in *Caulobacter crescentus*. Curr. Opin. Microbiol. *12*, 715–721.

Treuner-Lange, A., Aguiluz, K., van der Does, C., Gómez-Santos, N., Harms, A., Schumacher, D., Lenz, P., Hoppert, M., Kahnt, J., Muñoz-Dorado, J., *et al.* (2013). PomZ, a ParA-like protein, regulates Z-ring formation and cell division in *Myxococcus xanthus*. Mol. Microbiol. *87*, 235–253.

Tzeng, L., and Singer, M. (2005). DNA replication during sporulation in *Myxococcus xanthus* fruiting bodies. Proc. Natl. Acad. Sci. U.S.A. *102*, 14428–14433.

Tzeng, L., Ellis, T.N., and Singer, M. (2006). DNA replication during aggregation phase is essential for *Myxococcus xanthus* development. J. Bacteriol. *188*, 2774–2779.

Viollier, P.H., Thanbichler, M., McGrath, P.T., West, L., Meewan, M., McAdams, H.H., and Shapiro, L. (2004). Rapid and sequential movement of individual chromosomal loci to specific subcellular locations during bacterial DNA replication. Proc. Natl. Acad. Sci. U.S.A. *101*, 9257–9262.

Wang, X., Possoz, C., and Sherratt, D.J. (2005). Dancing around the divisome: asymmetric chromosome segregation in *Escherichia coli*. Genes Dev. *19*, 2367–2377.

Wang, X., Liu, X., Possoz, C., and Sherratt, D.J. (2006). The two *Escherichia coli* chromosome arms locate to separate cell halves. Genes Dev. *20*, 1727–1731.

Wang, X., Reyes-Lamothe, R., and Sherratt, D.J. (2008). Visualizing genetic loci and molecular machines in living bacteria. Biochem. Soc. Trans. *36*, 749–753.

Whitworth, D.E., ed. (2008). Myxobacteria. Multicellularity and differentiation (Washington, DC: ASM Press).

Willemse, J., Borst, J.W., de Waal, E., Bisseling, T., and van Wezel, G.P. (2011). Positive control of cell division: FtsZ is recruited by SsgB during sporulation of *Streptomyces*. Genes Dev. *25*, 89–99.

Woese, C.R. (1958). Comparison of the X-ray sensitivity of bacterial spores. J. Bacteriol. *75*, 5–8.

Zusman, D., and Rosenberg, E. (1968). Deoxyribonucleic acid synthesis during microcyst germination in *Myxococcus xanthus*. J. Bacteriol. *96*, 981–986.

Zusman, D., and Rosenberg, E. (1970). DNA cycle of *Myxococcus xanthus*. J. Mol. Biol. *49*, 609–619.

Zusman, D.R., Krotoski, D.M., and Cumsky, M. (1978). Chromosome replication in *Myxococcus xanthus*. J. Bacteriol. *133*, 122–129.

Social Interactions Mediated by Outer Membrane Exchange

5

Daniel Wall

Abstract

Myxobacteria are unusual because they exhibit complex social behaviours that involve cooperative cell–cell interactions to coordinate multicellular processes. The complexity of their social behaviours is mirrored by their genetic complexity; myxobacteria contain some of the largest known bacterial genomes. Other chapters in this book review cooperative behaviours of myxobacteria, include gliding motility and development. In this chapter, I will review our current understanding of a newly discovered social interaction whereby myxobacteria exchange their outer membrane (OM) proteins and lipids. The mechanism of transfer requires physical contact between aligned cells on hard surfaces. Transfer is mediated by OM fusion in which membrane contents laterally diffuse and are exchanged bidirectionally between cells. TraA and TraB are recently identified proteins that are required in donor and recipient cells for transfer to occur. OM exchange has phenotypic consequences that can alter cellular behaviours, including motility and development. Here I argue that OM exchange represents a new microbial platform whereby cells interact and communicate to coordinate multicellular activities.

Discovery path for OM exchange

One of the key scientific features of myxobacteria is their ability to communicate and coordinate cellular behaviours. Central to these behaviours is gliding motility, which powers and organizes swarm expansion, rippling, predation and fruiting body development on solid surfaces. *Myxococcus xanthus* has two separate motility systems or motors that drive gliding. They are called A (adventurous) and S (social) motility and are reviewed in Chapters 9 and 10. To study cooperative behaviours, genetic approaches have served as the primary discovery tool. In particular, the ability of mutants that are defective in social processes, such as swarming or development can, in some cases, be complemented extracellularly by other cells with a corresponding wild-type gene; this phenomenon has been used as a powerful instrument to identify gene products that function in cell–cell interactions/communication (Hagen *et al.*, 1978). In this regard, early work by Hodgkin and Kaiser found that a small subset of motility mutants could be complemented extracellularly when mixed with other motility mutants that encode the corresponding wild-type gene – a phenomenon they called stimulation, as the recipient mutant transiently gained the ability to glide or swarm (Fig. 5.1A) (Hodgkin and Kaiser, 1977). In a typical stimulation or extracellular complementation assay, two nonmotile mutants (A^-S^-) are mixed and placed on an agar surface. After several hours to a couple of days of incubation, 'stimulation' is detected as emergent flares derived from 'recipient cells' (Fig. 5.1A). As the 'donor cells' are A^-S^- and are non-stimulatable, they remain in their original inoculum spot, a process that can be visually tracked by labelling donor or recipient cells with green fluorescent protein (GFP) (Nudleman *et al.*, 2005). Stimulation is transient and involves only phenotypic changes to recipient cells; there are no genotypic changes. The original Hodgkin and Kaiser study found six stimulatable

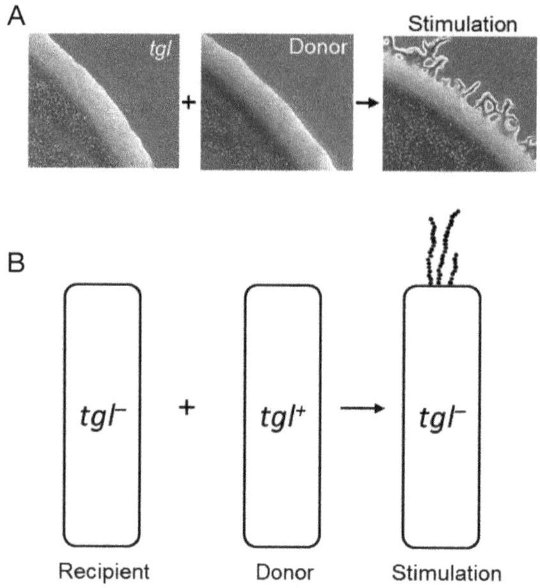

Figure 5.1 Stimulation of gliding motility and pili biogenesis in *M. xanthus*. (A) Left panel shows a nonmotile *tgl* mutant (A^-S^-), and the middle panel shows a nonmotile and nonstimulatable donor strain (A^-S^-) that is *tgl*⁺. After 1 day, the colonies have sharp edges; there is no swarming. Right panel shows a 1:1 mixture of the same strains, in which many S-motility flares are visible at the edge of the colony. Hence the *tgl* mutant was stimulated to swarm. (B) A cartoon depicting *tgl* stimulation. The donor strain lacks type IV pili (e.g. Δ*pilA* mutation) but is *tgl*⁺. The recipient strain is *tgl*⁻ and also lacks pili. Following physical contact with the donor strain, the *tgl*⁻ recipient transiently produces polar pili.

motility genes, five of which belong to A-motility (*cglB/C/D/E/F*) and one of which belongs to S-motility (*tgl*) (Hodgkin and Kaiser, 1977). Of the >150 known motility genes, no other stimulatable motility mutants have been reported (Hartzell *et al.*, 2008), although others may exist. Stimulation was originally interpreted by two non–mutually exclusive models: (i) myxobacteria cells exchange certain motility components and thus rescue missing functions associated with particular mutants or (ii) cells exchange signals through contacts that may restore motility. As described below, the former explanation has been shown to occur, and elements of the latter explanation also appear to be correct.

Through the use of electron microscopy, Kaiser subsequently reported that *tgl* stimulation of S-motility also resulted in the production of polar pili (Fig. 5.1B) (Kaiser, 1979). This early finding strongly implicated pili, later shown to be type IV pili, in S-motility (Wu and Kaiser, 1995). Type IV pili, which have since been found in many bacterial species, power surface motility, such as twitching or S-motility, by retracting extended pili that have adhered to surfaces, effectively pulling the cell forward (Skerker and Berg, 2001; Wall and Kaiser, 1999). The Kaiser result linked pili biogenesis and S-motility to *tgl* stimulation. Based on this finding, Wall *et al.* developed a pili shear assay that allowed *tgl* stimulation to be rapidly followed under different conditions (Wall and Kaiser, 1998; Wall *et al.*, 1998).

The next advance in understanding stimulation came with the identification of the *tgl* and *cglB* genes (Rodriguez-Soto and Kaiser, 1997b; Rodriguez and Spormann, 1999). Although these proteins are not similar, they both contain type II signal sequences, indicating and experimentally shown to be lipoproteins localized to the OM (Simunovic *et al.*, 2003). These results also suggested a common link, namely that stimulation might be specific to lipoproteins. That view persisted until the remaining four *cgl* genes were identified (Table 5.1) (Pathak and Wall, 2012). Similar to *tgl* and *cglB*, *cglC* and *cglD* gene products are lipoproteins, i.e. they encode type II signal sequences. However, *cglE* and *cglF* instead encode type I signal sequences (non-lipoproteins), thus revealing that stimulation is not limited to lipoproteins.

Key mechanistic insights into stimulation were found by tracking the whereabouts of the Tgl and CglB proteins between donor and recipient cells (Nudleman *et al.*, 2005). Strikingly, when a *tgl* deletion mutant was placed in physical contact with *tgl*⁺ donor cells for a period of time and then cell types were separated as the result of swarm expansion caused by stimulation or by FACS, the Tgl protein was found to transfer from donor to recipient cells. Remarkably, the Tgl lipoprotein accumulated to nearly equal levels in recipient cells as in donors. Similar results were obtained with CglB. These results thus provided a definitive answer to the question of how stimulation works. Namely, *tgl* or *cglB* mutants lack an essential protein for motility, and that defect is corrected when

Table 5.1 *tgl* and *cgl* genes

Gene	Locus tag	Amino acids	Signal sequence	Domains	Distribution
tgl	MXAN_3084	253	Type II	TPR/PilF	Widely
cglB	MXAN_3060	416	Type II	von Willebrand	Myxobacteria
cglC	MXAN_2538	172	Type II	Hypothetical	Myxobacteria
cglD	MXAN_0962	1130	Type II	von Willebrand	Widely
cglE	MXAN_4866	209	Type I	Hypothetical	Myxobacteria
cglF	MXAN_4868	89	Type I	Hypothetical	Myxobacteria

the corresponding Tgl or CglB protein is physically transferred to mutant recipients. 'Stimulation' thus restores gliding by providing the missing protein and its corresponding function. Moreover, because the recipient cells contain mutations and cannot produce functional Tgl or Cgl proteins, stimulation is transient, as the transferred proteins become diluted over time by cell division and protein turnover. To state this in another way, DNA is not transferred between cells and thus subsequent progenies of recipient cells retain the genetic and phenotypic properties (*cgl*$^-$ or *tgl*$^-$ and an absence of motility, respectively) of the parental strain.

In the case of Tgl, further mechanistic insight is known about stimulation. First, Tgl is an OM lipoprotein that is required for the assembly of the PilQ secretin (Kahnt *et al.*, 2010; Nudleman *et al.*, 2006; Rodriguez-Soto and Kaiser, 1997a; Simunovic *et al.*, 2003), which functions as a gated channel for the secretion and retraction of assembled type IV pili through the OM (Wall and Kaiser, 1999; Wall *et al.*, 1999). The secretin consists of 10–12 PilQ monomers (approximately a mega-Dalton size complex) that are stably assembled in the OM and are characteristically resistant to disassembly by heat and SDS detergent as judged by Western blot analysis (Nudleman *et al.*, 2006). *tgl* mutants are the only ones known in *M. xanthus*, other than particular *pilQ* point mutations, in which PilQ does not assemble into stable multimers (Nudleman *et al.*, 2006). Tgl thus functions as an assembly factor or 'pilotin' for the PilQ secretin, a function that is found in many type IV pili and other homologous systems, e.g. type II secretion systems, and in some cases these pilotins are orthologues of Tgl (Koo *et al.*, 2012). Thus, following Tgl transfer into *tgl* mutants, PilQ monomers are assembled into a stable multimeric secretin that is resistant to heat and SDS detergent (Fig. 5.2) (Nudleman *et al.*, 2005). In summary, the sequence of events that occurs during Tgl stimulation involves physical contact between donor and recipient cells → Tgl transfer → PilQ assembly → type IV pili assembly and retraction → S-motility (Fig. 5.2). In contrast, the mechanism of A-motility is recently described (Nan *et al.*, 2011; Sun *et al.*, 2011), and the function of the five Cgl proteins remains unknown.

Cis requirement in cargo proteins for transfer

As described in detail below, protein transfer requires physical contact between cells, a hard surface and cell motility; transfer does not occur through secretion into the medium or through diffusible OM vesicles. To gain insight into the mechanism of transfer, Wei *et al.* (2011) sought to define the *cis* requirements within Tgl and/or CglB that allowed transfer. To address this, a series of Tgl and Cgl fusions were made to the fluorescent reporter protein mCherry. Unfortunately, these constructs were not stably expressed in *M. xanthus*. In contrast, heterologous fusions between mCherry and type II signal sequences for OM (SS_{OM}-mCherry) and inner membrane (SS_{IM}-mCherry) localization were stably expressed (Wei *et al.*, 2011). By using different approaches, it was found that in live cells the SS_{OM}-mCherry reporter was transferred to recipient cells that were easily identified because they expressed cytoplasmically localized GFP. Transfer was efficient, as the OM of recipient cells exhibited similar fluorescence intensity as did donor cell OMs. In contrast, the SS_{IM}-mCherry and cytoplasmic GFP reporters did not transfer. Thus, OM lipoproteins transferred,

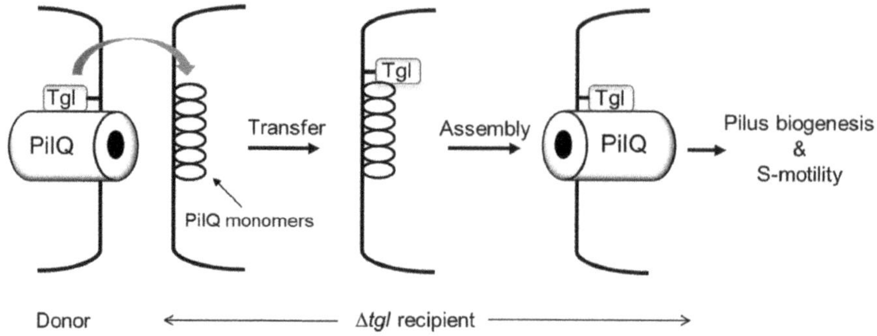

Figure 5.2 Model for Tgl stimulation in *M. xanthus*. Juxtaposed cell poles of a Tgl$^+$ donor and a Δ*tgl* recipient are shown (left). *tgl* mutants express PilQ monomers, but they fail to assemble into a stable multimeric channel in the OM (Nudleman *et al.*, 2006). Following physical contact, the Tgl protein is transferred from donor to recipient, where Tgl facilitates assembly of the PilQ secretin (Nudleman *et al.*, 2005). After PilQ assembly, type IV pili are polymerized at the cell pole to power S-motility.

whereas IM lipoproteins and cytoplasmic proteins did not transfer (Wei *et al.*, 2011). Importantly, these results showed that a type II signal sequence for OM localization was necessary and sufficient for heterologous protein transfer between live *M. xanthus* cells. Moreover, the finding that there was no other *cis* requirement for heterologous cargo protein transfer implies that many other OM lipoproteins are shuttled between *M. xanthus* cells. As the *M. xanthus* genome contains ~400 predicted lipoproteins, there are many candidate transfer proteins (Goldman *et al.*, 2006). We would expect, however, that some OM lipoproteins would not transfer because they are assembled into stable macromolecular complexes, e.g. bound to the cell wall or pilus apparatus, or might be sequestered away from a putative translocation site.

Recently, we reported that CglE and CglF proteins deviate from the above paradigm; they contain type I signal sequences *in lieu* of type II signal sequences (Table 5.1) (Pathak and Wall, 2012). These results therefore suggest that non-lipoproteins are also transferred. Both direct and indirect subcellular localization studies suggest that CglE and CglF localize to the OM (Kahnt *et al.*, 2010; Luciano *et al.*, 2011; Pathak and Wall, 2012). Assuming that these conclusions are correct, the reservoir of cargo proteins may be restricted to proteins associated with the OM; however, additional studies are needed to clarify this issue. Thus in additions to OM lipoproteins, peripheral and integral OM proteins are candidate cargo proteins. If integral OM proteins are transferred, there apparently are restrictions, as *pilQ* mutants cannot be stimulated (Hodgkin and Kaiser, 1979; Nudleman *et al.*, 2006), which suggests that the integral OM PilQ protein does not transfer. It remains an open question whether soluble periplasmic material is transferred and whether there is a gating function that restricts protein transfer. As shown in Table 5.1, the size of cargo proteins can apparently vary over a wide range, e.g. from 71 to >1100 amino acids (after cleavage of their signal sequences). Our findings thus suggest that a significant fraction of OM-associated and perhaps soluble periplasmic proteins are exchanged between myxobacteria (Wei *et al.*, 2011). We further predict that transfer proteins act in a variety of cellular processes and are not restricted to motility functions. To date, however, only (endogenous) motility proteins have been implicated in transfer.

Transfer requires cell–cell contact, a solid surface and cell motility

A simple explanation for the stimulation phenomena invokes the secretion of naked or OM-enclosed protein vesicles into the extracellular milieu and their subsequent uptake by recipient cells. This is, however, not the case. First, Tgl is not secreted (Rodriguez-Soto and Kaiser, 1997a). Second, as assayed by pilus assembly or by the reporter SS_{OM}-mCherry, no stimulation or transfer occurs in liquid whether cells are shaking or static, even after prolonged incubation (Wall and Kaiser, 1998; Wei *et al.*, 2011). Moreover, although *M. xanthus* is much more proficient at S-motility on soft agar than on hard agar (Shi and Zusman, 1993), Tgl stimulation, as measured by flare production or type IV pili assembly, or SS_{OM}-mCherry transfer, does not detectably occur on soft agar (Wall and Kaiser, 1998; Wei *et al.*, 2011). Instead, Tgl stimulation of type IV pili assembly/motility and SS_{OM}-mCherry transfer require a hard solid surface. These findings further argue that protein transfer does not occur through diffusion but instead requires direct and aligned cell–cell contacts (Wall and Kaiser, 1998; Wei *et al.*, 2011).

During the course of investigations, it was found that cell motility has a profound effect on protein transfer. First, when either donor or recipient cells possess A-motility, Tgl stimulation of type IV biogenesis is dramatically enhanced (Wall and Kaiser, 1998). Similarly, cell motility dramatically enhances heterologous SS_{OM}-mCherry transfer (Wei *et al.*, 2011). In this latter study, it was shown that either A- or S-motility allowed rapid and efficient SS_{OM}-mCherry transfer, as >90% of recipients acquired the reporter within 2 h. Similar transfer rates and efficiencies were found when the donor or recipient or both cell types were endowed with either motility system. In contrast, when both donor and recipient cells were nonmotile (A^-S^-), very little SS_{OM}-mCherry transfer occurred (<5%), even after extensive incubations (Fig. 5.3). These results show that cell motility plays a critical, but indirect, role in protein transfer. That is, there was no requirement for donor or recipient cells to be motile, rather motility was only required in one cell type. Moreover, motility enhancement of transfer was not specific to either A- or S-motility, as both systems worked similarly well (Wei *et al.*, 2011). To further investigate the apparent indirect role motility plays in protein transfer, the identical nonmotile and non-stimulatable strains (A^-S^-) described above were mixed 1:1:1 with a third-party non-fluorescent motile strain (A^+S^+). In this experiment, transfer from the nonmotile donor (SS_{OM}-mCherry) strain to the nonmotile recipient (GFP^+) strain was rapidly and efficiently restored (Fig. 5.3, indicated by dashed arrow). Although the transfer of SS_{OM}-mCherry from the donor to labelled recipients could occur via the third-party motile strain, this is not required, as similar results were obtained if the third-party strain was not competent for transfer (*traA* mutant, see below; X. Wei and D. Wall, unpublished). Interestingly, an intermediate SS_{OM}-mCherry transfer rate was seen when a nonmotile donor was mixed with a labelled nonmotile, but stimulatable, recipient (A^- *tgl*$^-$ *gfp*$^+$; Fig. 5.3). In this experiment, both strains were initially nonmotile, but over a 2-day period the recipient cells were 'stimulated' to produce type IV pili and thus slowly gained (weak) S-motility that facilitated protein transfer. As mentioned below, motility plays a role in transfer, presumably by aligning cells. Finally, the critical role that motility plays in transfer is not the result of creating local areas of high cell density, because concentrating cells as a result of growth (Fig. 5.3; cells were densely packed after 2 days of growth) or artificially by centrifugation (X. Wei and D. Wall, unpublished) does not enhance transfer.

The interpretation of the indirect role that gliding motility plays in facilitating protein transfer has been that it provides a mechanical means to align long rod-shaped cells (Wall and Kaiser, 1998; Wei *et al.*, 2011). In turn, cell alignment enriches end-to-end cell contacts between these rod-shaped *M. xanthus* cells, enabling critical cell–cell contacts to form. In the case of type IV pili and the protein components thereof, they are in fact located at cell poles (Bulyha *et al.*, 2009; Nudleman *et al.*, 2006). Thus, end-to-end cell contacts may allow close spacial proximity for protein departure and arrival between juxtaposed cells. As described below, cell poles also have high membrane curvature resulting in membrane stress, a condition that promotes membrane fusion (Martens and McMahon, 2008). It should be noted that side-by-side cell contacts have been implicated in myxobacterial cell–cell interactions and thus their involvement in transfer cannot be ruled out (Mauriello *et al.*, 2009). The mechanism of protein transfer is discussed in more detail below. After OM fusion, gliding motility is likely to facilitate the separation of cells, i.e. OM fission. Given that motility dramatically enhances transfer (Fig. 5.3), further experiments are needed to precisely elucidate its role in this process.

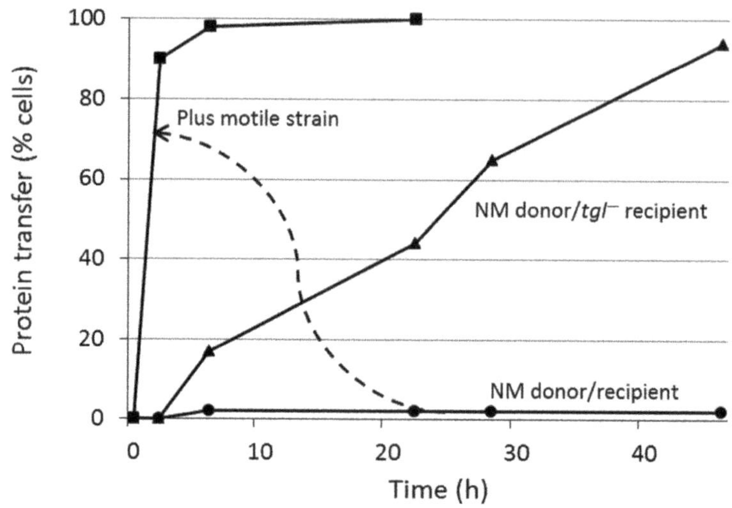

Figure 5.3 Cell motility is required for protein transfer between cells. The ability of cells to transfer proteins was assayed with the SS_{OM}-mCherry reporter. Recipient cells express GFP in the cytoplasm, and thus transfer is quantified as the per cent of green cells that fluoresce red from SS_{OM}-mCherry transfer. When both cell types are nonmotile (NM), almost no transfer occurs (<5%). In contrast when a third-party (unlabeled) motile cell is added, SS_{OM}-mCherry transfer occurs rapidly and efficiently to the NM GFP^{+} recipient (plus motile strain indicated by dashed arrow). When a nonmotile *tgl* recipient is mixed with a nonmotile donor, transfer slowly occurs and cells are slowly stimulated to move. Figure adapted from Wei *et al.* (2011).

OM lipids are transferred

Because OM protein transfer occurs efficiently and involves a form of bulk protein transfer, we hypothesized that OMs between cells fuse and thus lipids may also exchange. This hypothesis was tested and supported by the finding that lipophilic fluorescent dyes are indeed readily exchanged between cells (Pathak *et al.*, 2012a,b). Importantly, the transfer of lipophilic dyes, and hence membrane lipids, has the same stringent requirements for transfer as do OM lipoproteins. That is, dye transfer is concurrent with protein transfer and does not occur by diffusion into the milieu or by diffusible OM vesicles. Instead lipophilic dye transfer requires cell motility, a solid surface and specific cell–cell contacts, i.e. a structured biofilm. Conceptually we envisage fusion of the OM as consisting of the outer leaflet lipopolysaccharides and the inner leaflet phospholipids. The cell wall is predicted to act as a physical barrier and to preclude fusion between inner membranes. Consistent with this arrangement, only OM lipoproteins, and not IM lipoproteins, are transferred (Wei *et al.*, 2011). We conclude that the underlying mechanism for protein transfer involves the fusion of the OM between cells. The molecular details of this process remain largely unknown, although progress is being made (see below), and future efforts would be facilitated by detailed electron microscopy studies. In this respect, there could be mechanistic parallels to *Borrelia* species, as they have been observed to directly fuse their OMs, a process that is apparently mediated by cell motility (Kudryashev *et al.*, 2011).

Identification of transfer machinery proteins

To help understand the mechanism of lipoprotein transfer, we sought mutants that are universally defective in *cgl*/*tgl* stimulation and protein transfer (Pathak *et al.*, 2012a). Two mutants were identified, and the resulting proteins were named TraA and TraB for <u>tra</u>nsfer. Phenotypically, the defect in transfer is absolute; under no condition was stimulation or protein transfer detected in *traA* or *traB* mutants. Moreover, in a forward genetic screen, *traAB* mutants were repeatedly isolated (A. Dey and D. Wall, unpublished). From these efforts, we conclude that TraAB are major determinants for transfer and perhaps the only proteins that are directly involved in the transfer process.

Importantly, the *traAB* mutant defects also affect lipophilic dye transfer, thus cementing the interrelationship between protein and lipid transfer (Pathak *et al.*, 2012a). A further key finding was that TraAB are required in both donor and recipient cells. That is, no stimulation, protein or lipid transfer occurs when recipient cells contain *traA* or *traB* mutations (Pathak *et al.*, 2012a). These findings, in conjunction with the ability of *M. xanthus* cells to rapidly and homogeneously exchange lipoproteins (Nudleman *et al.*, 2005; Wei *et al.*, 2011), implies that transfer occurs bidirectionally between cells.

The *traA* (MXAN_6895) and *traB* (MXAN_6898) genes form a bicistronic operon, and their open reading frames overlap by four bases, which suggests that the relative ratio of these proteins is important and is regulated (Pathak *et al.*, 2012a). Bioinformatic analysis reveals that TraAB each contain type I signal sequences indicating cell envelope localization. The N-terminal region of TraA contains a distant PA14 domain, a domain that is broadly distributed throughout the tree of life and is implicated in glycan binding (Rigden *et al.*, 2004). However, the distant sequence relationship of the TraA PA14 domain to that of other family members leaves doubt whether TraA is a *bona fide* lectin. After the distant PA14 domain, TraA contains a cysteine-rich region that includes nine degenerative repeats, five of which are in tandem and each of which consists of 25 amino acids. This region is predicted to contain many disulphide bonds that may provide TraA with a rigid and stable structure. Finally, the C-terminal end of TraA contains a recently characterized MYXO-CTERM motif (GC–transmembrane helix–RRRR) predicted to function as a protein-sorting tag for cell surface localization (Pathak *et al.*, 2012a). Based on its domain architecture, we predict that TraA localizes to the cell surface and functions as a receptor/adhesion. This prediction has been experimentally supported, as TraAB overexpression results in cells that adhere to each other in end-to-end chains and side-by-side rafts (Pathak *et al.*, 2012a). Interestingly, the architectural features of TraA domains are similar to those of the *Saccharomyces cerevisiae* FLO1 and FLO5 cell surface adhesins, which are involved in kin recognition and flocculation in yeast (Rigden *et al.*, 2004; Smukalla *et al.*, 2008; Veelders *et al.*, 2010). Thus, *M. xanthus* and other microbes, including eukaryotes, appear to utilize PA14-containing proteins as extracellular signalling and recognition molecules to mediate social interactions.

The C-terminal region of TraB contains an OmpA-like domain, and thus it is likely to bind non-covalently to the cell wall in the periplasmic space (Pathak *et al.*, 2012a). The N-terminal region, the majority of TraB (~400 amino acids), has no ascribed function. It is, however, plausible that this region interacts with the OM and/or TraA on the cell surface. TraB could also play a role in proper localization of TraA to the cell surface. As described below, our working model is that TraAB functions as a cell surface adhesin that catalyses cell–cell fusion of the OM for protein exchange.

Synthetic CglF peptide restores A-motility

We recently described the identification of the *cglF* gene, which encodes a predicted open reading frame of 89 amino acids that includes a type I signal sequence (Pathak and Wall, 2012). After cleavage of the signal sequence, the CglF protein is predicted to contain 71 amino acids. We tested whether exogenous addition of CglF protein could restore the motility defect of a Δ*cglF* mutant. As CglF is unusually small, the CglF peptide was commercially synthesized, purified and added to a *cglF* mutant and shown to be functional in a stimulation assay. This result suggests that CglF, similar to CglB and Tgl, is indeed transferred between biofilm cells. Interestingly, the synthetic CglF peptide was also found to stimulate a *cglF*$^-$ *traA*$^-$ mutant, therefore demonstrating that exogenous CglF addition bypasses the requirement for TraA (Pathak and Wall, 2012). The observation that exogenous CglF peptide addition stimulates in a TraA-independent manner, whereas the use of live donors requires TraA in donors and recipients, suggests that the mechanism of CglF delivery fundamentally differs between these conditions. A possible interpretation of this result is that CglF is tightly bound on the cell surface, and thus live donors cannot transfer CglF unless the TraAB machinery is functional for OM fusion. In contrast, exogenous CglF addition bypasses TraAB function because the peptide can directly bind the OM of the *cglF* mutant.

OM exchange regulates cell behaviours

To date, *M. xanthus* is the only reported species in which cells fuse their OMs and exchanged proteins, which results in phenotypic changes to the cells. Microscopy studies suggest that *Borrelia* species also fuse their OMs, but the transfer of proteins and phenotypic changes have not been demonstrated (Kudryashev *et al.*, 2011). It should also be noted that many Gram-negative bacteria secrete diffusible OM vesicles that can fuse with other cells, including eukaryotic cells (Kulp and Kuehn, 2010), but this process is mechanistically different and is not directly regulated by producer cells. In any event, the phenomenon described here is likely to include many myxobacteria, as other species genomes encode TraAB orthologues and exhibit complex social behaviours (Pathak *et al.*, 2012a). Although other bacterial groups may fuse and exchange OM components (Kudryashev *et al.*, 2011), orthologues of TraAB are not readily found in other bacterial phyla.

A central and obvious question is why do myxobacteria fuse and exchange OM components? Experimentally, we have evidence that sheds light on this question, although our current understanding is incomplete; answering this question fully will require additional experimentation. As a general response, which is discussed in more detail below, OM exchange is likely to be related to the complex social interactions that myxobacteria exhibit. That is, it provides a mechanism to identify neighbouring cells, both friend and foe, and to coordinate behaviours that are dependent on cell–cell contact, without relying on diffusible signals, such as the quorum signals found in other bacteria.

As noted above, TraAB function is required for extracellular complementation or stimulation of certain motility mutants (Table 5.1). Given this, it is noteworthy that *traA* and *traB* mutations, in an otherwise wild-type strain, exhibit no overt defect in A- or S-motility, showing that OM transfer is not required for motility motor function (Pathak *et al.*, 2012a). Although the transient rescue of motility defects to certain mutants is an interesting phenomenon, its role in the interactions of wild-type cells, which should normally contain a full complement of motility proteins, is unknown. However, the exchange of OM lipids and proteins involves significant cellular resource sharing, and therefore this process must involve physiological consequences. Insights into the physiological significance of OM exchange were found in strain mixing experiments (Pathak *et al.*, 2012a,b). Here, it was shown that OM exchange can regulate swarming and development behaviours between strains. An illustration of this is shown in Fig. 5.4. By a Tra-dependent mechanism, swarm expansion of a motile strain is blocked when cells are mixed at a 1:1 ratio with a nonmotile strain. In contrast, when identical mixing experiments were conducted between isogenic *traA*$^-$ strains, swarm expansion was restored (Fig. 5.4A). As was found for lipoprotein and lipid transfer, the 'relief of swarm inhibition' occurred when a *tra* mutation was introduced into either the nonmotile or motile strains. Time-lapse video microscopy revealed that during the first day of mixing the motile strains exhibit similar behaviors with respect to cellular speeds, reversal frequencies and per cent of cells moving, and thus motility is not physically impaired. However, by day 2 these strain mixtures exhibit drastically different behaviors. That is, mixtures containing *tra*$^+$ cells exhibit a nearly complete absence of cell movement. In sharp contrast, isogenic strain mixtures with a *tra* mutation in either the motile or nonmotile strain exhibit robust motility (Fig. 5.4A). From these studies, it was concluded that nonmotile cells produce a time-dependent physiological signal that is transferred via OM exchange to motile cells that inhibits their motility (Pathak *et al.*, 2012b). The nature of that signal is unknown and is under investigation.

A hallmark of myxobacteria behaviour is their ability to form multicellular fruiting bodies and to sporulate in response to starvation. To test for a role in development, a *traA*$^-$ mutation was introduced into a wild-type strain. When a genetically and physiologically homogenous *traA*$^-$ culture was starved under standard laboratory conditions,

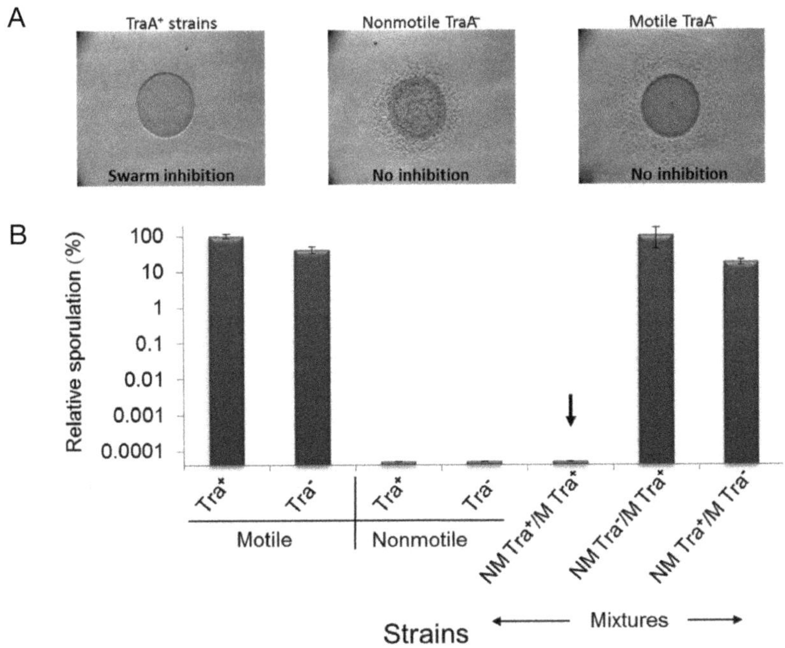

Figure 5.4 OM exchange regulates multicellular behaviours. (A) Motile and nonmotile strains were mixed at a 1:1 ratio on a hard agar surface and incubated for 4 days. The left panel, where both strains are Tra$^+$, shows swarming inhibition. The middle and right panels show swarm expansion in the presence of isogenic TraA$^-$ strain mixtures. (B) TraA-dependent inhibition of development. The ability of the motile (M) strain to sporulate was arbitrarily set at 100%; a nonmotile (NM) strain does not sporulate. The sporulation efficiencies of these strains do not significantly change with the presence of a *traA*$^-$ mutation. These motile and nonmotile strains were then mixed at a 1:1 ratio, and sporulation of the motile strain was blocked (arrow) in a Tra$^+$-dependent manner. Development was analysed on TPM starvation agar, and the assay for viable spores was carried out after 5 days. Figures adapted from Pathak *et al.* (2012).

no overt defect in fruiting body formation or sporulation was found. In contrast, in strain mixing experiments, as described above, dramatic differences were observed in the development outcome that were Tra dependent (Pathak *et al.*, 2012a). These results are summarized in Fig. 5.4B. First, as mentioned, the *traA* mutation did not significantly alter the ability of a motile strain to sporulate. Second, as development is coupled to motility (Kim and Kaiser, 1990), nonmotile strains cannot fruit or sporulate, and a *traA* mutation does not alter this defect. Importantly, when a motile strain was mixed at a 1:1 ratio with a non-motile strain no viable spores were detected; the nonmotile strain blocked the development of the motile strain. In sharp contrast, when isogenic strains contained a *traA* mutation in either the motile or nonmotile strain, the ability of the motile strain to sporulate was restored to control levels, i.e. similar to no strain mixing (Fig. 5.4B). Thus, like swarm inhibition, a nonmotile strain blocks development of a motile strain by a Tra-dependent mechanism. These results were again interpreted to mean that the non-motile strain produces a signal that is transferred via TraAB to the motile strain that blocks its development (Pathak *et al.*, 2012a,b).

From the above studies, we hypothesize that OM exchange serves as a cell–cell communication mechanism. In a biological context, cellular communication mediated by direct cell contact offers advantages over signalling by diffusible molecules. Cell contact-mediated signalling allows cells to definitively identify neighbouring cells and to directly coordinate cellular behaviours. Problems associated with diffusible signals being sequestered, degraded, retarded, diluted or mixed with other signals are avoided. When cells form multicellular tissues, such as myxobacteria, direct cell–cell contact is likely to represent a key signalling mechanism, which has been established for developmental signalling (Kaiser *et al.*, 2010; Konovalova *et al.*, 2010).

Proposed mechanism of OM exchange

A working and testable model for the mechanism of cell contact-dependent exchange is outlined in Fig. 5.5. First, the model proposes that TraA serves as a cell surface receptor/adhesin because TraAB overexpression increases the adhesive properties of cells and because its domain architecture is consistent with this proposition (Pathak *et al.*, 2012a). We further suggest, based on parallels to other family members, that the distant PA14 domain within TraA may be the actual receptor that binds adjoining cells (Rigden *et al.*, 2004; Veelders *et al.*, 2010). Similar to the FLO1 system, a key component of this model involves reciprocal TraA binding by both cells. We also have evidence that TraA serves as a molecular determinant for recognizing other cells (D. Pathak, X. Wei, A. Arup and D. Wall, unpublished). In this context, TraA may thus function as a homophilic receptor, which necessitates reciprocal binding. The function of TraB is less clear, but as mentioned above, it could interact with TraA and/or facilitate the subcellular localization of TraA. In Fig. 5.5, a 'donor' cell was arbitrarily assigned, and its OM (mCherry) lipoproteins are symbolized as black lollipops. Upon aligned cell–cell contact, mediated by cell motility, the receptors become engaged and temporarily hold adjoining cells together. Next, we propose that tight cell–cell binding facilitates OM fusion. The molecular details of fusion are unknown, but TraAB are candidate fusagen proteins. Higher membrane curvatures at cell poles may also facilitate OM fusion. Although membrane fusion is poorly understood in bacteria, considerable information is known from eukaryotic systems (Martens and McMahon, 2008; Wickner and Schekman, 2008). From these systems it is known that there are numerous proteins and lipids that facilitate membrane fusion between different biological particles or organelles or between cells. Thus, the identification of a fusagen is not restricted by sequence conservation, i.e. the lack of homology between TraAB and known fusagens does not argue against their classification as such. Second, an underlying principle of fusion is that the local membrane environment destabilizes the juxtaposed lipid bilayers, resulting in lipid molecules spontaneously flipping between bilayers, which lead to hemifusion followed by full membrane fusion. In this context, TraAB may help bring OMs into close proximity and may help destabilize lipids for fusion. Cell motility could also create local mechanical stress on membranes once the cells are adhered by their TraA receptors. Upon fusion, OM phospholipids, lipopolysaccharides and proteins exchange bidirectionally – a process that is probably the result of lateral diffusion. Evidence clearly suggests that the inner membranes do not fuse, as inner membrane lipoproteins and cytoplasmic proteins do not transfer (Wei *et al.*, 2011). After fusion, the OMs undergo physical separation, which is presumably driven by gliding motility. As protein transfer is robust and rapid, occurring on a timescale of several minutes to an hour, we expect exchange to occur

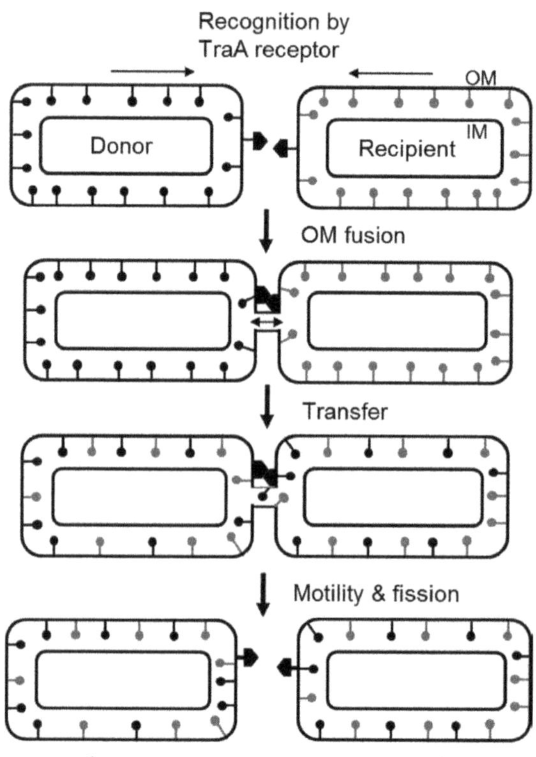

Figure 5.5 OM exchange working model. See text for details.

repeatedly within a swarm during the lifespan (~5 h) of logarithmically growing cells.

Physiological and ecological perspectives of OM exchange

As mentioned above, the biology of myxobacteria is rather unusual. Like many bacteria, they live as single cells that undergo binary fission to yield daughter cells. Their vegetative life, as understood in a laboratory setting, is social – the cells swarm in tightly associated packs, although at times single cells venture out and briefly separate themselves from the community. In response to starvation, they aggregate into mounds consisting of tens of thousands of cells and build multicellular fruiting bodies whose structural complexity varies between species (Shimkets *et al.*, 2006), a behaviour that is unparalleled in the bacterial kingdom. Their developmental programme touches on a fundamental biological and evolutionary question of how single cells coalesce and cooperate to build a multicellular structure or tissue. Central to this process is an appreciation that myxobacteria do not exist as prototypic eukaryotic cells, where sibling cells are derived from clonal expansion of a fertilized egg within a privileged environment. Instead, myxobacteria exist in heterogeneous populations that consist of assorted microbial species and various myxobacterial species and subspecies (genetically distinct strains). Thus the question arises, how do myxobacteria recognize their kin, or at least related strains, to build multicellular structures? Work conducted by Velicer and colleagues suggest that this selective process is not absolute, meaning environmental fruiting bodies do not consist simply of siblings (Kraemer and Velicer, 2011). Nevertheless, their work also indicates that a form of kin selection occurs during development, as fruiting bodies are enriched for strains that are genetically and phenotypically similar. Cell enrichment may also occur during the colonization of a rich nutrient source. For instance, when *M. xanthus* colonize a dung pellet, they feast off enteric bacteria and the population grows by clonal expansion. Environmental fruiting bodies or populations are not, however, simply derived from clonal expansion (Kraemer and Velicer, 2011; Vos and Velicer, 2006).

Another noted property of myxobacteria is that they are highly antagonistic towards other strains (Smith and Dworkin, 1994; Vos and Velicer, 2009). Given the heterogeneous and antagonistic nature of myxobacteria populations, how cells recognize themselves or closely related kin to elicit cooperative behaviours is an intriguing problem. A comprehensive answer to this question is unknown; however, we hypothesize that OM exchange represents one layer of strain recognition to allow single cells to coalesce into a more homogenous population to build a tissue. In unpublished work, we have found that TraA serves as a molecular determinant for self/non-self-recognition, which provides a mechanism for kin selection (D. Pathak, W. Wei, A. Dey and D. Wall). Future work will need to elucidate if this recognition system for cell–cell OM exchange plays a role in cell recognition of partnering cells to assemble multicellular structures. Other microbes, such as *Dictyostelium* species, are confronted with similar problems of kin recognition during multicellular development and have devised molecular mechanisms for self-recognition (Strassmann *et al.*, 2011).

OM exchange may also represent a form of cooperative behaviour whereby cells within recognition groups share resources. Selective recognition and transfer would ensure that only group members obtain population resources and non-group members or 'cheater' cells do not. In contrast, resources that are secreted into the environment can be exploited by a variety of cells. OM exchange may thus represent an extension of documented (unregulated) resource sharing whereby *M. xanthus* cells cooperatively secrete hydrolytic enzymes and antibiotics into the extracellular milieu to kill and feed on prey as microbial 'wolf-packs' (Berleman and Kirby, 2009; Rosenberg *et al.*, 1977; Xiao *et al.*, 2011). The ability of cells to share resources also provides a mechanism to reduce physiological heterogeneity across a population. In heterogeneous populations some cells are less fit and the ability of the community to coordinate behaviours is complicated by physiological noise among cells. Thus, we hypothesize that the ability to exchange and share the OM proteome allows the population to establish OM homeostasis and for some individuals to gain fitness. In turn, homeostasis may increase population fitness by normalizing intercellular signal output and reception leading to improved communication to coordinate community behaviours.

Acknowledgements

I thank the National Science Foundation (grant MCB-848141) and the National Institutes of Health (grant GM101449) for funding to support this research.

References

Berleman, J.E., and Kirby, J.R. (2009). Deciphering the hunting strategy of a bacterial wolfpack. FEMS Microbiol. Rev. 33, 942–957.

Bulyha, I., Schmidt, C., Lenz, P., Jakovljevic, V., Hone, A., Maier, B., Hoppert, M., and Sogaard-Andersen, L. (2009). Regulation of the type IV pili molecular machine by dynamic localization of two motor proteins. Mol. Microbiol. 74, 691–706.

Goldman, B.S., Nierman, W.C., Kaiser, D., Slater, S.C., Durkin, A.S., Eisen, J.A., Ronning, C.M., Barbazuk, W.B., Blanchard, M., Field, C., *et al.* (2006). Evolution of sensory complexity recorded in a myxobacterial genome. Proc. Natl. Acad. Sci. U.S.A. *103*, 15200–15205.

Hagen, D.C., Bretscher, A.P., and Kaiser, D. (1978). Synergism between morphogenetic mutants of *Myxococcus xanthus*. Dev. Biol. *64*, 284–296.

Hartzell, P., Shi, W., and Youderian, P. (2008). Gliding motility in *Myxococcus xanthus*. In Myxobacteria: Multicellularity and Differentiation, D.E. Whitworth, ed. (Washington, DC: ASM Press), pp. 103–122.

Hodgkin, J., and Kaiser, D. (1977). Cell-to-cell stimulation of movement in nonmotile mutants of *Myxococcus*. Proc. Natl. Acad. Sci. U.S.A. *74*, 2938–2942.

Hodgkin, J., and Kaiser, J. (1979). Genetics of gliding motility in *Myxococcus xanthus* (Myxobacterales): Two gene systems control movement. Mol. Gen. Genet. *171*, 177–191.

Kahnt, J., Aguiluz, K., Koch, J., Treuner-Lange, A., Konovalova, A., Huntley, S., Hoppert, M., Sogaard-Andersen, L., and Hedderich, R. (2010). Profiling the outer membrane proteome during growth and development of the social bacterium *Myxococcus xanthus* by selective biotinylation and analyses of outer membrane vesicles. J. Proteome Res. *9*, 5197–5208.

Kaiser, D. (1979). Social gliding is correlated with the presence of pili in *Myxococcus xanthus*. Proc. Natl. Acad. Sci. U.S.A. *76*, 5952–5956.

Kaiser, D., Robinson, M., and Kroos, L. (2010). Myxobacteria, polarity, and multicellular morphogenesis. Cold Spring Harb. Perspect. Biol. *2*, a000380.

Kim, S.K., and Kaiser, D. (1990). Cell motility is required for the transmission of C-factor, an intercellular signal that coordinates fruiting body morphogenesis of *Myxococcus xanthus*. Genes Dev. *4*, 896–904.

Konovalova, A., Petters, T., and Sogaard-Andersen, L. (2010). Extracellular biology of *Myxococcus xanthus*. FEMS Microbiol. Rev. *34*, 89–106.

Koo, J., Burrows, L.L., and Howell, P.L. (2012). Decoding the roles of pilotins and accessory proteins in secretin escort services. FEMS Microbiol. Lett. *328*, 1–12.

Kraemer, S.A., and Velicer, G.J. (2011). Endemic social diversity within natural kin groups of a cooperative bacterium. Proc. Natl. Acad. Sci. U.S.A. *108* (Suppl. 2), 10823–10830.

Kudryashev, M., Cyrklaff, M., Alex, B., Lemgruber, L., Baumeister, W., Wallich, R., and Frischknecht, F. (2011). Evidence of direct cell–cell fusion in *Borrelia* by cryogenic electron tomography. Cell Microbiol. *13*, 731–741.

Kulp, A., and Kuehn, M.J. (2010). Biological functions and biogenesis of secreted bacterial outer membrane vesicles. Annu. Rev. Microbiol. *64*, 163–184.

Luciano, J., Agrebi, R., Le Gall, A.V., Wartel, M., Fiegna, F., Ducret, A., Brochier-Armanet, C., and Mignot, T. (2011). Emergence and modular evolution of a novel motility machinery in bacteria. PLoS Genet. *7*, e1002268.

Martens, S., and McMahon, H.T. (2008). Mechanisms of membrane fusion: disparate players and common principles. Nat. Rev. Mol. Cell Biol. *9*, 543–556.

Mauriello, E.M., Astling, D.P., Sliusarenko, O., and Zusman, D.R. (2009). Localization of a bacterial cytoplasmic receptor is dynamic and changes with cell–cell contacts. Proc. Natl. Acad. Sci. U.S.A. *106*, 4852–4857.

Nan, B., Chen, J., Neu, J.C., Berry, R.M., Oster, G., and Zusman, D.R. (2011). Myxobacteria gliding motility requires cytoskeleton rotation powered by proton motive force. Proc. Natl. Acad. Sci. U.S.A. *108*, 2498–2503.

Nudleman, E., Wall, D., and Kaiser, D. (2005). Cell-to-cell transfer of bacterial outer membrane lipoproteins. Science *309*, 125–127.

Nudleman, E., Wall, D., and Kaiser, D. (2006). Polar assembly of the type IV pilus secretin in *Myxococcus xanthus*. Mol. Microbiol. *60*, 16–29.

Pathak, D.T., and Wall, D. (2012). Identification of the *cglC*, *cglD*, *cglE*, and *cglF* genes and their role in cell contact-dependent gliding motility in *Myxococcus xanthus*. J. Bacteriol. *194*, 1940–1949.

Pathak, D.T., Wei, X., Bucuvalas, A., Haft, D.H., Gerloff, D.L., and Wall, D. (2012a). Cell contact-dependent outer membrane exchange in myxobacteria: genetic determinants and mechanism. PLoS Genet. *8*, e1002626.

Pathak, D.T., Wei, X., and Wall, D. (2012b). Myxobacterial tools for social interactions. Res. Microbiol. *163*, 579–591.

Rigden, D.J., Mello, L.V., and Galperin, M.Y. (2004). The PA14 domain, a conserved all-beta domain in bacterial toxins, enzymes, adhesins and signaling molecules. Trends Biochem. Sci. *29*, 335–339.

Rodriguez, A.M., and Spormann, A.M. (1999). Genetic and molecular analysis of *cglB*, a gene essential for single-cell gliding in *Myxococcus xanthus*. J. Bacteriol. *181*, 4381–4390.

Rodriguez-Soto, J.P., and Kaiser, D. (1997a). Identification and localization of the Tgl protein, which is required for *Myxococcus xanthus* social motility. J. Bacteriol. *179*, 4372–4381.

Rodriguez-Soto, J.P., and Kaiser, D. (1997b). The *tgl* gene: social motility and stimulation in *Myxococcus xanthus*. J. Bacteriol. *179*, 4361–4371.

Rosenberg, E., Keller, K.H., and Dworkin, M. (1977). Cell density-dependent growth of *Myxococcus xanthus* on casein. J. Bacteriol. *129*, 770–777.

Shi, W., and Zusman, D.R. (1993). The two motility systems of *Myxococcus xanthus* show different selective advantages on various surfaces. Proc. Natl. Acad. Sci. U.S.A. *90*, 3378–3382.

Shimkets, L.J., Dworkin, M., and Reichenbach, H. (2006). The myxobacteria. In The Prokaryotes, M. Dworkin, S. Falkow, E. Rosenberg, K.H. Schleifer, and E. Stackebrandt, eds. (Heidelberg: Springer Verlag), pp. 31–115.

Simunovic, V., Gherardini, F.C., and Shimkets, L.J. (2003). Membrane localization of motility, signaling, and polyketide synthetase proteins in *Myxococcus xanthus*. J. Bacteriol. *185*, 5066–5075.

Skerker, J.M., and Berg, H.C. (2001). Direct observation of extension and retraction of type IV pili. Proc. Natl. Acad. Sci. U.S.A. *98*, 6901–6904.

Smith, D.R., and Dworkin, M. (1994). Territorial interactions between two *Myxococcus* species. J. Bacteriol. *176*, 1201–1205.

Smukalla, S., Caldara, M., Pochet, N., Beauvais, A., Guadagnini, S., Yan, C., Vinces, M.D., Jansen, A., Prevost, M.C., Latge, J.P., *et al.* (2008). *FLO1* is a variable green beard gene that drives biofilm-like cooperation in budding yeast. Cell *135*, 726–737.

Strassmann, J.E., Gilbert, O.M., and Queller, D.C. (2011). Kin discrimination and cooperation in microbes. Annu. Rev. Microbiol. *65*, 349–367.

Sun, M., Wartel, M., Cascales, E., Shaevitz, J.W., and Mignot, T. (2011). Motor-driven intracellular transport powers bacterial gliding motility. Proc. Natl. Acad. Sci. U.S.A. *108*, 7559–7564.

Veelders, M., Bruckner, S., Ott, D., Unverzagt, C., Mosch, H.U., and Essen, L.O. (2010). Structural basis of flocculin-mediated social behavior in yeast. Proc. Natl. Acad. Sci. U.S.A. *107*, 22511–22516.

Vos, M., and Velicer, G.J. (2006). Genetic population structure of the soil bacterium *Myxococcus xanthus* at the centimeter scale. Appl. Environ. Microbiol. *72*, 3615–3625.

Vos, M., and Velicer, G.J. (2009). Social conflict in centimeter-and global-scale populations of the bacterium *Myxococcus xanthus*. Curr. Biol. *19*, 1763–1767.

Wall, D., and Kaiser, D. (1998). Alignment enhances the cell-to-cell transfer of pilus phenotype. Proc. Natl. Acad. Sci. U.S.A. *95*, 3054–3058.

Wall, D., and Kaiser, D. (1999). Type IV pili and cell motility. Mol. Microbiol. *32*, 1–10.

Wall, D., Wu, S.S., and Kaiser, D. (1998). Contact stimulation of Tgl and type IV pili in *Myxococcus xanthus*. J. Bacteriol. *180*, 759–761.

Wall, D., Kolenbrander, P.E., and Kaiser, D. (1999). The *Myxococcus xanthus pilQ* (*sglA*) gene encodes a secretin homolog required for type IV pilus biogenesis, social motility, and development. J. Bacteriol. *181*, 24–33.

Wei, X., Pathak, D.T., and Wall, D. (2011). Heterologous protein transfer within structured myxobacteria biofilms. Mol. Microbiol. *81*, 315–326.

Wickner, W., and Schekman, R. (2008). Membrane fusion. Nat. Struct. Mol. Biol. *15*, 658–664.

Wu, S.S., and Kaiser, D. (1995). Genetic and functional evidence that Type IV pili are required for social gliding motility in *Myxococcus xanthus*. Mol. Microbiol. *18*, 547–558.

Xiao, Y., Wei, X., Ebright, R., and Wall, D. (2011). Antibiotic production by myxobacteria plays a role in predation. J. Bacteriol. *193*, 4626–4633.

Developmental Gene Regulation

6

Ramya Rajagopalan, Zaara Sarwar, Anthony G. Garza and Lee Kroos

Abstract

Starvation induces *Myxococcus xanthus* cells to glide into aggregates and form multicellular fruiting bodies in which some cells differentiate from rods to spores. The gene regulatory network controlling this developmental process involves three interconnected modules. The first module is a cascade of eukaryotic-like enhancer binding proteins that activate early genes in response to starvation, although products of some of those genes function later in development. The second module depends on the first, linking it indirectly to starvation, but other pathways directly link the second module to starvation, and govern accumulation of MrpC and its truncated form MrpC2, which directly activate transcription of *fruA*, the gene encoding the key component of the third module. Activity of FruA is governed by short-range C-signalling between cells, which is believed to increase as cells become aligned in aggregates. FruA and MrpC2 bind cooperatively to promoter regions of several late genes, subjecting them to control by the Mrp and FruA modules. Among these late genes is the *dev* operon, whose products somehow control expression of genes required for spore formation. Feedback loops within modules and between modules reinforce developmental progression. Interestingly, some of the key regulators appear to have been acquired by lateral gene transfer and are not conserved in distantly related mxyobacteria that form fruiting bodies.

Introduction

This chapter focuses on recent progress towards understanding the gene regulatory network (GRN) controlling *Myxococcus xanthus* development. The *M. xanthus* GRN relies heavily on eukaryotic-like transcription factors called enhancer binding proteins (EBPs) that enable σ^{54} RNA polymerase (RNAP) to activate transcription. We first review the features of the σ^{54} regulatory system. It has been speculated that signal-responsive EBPs enable *M. xanthus* to build a multicellular fruiting body of defined size and shape (Kroos, 2007). In contrast, the endosporulation process of *Bacillus subtilis* does not rely on EBPs (Kroos, 2007). Rather, alternative σ factors are organized into cascades coordinated by signal transduction pathways between the mother cell and forespore compartments. While alternative σ factors surely play important roles in *M. xanthus* development (*sigB-F* mutants exhibit defects), no progress has been reported on these or on most of the 38 predicted extracytoplasmic function (ECF) σ factors (exceptions being CarQ and CorE; see Chapter 7) in terms of the genes they transcribe, since the last time this topic was reviewed (Kroos and Inouye, 2008). So except for σ^{54}, our knowledge of σ factor function in the GRN is conspicuously lacking, while our knowledge of how EBPs and two other key transcription factors, MrpC and FruA, function in the GRN has advanced significantly. These advances led to a view of the GRN in terms of modular design, with starvation, intracellular ppGpp, and extracellular A- and C-signals providing input into three regulatory modules designated the EBP cascade, Mrp, and FruA (Kaiser *et al.*, 2010). We update this view in light of recent progress and we emphasize autoregulatory and feedback loops, connections between modules, and cooperation between MrpC and FruA in controlling late gene expression (Fig. 6.1). Much remains to be learned about the GRN, so we attempt to identify some of the most interesting and important questions.

The σ^{54} regulatory system

σ factors control the specificity of transcription by binding to RNAP and directing it to promoter sequences upstream of certain target genes (Wösten, 1998). Two families of bacterial σ factors have been identified: σ^{70} and σ^{54}. Bacteria typically have many genes that code for σ^{70} family members (*sigA-G* and 38 predicted ECF σ factors in *M. xanthus*), but only one gene (*rpoN*) coding for σ^{54}. σ^{54} mediates transcription of genes that have a variety of different functions, including nitrogen assimilation, flagellar biogenesis, and production of extracellular alginate (Barrios *et al.*, 1999). In *M. xanthus*, σ^{54} regulates genes that control many aspects of fruiting body development (Kroos and Inouye, 2008). In addition, σ^{54}-mediated gene regulation has a unique role in *M. xanthus*: it is essential for growth (Keseler and Kaiser, 1997).

σ^{54} directs RNAP to promoters with conserved sequences in their −12 and −24 regions. The hallmarks of these σ^{54} promoters are a conserved GC dinucleotide in the −12 region, a conserved GG dinucleotide in the −24 region, and the strict

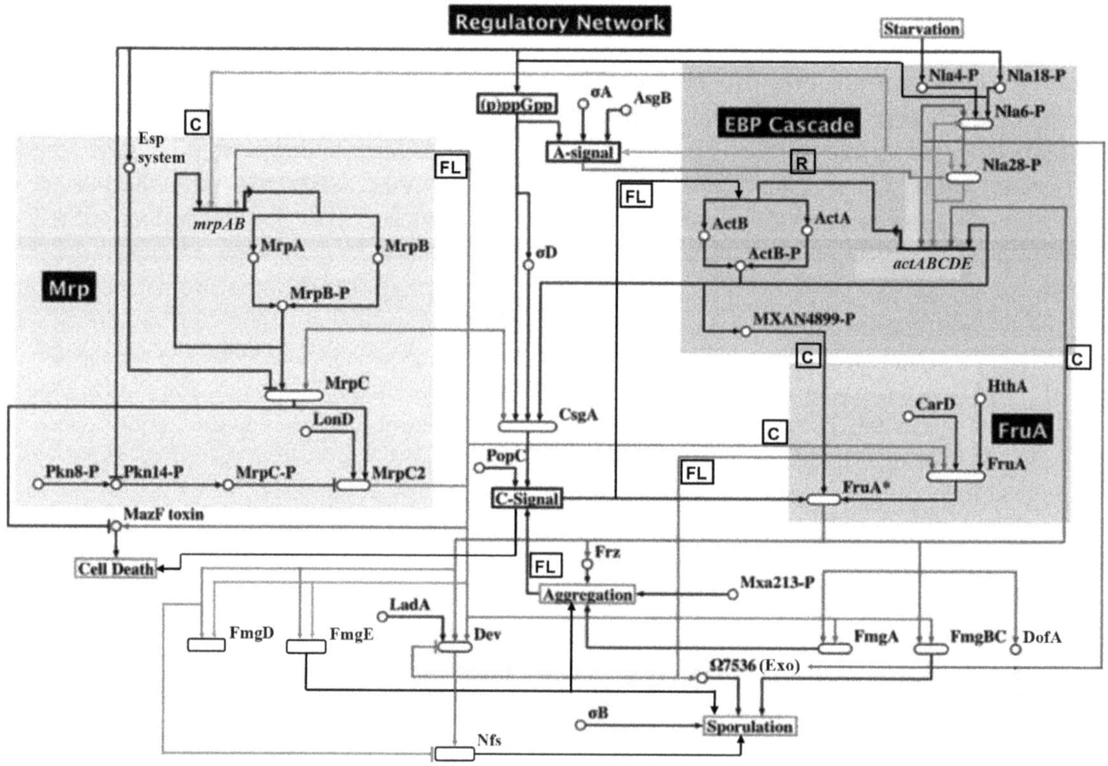

Figure 6.1 Gene regulatory network during *M. xanthus* development. Arrows indicate positive regulation and lines with barred ends indicate negative regulation, which in most cases may be at the transcriptional or post-transcriptional level. In two cases, operons are depicted (*mrpAB* and *actABCDE*) and the regulation is at the transcriptional level, although it may be direct or indirect. Some of the highly branched lines are coloured to make them easier to follow. Circles and ellipses indicate proteins or genes (no distinction is intended). Background colours distinguish the EBP cascade, Mrp, and FruA regulatory modules. Connections (C) between modules, feedback loops (FL), and reciprocal regulation (R) are indicated. The EBP cascade has three positive autoregulatory loops involving Nla6, Nla28, and ActB. The Mrp module has two positive autoregulatory loops involving MrpB and MrpC. Products of the *dev* operon (Dev) negatively autoregulate. An earlier version of this figure was published in Cold Spring Harb. Perspect. Biol. *2*, a000380 (Kaiser *et al.*, 2010). A colour version of this figure is located in the plate section at the back of the book.

requirement for a 4-bp spacer between these regions (Barrios *et al.*, 1999). These features distinguish σ^{54} promoters from σ^{70} promoters, which have conserved sequences in their −10 and −35 regions, and seem to have a greater tolerance for variations in spacer length (Barrios *et al.*, 1999). In addition to the σ^{54} RNAP binding sites in their −12 and −24 regions, σ^{54} promoter regions contain tandem repeat sequences known as enhancer elements. Enhancer elements are typically found 70–150 bp upstream of the −24 region (Xu and Hoover, 2001) and they are the binding sites of transcriptional activators known as EBPs. EBPs are crucial to the σ^{54} regulatory system because EBP-catalysed ATP hydrolysis opens the σ^{54} RNAP promoter complex so that transcription can initiate (Popham *et al.*, 1989; Sasse-Dwight and Gralla, 1990; Wedel and Kustu, 1995). The ATPase activity of each EBP is controlled by a specific intracellular or extracellular signal. Some EBPs bind directly to a signalling molecule. However, most EBPs interact with a signal transduction partner that detects a signal (Shingler, 2011). Because σ^{54} RNAP requires the energy from EBP-catalysed ATP hydrolysis to initiate transcription and the EBP's ATPase activity is controlled by signal input, σ^{54}-regulated genes can be turned off or expressed at a particular rate depending on whether the appropriate signal is present. This feature makes EBPs ideal regulators in multicellular systems such as *M. xanthus* fruiting body development, since cells must link information about a variety of signals to the proper temporal expression of developmental genes.

Domain architecture of *Myxococcus xanthus* EBPs

M. xanthus has an abundance of EBPs (53) compared to other bacterial species such as *E. coli*, which has only 12 (Blattner *et al.*, 1997; Goldman *et al.*, 2006). These EBPs have a common domain organization with a central AAA-ATPase domain responsible for the ATP hydrolysis and interaction with σ^{54}, a C-terminal helix–turn–helix (HTH) DNA-binding domain, and an N-terminal sensory domain that regulates the ATPase activity of the central domain in response to a particular signal (Studholme and Dixon, 2003). A variety of different N-terminal sensory domains have been identified in *M. xanthus* EBPs (Goldman *et al.*, 2006) (Fig. 6.2). Most of the EBPs have an N-terminal response regulator receiver module that mediates a specific interaction with a histidine protein kinase (HPK) sensor that detects an intracellular or extracellular signal (Laub and Goulian, 2007; Diodati *et al.*, 2008). Other EBPs appear to have an N-terminal forkhead-associated (FHA) domain that mediates an interaction with a protein in a serine/threonine protein kinase (STPK) signal transduction pathway (Durocher *et al.*, 2000), a small molecule-binding domain such as a GAF domain (Anantharaman *et al.*, 2001), or an RtcR domain (Genschik *et al.*, 1998). There are five EBPs whose N-terminal region lacks a domain of known function and three of these EBPs also lack a C-terminal DNA-binding domain. This finding suggests that the last three EBPs activate transcription in an enhancer-independent manner similar to *H. pylori* FlgR (Brahmachary *et al.*, 2004).

EBPs and developmental gene expression

Changes in gene expression are an integral part of the behavioural and morphological events that occur during *M. xanthus* fruiting body development (Kroos and Inouye, 2008; Kaiser *et al.*, 2010). A recent DNA microarray analysis showed that a large number of genes are developmentally regulated; at least 2000 *M. xanthus* genes (about 30% of the genome) are differentially expressed in developing cells (Giglio *et al.*, 2011). These developmental genes are activated sequentially and, in the case of several genes involved in sporulation, expression is spatially localized to the fruiting body structure (Sager and Kaiser, 1993; Julien *et al.*, 2000; Muller *et al.*, 2010).

EBPs and the σ^{54} regulatory system have been linked to expression of many *M. xanthus* developmental genes; over 200 developmentally regulated operons with known σ^{54} promoters or putative σ^{54} promoters have been identified (Romeo and Zusman, 1991; Keseler and Kaiser, 1995; Wu and Kaiser, 1997; Garza *et al.*, 1998, 2000; Gronewold and Kaiser, 2001, 2007; Nariya and Inouye, 2005; Giglio *et al.*, 2011; Li, Murphy, Snyder, and Garza,

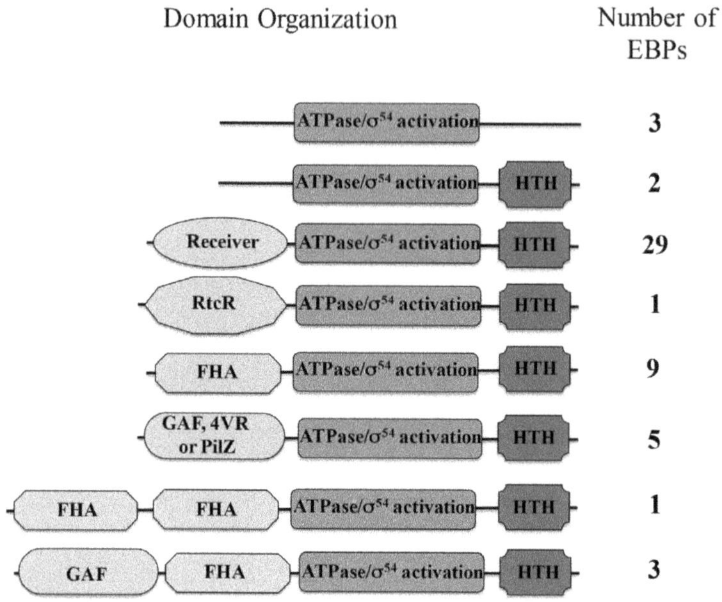

Figure 6.2 Domain organization of *M. xanthus* EBPs. The number of EBPs that are predicted to have a particular domain organization is shown in bold type. Abbreviations: FHA, forkhead-associated domain; GAF, c**G**MP-specific phosphodiesterase, **a**denylyl cyclase, and **F**hlA domain; HTH, helix–turn–helix DNA-binding domain; RtcR, regulator of RNA terminal phosphate cyclase domain; 4VR, 4-vinyl reductase domain.

unpublished). These σ^{54}-dependent and putative σ^{54}-dependent operons represent about 28% of the developmentally regulated operons identified using DNA microarrays (Giglio *et al.*, 2011). Most of these operons are activated in the early stages of fruiting body development, suggesting that EBPs and the σ^{54} regulatory system play a prominent role in early developmental gene expression.

The vast majority of the 53 *M. xanthus* EBP genes have been characterized via mutational analysis (Wu and Kaiser, 1997; Gorski and Kaiser, 1998, 2002; Guo *et al.*, 2000; Hager *et al.*, 2001; Sun and Shi, 2001b; Caberoy *et al.*, 2003; Kirby and Zusman, 2003; Jakobsen *et al.*, 2004; Jelsbak *et al.*, 2005; Giglio *et al.*, 2010). About 20 EBP gene mutations produce obvious defects in fruiting body development, suggesting that *M. xanthus* uses a large number of EBPs to regulate expression of its developmental genes. Ten of these EBP gene mutations produce strong sporulation defects and some these mutations also produce aggregation defects, as summarized in Table 6.1. The *nla24* and MXAN4196 EBP gene mutations affect gliding motility (Caberoy *et al.*, 2003; Lancero *et al.*, 2004; Giglio *et al.*, 2010), which is required for cells to move into aggregation centres and to build fruiting bodies. Two motility systems control *M. xanthus* gliding motility on solid surfaces: the adventurous (A) and the social (S) systems (Hodgkin and Kaiser, 1979a,b) (see Chapters 9 and 10, respectively). The *nla24* mutation causes defects in A- and S-motility, whereas the MXAN4196 mutation only affects the A-motility system. The eight remaining EBP gene mutations have no appreciable effect on gliding motility. These mutations do, however, affect the morphological changes associated with fruiting body development in a variety of different ways. Furthermore, these mutations affect a number of crucial developmental processes such as the accumulation of the intracellular starvation signal (p)ppGpp, which initiates the transition into development (Singer and Kaiser, 1995; Harris *et al.*, 1998), and the accumulation of the cell density signal known as A-signal (Kuspa *et al.*, 1992), which triggers aggregation (Kuspa *et al.*, 1986). These findings and the prevalence of σ^{54}-like promoters upstream of developmental operons led to the proposal that *M. xanthus* uses a series of EBPs during the formation of multicellular fruiting

Table 6.1 Phenotypes of EBP gene mutants

Gene	MXAN	Motility[a]	Aggregation[b]	Sporulation	Change in developmental gene expression pattern[c]	Function/putative function of EBP gene
nla4	2516	A^+ S^+	–	–	1 h	ppGpp synthesis
nla18	3692	A^+ S^+	±	–	1 h	ppGpp synthesis
nla6	4042	A^+ S^+	±	–	1 h	A-signal production/sporulation
nla28	1167	A^+ S^+	±	–	1 h	A-signal production/sporulation
sasR	1245	A^+ S^+	–	–	1–2 h	A-signal response
mrpB	5124	A^+ S^+	–	–	NDA	Regulation of mrpC
actB	3214	A^+ S^+	±	–	6 h	C-signal production/sporulation
MXAN4899	4899	A^+ S^+	±	–	6–12 h	Regulation of FruA
nla24	7440	A^- S^-	–	–	NDA	Motility
MXAN4196	4196	A^- S^+	–	–	NDA	Motility

[a]A, adventurous; S, social.
[b]–, failed to produce normal looking fruiting bodies; ±, produced normal-looking fruiting bodies but aggregation was delayed.
[c]NDA, no data available.

bodies in much the same way that *B. subtilis* uses a cascade of sigma factors during sporulation (Keseler and Kaiser, 1995; Gorski and Kaiser, 1998; Giglio *et al.*, 2011).

The early-functioning EBPs form a regulatory cascade

Nla4, Nla6, Nla18, Nla28, ActB, and MXAN4899 were selected to test the idea that *M. xanthus* developmental EBPs form a regulatory cascade. Phenotypic analyses and expression studies placed these EBPs early in the developmental pathway. Fig. 6.3 indicates the approximate time during development that EBPs in the cascade begin to function. Nla4 and Nla18 are important for gene expression during vegetative growth and development (Diodati *et al.*, 2006; Ossa *et al.*, 2007; Giglio *et al.*, 2011), a property they share with the (p)ppGpp synthetase RelA (Singer and Kaiser, 1995; Harris *et al.*, 1998). Indeed, Nla4 and Nla18 are important for the starvation-induced accumulation of (p)ppGpp, suggesting that they regulate the transition from vegetative growth to development (Diodati *et al.*, 2006; Ossa *et al.*, 2007) (Fig. 6.1). Accumulation of the cell density (quorum) signal known as A-signal is the hallmark of the pre-aggregation stage of development, which follows the vegetative growth to development transition. Based on extracellular complementation studies with the *nla6* and *nla28* mutants, and an A-signal-defective mutant, Nla6 and Nla28 have been linked to the accumulation of A-signal (Caberoy *et al.*, 2003) (Fig. 6.1 shows an arrow only from Nla28 to A-signal since the Nla6 effect may be indirect *via* Nla28, see below). This finding indicates that Nla6 and Nla28 begin functioning during the pre-aggregation stage of development, which is consistent the results of DNA microarray studies indicating that Nla6 and Nla28 become important for gene expression as cells make the transition into pre-aggregation (Giglio *et al.*, 2011) (Fig. 6.3). ActB and MXAN4899, which are known to be important for switching on sporulation (Gorski *et al.*, 2000; Gronewold and Kaiser, 2002; Jelsbak *et al.*, 2005), are the last of these EBPs to begin performing their developmental functions. In particular, ActB starts to be crucial for gene expression when cells are making the transition into aggregation (Gronewold and Kaiser, 2002), whereas MXAN4899 starts to be crucial for gene expression when aggregation is well under way (Jelsbak *et al.*, 2005) (Fig. 6.3). Collectively, these results indicate that pairs of

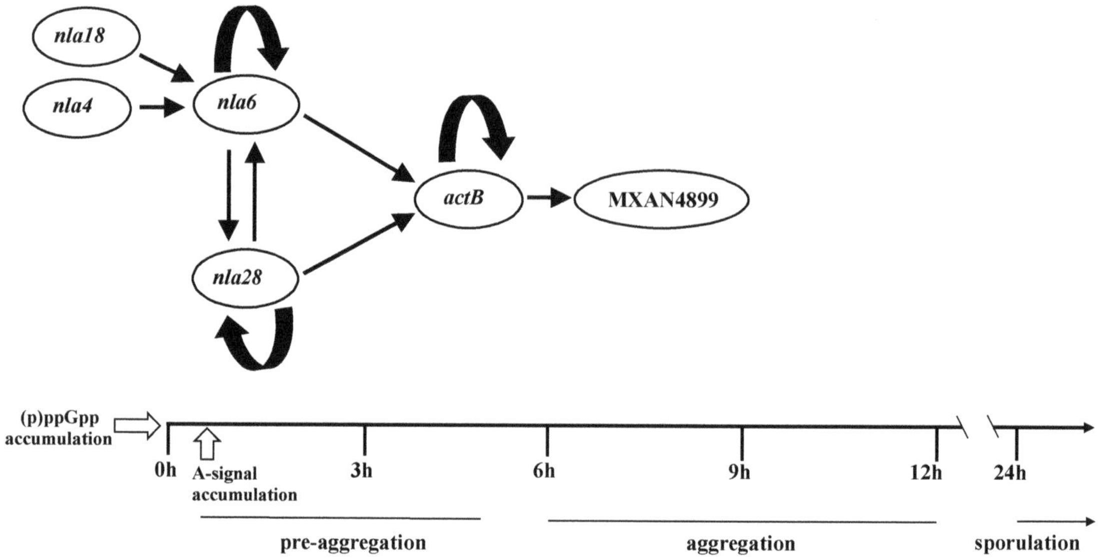

Figure 6.3 Model for the EBP transcriptional cascade. The approximate duration of the pre-aggregation (1–5h), aggregation (6–12h) and sporulation (24–120h) stages of fruiting body development are shown. The white arrows indicate that (p)ppGpp accumulation is required to make the transition from growth to development and that A-signal is required in the early part of pre-aggregation. The products of the EBP genes shown here have a specific operational sequence as indicated by their placement above the developmental timeline. The straight black arrows represent direct transcriptional regulation and the curved black arrows represent autoregulation. This figure was first published in Proc. Natl. Acad. Sci. U.S.A. *108*, E431-E439 (Giglio *et al.*, 2011).

the early functioning EBPs operate in sequential stages of fruiting body development: Nla4 and Nla18 regulate the vegetative growth to fruiting body development transition, Nla6 and Nla28 begin to activate developmental genes when cells enter the pre-aggregation stage, and ActB and MXAN4899 are crucial for expression of developmental genes when cells enter and are well into the aggregation stage, respectively (Fig. 6.3).

Using expression studies and *in vitro* promoter binding assays, Giglio *et al.* (2011) showed that the early functioning EBPs do indeed form a transcriptional cascade. While this is evident in Fig. 6.1, the many outputs from the EBP cascade make the connections within the cascade less obvious. Those connections are more clearly seen in Fig. 6.3. In this cascade, the transition state regulators Nla4 and Nla18 directly activate transcription of the pre-aggregation stage EBP gene *nla6*, Nla6 directly activates transcription of *nla28* and the two pre-aggregation stage EBPs Nla6 and Nla28 directly regulate transcription of the aggregation stage EBP gene *actB*, and ActB directly regulates the MXAN4899 gene. Thus, the regulatory cascade is designed such that EBP pairs functioning at one stage of development directly regulate transcription of an EBP gene important for the next developmental stage. Interestingly, the cascade uses two EBPs to directly regulate developmental expression of EBP genes that are positioned at stage-to-stage transition points in early development (i.e. Nla4 and Nla18 activate *nla6*, and Nla6 and Nla28 activate *actB*). By using this novel system of EBP co-regulation, expression of EBP genes that modulate stage-to-stage transitions is responsive to multiple signal transduction pathways. Presumably, these signals provide information indicating that the developmental process is ready to advance from one stage to the next. Three of the EBPs positively autoregulate, reinforcing forward progression of the cascade, perhaps in a subpopulation of the starving cells since positive autoregulation can produce a bistable switch, leading to different cell fates (Dubnau and Losick, 2006; Smits *et al.*, 2006; Ray *et al.*, 2011).

Inputs and outputs of the EBP regulatory cascade

Nla28 is the only member of the EBP regulatory cascade whose signal transduction partner has been defined experimentally (Sarwar and Garza, 2012). Nla28 has an N-terminal receiver domain, suggesting that it functions as a response regulator in a two-component signal transduction system (Goldman *et al.*, 2006). In two-component systems that contain EBPs such as Nla28, the EBP activates gene expression in response to phosphorylation by an HPK that detects a particular signal. Nla28S was identified as a strong candidate for the HPK partner of Nla28 based on data suggesting that *nla28S* and *nla28* are part of the same developmentally regulated operon (Goldman *et al.*, 2006; Giglio *et al.*, 2011; Sarwar and Garza, 2012). This proposal was tested using *in vitro* assays and the results support the idea that Nla28S and Nla28 form a two-component signal transduction circuit: the cytoplasmic domain of Nla28S has the signature biochemical properties of an HPK, it is capable of transferring a phosphoryl group to Nla28, and the phosphotransfer between the two proteins is specific (Sarwar and Garza, 2012). The developmental phenotypes of *nla28S* and *nla28* mutants suggest that the primary function of the Nla28S/Nla28 signal transduction circuit is to regulate sporulation (Caberoy *et al.*, 2003; Sarwar and Garza, 2012). Indeed, most of the developmental genes that are directly regulated by Nla28, including *nla6*, *actB*, MXAN0881, and MXAN0989 are primarily involved in sporulation (Gorski and Kaiser, 1998; Caberoy *et al.*, 2003; Giglio *et al.*, 2011) (Li, Murphy, Snyder, and Garza, unpublished). This is also true for Nla6, which directly regulates *nla28*, *actB*, and *exo* (Ω7536), a locus that is essential for the shape change associated with sporulation (Gorski and Kaiser, 1998; Licking *et al.*, 2000; Caberoy *et al.*, 2003; Giglio *et al.*, 2011) (Fig. 6.1). In addition to its sporulation function, Nla28 directly regulates the *mrpAB* operon (Li, Murphy, Snyder, and Garza, unpublished), which suggests that Nla28 serves as a bridge between the EBP cascade and the Mrp regulatory module (Fig. 6.1).

The signals that activate Nla28S/Nla28 and the other EBP signal transduction circuits associated with the regulatory cascade have yet to be defined. However, a number of early developmental cues are known and have the potential to serve as input signals, including the levels of nutrients in the environment, the level of the intracellular starvation signal (p)ppGpp, and the level of the extracellular quorum-indicating A-signal. Interestingly, A-signal is known to be important for full expression of the *nla28S*/*nla28* operon during development (Sarwar and Garza, 2012), suggesting that A-signalling regulates the availability of the Nla28S/Nla28 signal transduction circuit in developing cells. As mentioned above, Nla28 was linked to A-signal production using co-development assays (Caberoy *et al.*, 2003). Thus, A-signal and the Nla28S/Nla28 signal transduction circuit may have a reciprocal relationship: expression of the genes for the Nla28S/Nla28 signal transduction circuit is modulated by A-signal and the Nla28S/Nla28 signal transduction circuit may modulate A-signal production. These findings raise the following questions: does the Nla28S/Nla28 signal transduction circuit detect A-signal and does it respond by regulating genes that are important for A-signal production? The challenge for the future will be to answer these questions and to determine the inputs and outputs of the other EBP signal transduction circuits that function in early development.

The Mrp and FruA modules

The Mrp module, which consists primarily of the *mrpAB* operon and the adjacent downstream *mrpC* gene, seems to depend partly on the EBP cascade and begins to function slightly later. The FruA module depends on MrpC and C-signalling. The inputs and outputs of these two modules are discussed in the following sections.

Inputs into the Mrp module

As noted above, Nla28 (presumably in its phosphorylated form) activates *mrpAB* transcription, linking the EBP cascade to the Mrp module (Li, Murphy, Snyder, and Garza, unpublished) (Fig. 6.1). This could explain the observation that an *asgA* mutation impairs expression of an *mrpAB–lacZ* fusion (Sun and Shi, 2001a), since reciprocal regulation between the A-signal and the Nla28S/Nla28 signal transduction circuit was discovered recently (Sarwar and Garza, 2012). A mutation in *relA*, the gene that codes for (p)ppGpp synthetase (Singer and Kaiser, 1995; Harris *et al.*, 1998), also impairs *mrpAB-lacZ* expression (Sun and Shi, 2001a), but this too could be an indirect consequence of defective A-signalling, since intracellular (p)ppGpp regulates production of extracellular A-signal (Harris *et al.*, 1998). Hence, phosphorylated Nla28 appears to be the main input into the Mrp module from the EBP cascade, directly activating *mrpAB* transcription and integrating (p)ppGpp- and A-signalling.

A better understanding of MrpA and MrpB function is needed. MrpA is a predicted HPK essential for sporulation but dispensable for aggregation, whereas its presumed substrate MrpB is necessary for both aggregation and sporulation (Sun and Shi, 2001b). These observations and other experiments suggest that MrpA is not the only enzyme capable of phosphorylating MrpB, but that MrpA must dephosphorylate MrpB to permit sporulation. The signal(s) to which MrpA and the other putative MrpB kinase respond are unknown. MrpB is a predicted EBP and it positively autoregulates, perhaps directly from a putative σ^{54} promoter upstream of *mrpA*. This is one of several positive autoregulatory or feedback loops within the Mrp module (Fig. 6.1, see below). MrpB is necessary for expression of *mrpC* (Sun and Shi, 2001b), presumably activating transcription from its apparent σ^{54} promoter, although this remains to be established.

MrpC is not similar to EBPs, yet it has been reported to positively autoregulate and to positively regulate *mrpAB* (Sun and Shi, 2001b), which could be indirect effects on the putative σ^{54} promoters. In any case, the positive autoregulatory and feedback loops reinforce expression within the Mrp module (Fig. 6.1). As noted above for the EBP cascade, this network architecture can produce a bistable switch in a population, leading to different cell fates (Dubnau and Losick, 2006; Smits *et al.*, 2006; Ray *et al.*, 2011). MrpC is similar to the cyclic AMP receptor protein (CRP) family of transcription factors and is necessary for both aggregation and sporulation (Sun and Shi, 2001b).

At least three inputs regulate MrpC post-translationally (Fig. 6.1). The elaborate Esp system is composed of two hybrid HPKs, two STPKs, and a putative transport protein that delay development by promoting proteolytic degradation of MrpC (Cho and Zusman, 1999; Stein *et al.*, 2006; Higgs *et al.*, 2008; Schramm *et al.*, 2012) (see Chapter 7 for mechanistic details). Starvation inhibits the effect of the Esp system on MrpC but the precise input signal(s) and output protease(s) involved remain to be identified. Starvation also inhibits phosphorylation of MrpC. Under vegetative growth conditions, *mrpC* is expressed at a basal level, and its product is thought to be present in a phosphorylated state (MrpC-P) (Nariya and Inouye, 2005). MrpC-P shows reduced DNA-binding affinity compared with unphosphorylated MrpC (Nariya and Inouye, 2006). Phosphorylation of MrpC is carried out by an STPK, Pkn14, which in turn is phosphorylated by the STPK Pkn8 (Nariya and Inouye, 2005, 2006). It has been reported that deletion of either *pkn8* or *pkn14* speeds up the developmental programme. MrpC accumulation is much higher in a *pkn8* or a *pkn14* mutant. Expression of *pkn14* decreases during development, but the mechanism of regulation is unknown. Shortly after MrpC begins to accumulate during development, an N-terminally truncated form called MrpC2 is observed. The LonD (BsgA) protease is necessary to produce MrpC2 but whether MrpC2 results from cleavage by LonD or another protease, or from an indirect effect of LonD that causes an alternative translation start codon to be used is uncertain, as is whether the MrpC2 produced *in vivo* lacks 25 or 32 residues from the N-terminal part of MrpC (Ueki and Inouye, 2003). In either case, MrpC2 would lack the site in MrpC thought to be phosphorylated by Pkn14 (Nariya and Inouye, 2006). As a result, MrpC2 can evade deactivation by phosphorylation, which might be important during the transition from growth to development. MrpC2 produced in *E. coli* and purified for *in vitro* studies lacks 25 residues (Ueki and Inouye, 2003; Nariya and Inouye, 2006; Mittal and Kroos, 2009a). Purified MrpC2 was reported to bind to target DNA sequences better than purified MrpC (Nariya and Inouye, 2006). For this reason and for simplicity, outputs from the Mrp module are depicted in Fig. 6.1

only from MprC2, although MrpC appears to be more abundant and presumably contributes to the outputs (Nariya and Inouye, 2006; Higgs *et al.*, 2008; Lee *et al.*, 2012). Recently, MrpC and MrpC2 were shown to accumulate preferentially in aggregated cells (i.e. cells that sediment at 50×g) compared with non-aggregated cells during development (Lee *et al.*, 2012). Whether the positive autoregulatory and feedback loops within the Mrp module and/or the inputs described above that regulate MrpC post-translationally account for preferential accumulation in certain cells, affecting their fate, is an interesting possibility that remains to be explored (see Chapter 3).

Outputs from the Mrp module

In the *M. xanthus* strain DZF1, MrpC was shown to interact both *in vivo* and *in vitro* with a developmentally regulated endoribonuclease, MazF (Nariya and Inouye, 2008). MrpC also affects transcription of the *mazF* gene, and this is likely a direct effect since MrpC binds to the *mazF* promoter region. A *mazF* deletion has the effect of eliminating cellular death during development, and causes a reduction in spore numbers. A model was proposed whereby a novel toxin–antitoxin system consisting of MazF and MrpC regulates programmed cell death, thus contributing to one of three cell fates during the developmental process. Of the remaining cells in a developing population, some form spores within fruiting bodies, while others remain outside fruiting bodies as perhipheral rods (O'Connor and Zusman, 1991). However, deletion of the *mazF* gene in two wild-type *M. xanthus* backgrounds, DK1622 and DZ2, has very little or no effect on development (Lee *et al.*, 2012). The strain DZF1 is defective in social motility, and the MazF-related developmental effects were proposed to be an adaptation specific to that strain. Hence, MazF is dispensable for cell lysis during development of wild-type *M. xanthus.* This reopens the question of whether cell lysis is important for development (O'Connor and Zusman, 1988) and if so, the question of how lysis occurs in wild-type strains.

MrpC and/or MrpC2 (abbreviated MrpC/2 hereafter) activates transcription of *fruA* by binding to two sites upstream of the promoter (Ueki and Inouye, 2003). This key output of the Mrp module is an essential input into the FruA module (designated as a connection between modules in Fig. 6.1). MrpC/2 also directly regulates transcription of some C-signal-dependent genes in combination with FruA, which will be described below.

MrpC/2 directly or indirectly regulates the expression of other genes during the early to middle period of development, only one of which is shown in Fig. 6.1, *csgA*. The product of *csgA*, called p25, is cleaved at the cell surface to produce p17, which acts as the C-signal (Shimkets *et al.*, 1983; Kim and Kaiser, 1990a; Shimkets and Rafiee, 1990; Kruse *et al.*, 2001; Lobedanz and Sogaard-Andersen, 2003; Rolbetzki *et al.*, 2008; Konovalova *et al.*, 2012). Although the receptor for p17 is unknown, transmission of the C-signal requires cells to move into alignment (Kroos *et al.*, 1988; Kim and Kaiser, 1990b,c), suggesting that C-signalling requires contact between cells. C-signalling is required for formation of stable aggregates during development (Shimkets *et al.*, 1983; Curtis *et al.*, 2007) and cell alignment within mounds has been proposed to trigger C-signal-dependent gene expression leading to sporulation (Kim and Kaiser, 1991; Li *et al.*, 1992; Sager and Kaiser, 1993; Julien *et al.*, 2000; Kruse *et al.*, 2001). An *mrpC* mutant fails to accumulate p17 normally and fails to rescue development of a *csgA* mutant when the two strains are mixed (Sun and Shi, 2001a). It is not known whether MrpC/2 affects *csgA* transcription (as depicted for MrpC2 in Fig. 6.1) or a subsequent step in p17 production. A mutation in *mrpC* greatly reduces expression of the A-signal-independent *sdeK* gene, as well as the A-signal-dependent *spi* gene (Kuspa *et al.*, 1986; Sun and Shi, 2001a). Although *spi* does not appear to be required for development (Kroos *et al.*, 1986), *sdeK*, which codes for an HPK, is essential for development (Kroos *et al.*, 1986, 1990; Garza *et al.*, 1998; Pollack and Singer, 2001). Chromatin immunoprecipitation with MrpC antibodies 24 h after the induction of development followed by sequencing (ChIP-seq) has revealed a large number of binding sites on the *M. xanthus* chromosome, suggesting that MrpC directly regulates many developmental genes (Robinson, Son and Kroos, unpublished).

Inputs into the FruA module

The FruA module appears to respond to several other inputs. As noted above, MrpC appears to activate *fruA* transcription (Ueki and Inouye, 2003), connecting the Mrp module to the FruA module (Fig. 6.1). Other positive inputs into *fruA* transcription, which may be direct or indirect, are mediated by the eukaryotic-like transcription factor CarD (Elias-Arnanz *et al.*, 2008), the putative DNA-binding protein HthA (Nielsen *et al.*, 2004), and DevT (forming a feedback loop described below) (Boysen *et al.*, 2002). Expression of a *fruA–lacZ* fusion was reported to be severely reduced in signalling-defective *asgA*, *asgB*, *bsgA*, *csgA*, *dsgA*, and *esg* mutants (Ogawa *et al.*, 1996), but these likely indirect effects are not included in Fig. 6.1.

FruA appears to be regulated post-translationally. It is similar to response-regulators of two-component His-Asp phosphorelay systems (Ogawa *et al.*, 1996; Ellehauge *et al.*, 1998). In contrast to the reported effect of a *csgA* mutation on *fruA-lacZ* expression (Ogawa *et al.*, 1996), FruA was subsequently reported to accumulate normally in a *csgA* mutant, and some evidence suggested that C-signalling results in phosphorylation of FruA (Ellehauge *et al.*, 1998). The last member of the EBP cascade, MXAN4899, was also proposed to cause phosphorylation of FruA (Jelsbak *et al.*, 2005). However, the CheY-like receiver domain of FruA lacks key aspartate residues typically required for phosphorylation, a cognate HPK for FruA has not been found, and the DNA binding-ability of recombinant FruA was not altered by phosphatase treatment, mutation of the putative phosphorylation site of FruA from aspartate to glutamate, or the addition of small molecule phosphodonors (Mittal and Kroos, 2009a). Therefore, C-signalling and MXAN4899 might lead to activation of FruA by means other than phosphorylation, and for this reason active FruA is designated FruA* in Fig. 6.1. Elucidating the mechanism of FruA activation is an important challenge.

Outputs from the FruA module

Outputs from the FruA module include modulation of the Frz chemosensory system to bring about aggregation, and activation of target genes, some of which play a role in aggregation or sporulation (Fig. 6.1). Activated FruA* presumably accounts for the observed effect of C-signal in promoting the methylation of FrzCD, which stimulates aggregation behaviour in the developing population rather than cell reversals (Sogaard-Andersen and Kaiser, 1996). FrzCD is homologous to the methyl-accepting chemotaxis proteins and it exhibits a distinct pattern of methylation during vegetative growth and development (McBride and Zusman, 1993). It is methylated during vegetative growth and becomes demethylated two hours after starvation. By nine hours into development however, most of the FrzCD is converted back to the methylated form. This is disrupted in a *fruA* mutant; the methylation pattern largely remains what it was at two hours (Sogaard-Andersen and Kaiser, 1996). In wild type, aggregation brings cells into contact, enhancing C-signalling, which leads to more activated FruA*, forming a positive feedback loop (Kroos *et al.*, 1988; Kim and Kaiser, 1990b,c, 1991; Li *et al.*, 1992; Kruse *et al.*, 2001) (Fig. 6.1).

FruA* is also involved in a second positive feedback loop. Early expression of the *act* operon depends on FruA (Gronewold and Kaiser, 2007), connecting the FruA module to the EBP cascade (Fig. 6.1). The later more rapid phase of *act* expression depends on *actA* and *csgA*, leading to the idea that C-signalling feeds back on expression of the *act* operon *via* ActA phosphorylation of ActB (Gronewold and Kaiser, 2007) (Fig. 6.1). The many connections and feedback loops involving Act and MXAN4899 in the EBP cascade, the FruA module, C-signalling, and aggregation, presumably ensure that FruA* rises in aggregated cells, as appears to be the case (Lee *et al.*, 2012). Moreover, MrpC/2, which accumulate preferentially in aggregated cells (Lee *et al.*, 2012) provide input from the Mrp module by directly activating *fruA* transcription (Ueki and Inouye, 2003) and positively regulating C-signal production (Sun and Shi, 2001a).

Most target genes of the FruA module depend on C-signalling, which appears to somehow activate FruA as mentioned above, but there are two notable exceptions. Expression from *tps*, the gene coding for a major spore coat protein, Protein S, requires the presence of a functional *fruA* gene

(Ogawa *et al.*, 1996), but a *csgA* mutation has no effect (Kroos and Kaiser, 1987). The effect of FruA on *tps* expression may be indirect since purified FruA does not bind to the promoter region of *tps* (Ueki and Inouye, 2005a). Expression from *dofA* is likewise C-signal-independent (Horiuchi *et al.*, 2002a,b), but the purified C-terminal domain of FruA binds to two sites in the promoter region of *dofA* (Ueki and Inouye, 2005a). This suggests that even in the absence of C-signalling, enough FruA* is produced to activate transcription of *dofA*. The other target genes of the FruA module that have been studied depend at least in part on C-signalling for their expression, and these will be described in the next section.

Regulation of C-signal-dependent genes

Genes expressed after the early aggregation stage of development at about 6 h post-starvation exhibit two patterns of dependence on C-signalling (Kroos and Kaiser, 1987). Some depend partially on C-signalling and others depend absolutely, based on their expression in a *csgA* mutant. Many years of effort towards understanding the regulation of these genes culminated in the discovery that MrpC and FruA bind cooperatively in their promoter regions (Mittal and Kroos, 2009a). Different arrangements of binding sites with different affinities influence the timing of promoter activity and its dependence on C-signalling (Mittal and Kroos, 2009a; Mittal and Kroos, 2009b; Lee *et al.*, 2011; Son *et al.*, 2011). In this way, both the MrpC module and the FruA module appear to regulate the transcription of many genes expressed late in development.

Many C-signal-dependent genes were identified as *lacZ* fusions created by transposition of Tn*5 lac* into wild-type cells followed by transduction into a *csgA* mutant (Kroos and Kaiser, 1987). If *lacZ* expression was reduced or abolished in the *csgA* mutant compared with wild type, the gene was potentially C-signal-dependent, contingent upon showing in a subsequent experiment that *lacZ* expression in the *csgA* mutant could be restored by supplying extracellular C-signal. Typically this was done by co-developing the *csgA* mutant containing the *lacZ* fusion in a 1:1 mixture with wild-type cells lacking the *lacZ* fusion (Kroos and Kaiser, 1987; Fisseha *et al.*, 1996, 1999; Brandner and Kroos, 1998; Loconto *et al.*, 2005), although in some experiments purified C-signal was added to the *lacZ* fusion-containing *csgA* mutant (Kim and Kaiser, 1991). Genes that depended partially on C-signalling tended to be induced earlier in development at about 6–10 h post starvation, while genes that depended absolutely on C-signalling were induced after 10 h (Kroos and Kaiser, 1987).

Two Tn*5 lac* fusions that depended partially on C-signalling and two that depended absolutely were selected for studies aimed to understand their regulation. This involved cloning DNA upstream of the Tn*5 lac* insertion, mapping the transcriptional start site, and performing mutational analysis to define the boundaries of the promoter region and elucidate the important *cis*-regulatory elements (Fisseha *et al.*, 1996, 1999; Brandner and Kroos, 1998; Viswanathan and Kroos, 2003; Yoder and Kroos, 2004a,b; Loconto *et al.*, 2005; Viswanathan *et al.*, 2006). In all four cases, *cis*-regulatory elements were located immediately upstream of the promoter. In the case of the promoter at the Ω4406 locus, additional *cis*-regulatory elements were found slightly farther upstream of the promoter and downstream of it (Viswanathan *et al.*, 2006). This work set the stage for identifying transcription factors that bind to the *cis*-regulatory elements.

FruA was the obvious candidate to be a direct regulator of C-signal-dependent genes since FruA had been shown to mediate responses to C-signalling (Sogaard-Andersen *et al.*, 1996; Ellehauge *et al.*, 1998) and its C-terminal domain (FruA-DBD) had been shown to bind DNA (Ueki and Inouye, 2005a,b). Indeed, the FruA-DBD was used to isolate the *fdgA* promoter region from a genomic library of DNA fragments and expression of *fdgA* from one of two developmental promoters was subsequently shown to be reduced in a *csgA* mutant (Ueki and Inouye, 2005b). Not surprisingly, FruA-DBD bound upstream of the promoter identified by Tn*5 lac* Ω4400, one of *lacZ* fusion mentioned above that depends partially on C-signalling (Yoder-Himes and Kroos, 2006). Mutations between −86 and −77 greatly reduced binding of FruA-DBD. These mutations had previously been shown to reduce promoter activity and in the one case tested, to eliminate dependence on C-signalling (Yoder and Kroos, 2004a). Taken together, the two studies provided strong evidence that FruA is a direct activator of C-signal-dependent genes.

Mutational analysis of the promoter region at the Ω4400 locus implied that a second transcription factor in addition to FruA was required for developmental promoter activity. Mutations between −63 and −46 greatly impaired promoter activity but had no effect on FruA-DBD binding (Yoder and Kroos, 2004a; Yoder-Himes and Kroos, 2006). DNA spanning this *cis*-regulatory element was used to isolate the second transcription factor from extracts of developing *M. xanthus* cells that had been subjected to ammonium sulphate fractionation (Mittal and Kroos, 2009a). Surprisingly, the second transcription factor was MrpC/2. This suggested that MrpC/2 not only activated transcription of *fruA* (Ueki and Inouye, 2003), it participated with FruA to activate C-signal-dependent genes (Mittal and Kroos, 2009a) (Fig. 6.1). Even more surprising was the discovery that MrpC/2 occupancy at the Ω4400 promoter region required FruA *in vivo* (as determined by chromatin immunoprecipitation analysis of a *fruA* mutant), which led to the finding that MrpC2 and FruA bind cooperatively to the promoter region *in vitro* (using EMSAs and DNase I footprinting) (Mittal and Kroos, 2009a). Never before had a response regulator (FruA) been shown to bind DNA cooperatively with a CRP-family transcription factor (MrpC). The gene at the Ω4400 locus was named *fmgA* for FruA- and MrpC-regulated gene A. The arrangement of FruA and MrpC2 binding in the *fmgA* promoter region is depicted in Fig. 6.4. The detailed mechanism by which these two transcription factors activate *fmgA* transcription is unknown, as is the σ factor responsible for recognition of the *fmgA* promoter. Given their proximity to the promoter, presumably FruA and/or MrpC/2 interact directly with RNAP holoenzyme, recruiting it to the promoter and/or stimulating a subsequent step in transcription initiation.

The discovery that FruA and MrpC/2 bind cooperatively to the *fmgA* promoter region begged the question of whether other C-signal-dependent genes are regulated similarly. Indeed, cooperative binding of these two transcription factors appears to regulate genes at each of the other three C-signal-dependent loci identified by the Tn*5* *lac* insertions mentioned above (Mittal and Kroos, 2009b; Lee *et al.*, 2011; Son *et al.*, 2011). Hence, the genes at these loci were named *fmg* genes (Fig. 6.1). Table 6.2 lists the Tn*5* *lac* locus, *fmg*, and MXAN designations, as well as functional annotation and mutant phenotypes. A fifth locus, *dev*, will be described below. It is worth noting that three of the Tn*5* *lac* insertions (4400, 4499 and 4406) cause subtle developmental defects that required more careful examination of aggregation and sporulation than used in the initial screen of a large number of insertions (Kroos *et al.*, 1986).

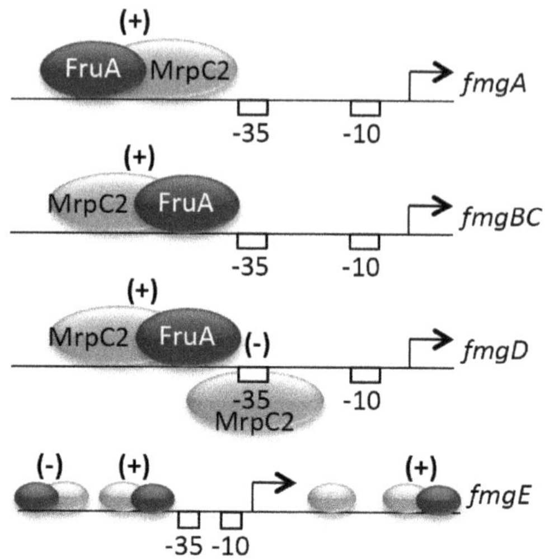

Figure 6.4 Arrangements of FruA and MrpC2 binding sites in *fmg* promoter regions. The positions of FruA-His_6 and His_{10}-MrpC2 binding in vitro are shown, based on DNase I footprinting for *fmgA* (Mittal and Kroos, 2009a) or on the effects of mutations in DNA probes in electrophoretic mobility shift assays for *fmgBC*, *fmgD*, and *fmgE* (Mittal and Kroos, 2009b; Lee *et al.*, 2011; Son *et al.*, 2011). (+) and (–) indicate positive and negative effects on promoter activity, as inferred from expression of *lacZ* fusions to promoter regions with mutations in transcription factor binding sites. A larger region is shown for *fmgE* (–120 to +120).

Fig. 6.4 shows the arrangements of FruA and MrpC2 binding in the *fmg* promoter regions. Interestingly, FruA and MrpC2 bind in a different arrangement in the *fmgBC* promoter region

Table 6.2 C-signal-dependent genes directly regulated by FruA and MrpC/2

Tn*5 lac* locus	Gene(s)	MXAN	Predicted function(s)	Mutant phenotype
4400	*fmgA*	2884	ATP/GTP-binding protein	Aggregation delayed by 8 h
4499	*fmgBC*	4126–7	Cytochrome P450 system	Sporulation reduced 5-fold
4403	*fmgD*	1501	Serine protease	No defect
4406	*fmgE*	3464	Unknown	Agggregation delayed/reduced Sporulation reduced 8-fold
4414	*dev*	7266–59	CRISPR-Cas system	Sporulation reduced 100-fold

than in the *fmgA* promoter region (Mittal and Kroos, 2009a; Mittal and Kroos, 2009b). FruA is proximal to the *fmgBC* promoter, whereas MrpC2 is proximal to the *fmgA* promoter. In both cases, the pair of transcription factors bind immediately upstream of the promoter, so the contacts with RNAP are likely different, yet transcription from both promoters depends partially on C-signalling (Kroos and Kaiser, 1987). Neither promoter is active in a *fruA* mutant (Yoder-Himes and Kroos, 2006; Mittal and Kroos, 2009b). Apparently, enough active FruA is produced in the absence of C-signalling to allow partial occupancy of cooperatively binding FruA and MrpC/2 at their sites in the *fmgA* and *fmgBC* promoter regions, accounting for the reduced expression observed in a *csgA* mutant relative to wild type. Indeed, FruA was found to be associated with the *fmgA* promoter region in a *csgA* mutant using ChIP analysis (Yoder-Himes and Kroos, 2006).

What about genes that depend absolutely on C-signalling? Both genes studied exhibit more complex arrangements of FruA and MrpC2 binding than the two partially C-signal-dependent genes. In the case of *fmgD*, a second MrpC2 binding site overlaps the FruA binding site and the promoter −35 region (Lee *et al.*, 2011) (Fig. 6.4). The effects of mutations on promoter activity *in vivo* and on transcription factor binding *in vitro* support a model in which binding of MrpC/2 to the downstream site represses *fmgD* transcription until C-signalling causes the concentration of active FruA to increase sufficiently to outcompete the downstream MrpC/2 for cooperative binding with the upstream MrpC/2. If a higher concentration of active FruA is needed for binding to the *fmgD* promoter region than for binding to the *fmgA* and *fmgBC* promoter regions, it would explain why *fmgD* transcription begins later during development and is more dependent on C-signalling than transcription of *fmgA* and *fmgBC* (Kroos and Kaiser, 1987).

In the case of *fmgE*, three sites with differential affinity for cooperative binding by FruA and MrpC2 regulate promoter activity (Son *et al.*, 2011). The highest-affinity site, centred at about +100, correlates with positive regulation (Fig. 6.4), so it was proposed to recruit FruA and MrpC/2 to the vicinity of the *fmgE* promoter. The medium-affinity site, centred at about −100, correlates with negative regulation, so it was proposed to compete with the lowest-affinity site, centred just upstream of the promoter, for binding of FruA and MrpC/2. Occupancy of the lowest-affinity site is required to activate the promoter but presumably would only occur when C-signalling produces a high concentration of active FruA, explaining the late, absolutely C-signal-dependent transcription of *fmgE* (Kroos and Kaiser, 1987).

Regulation of the *dev* operon depends on C-signalling and three transcription factors

Expression of the *dev* operon depends partially on C-signalling (Kroos and Kaiser, 1987) and also involves cooperative binding of FruA and MrpC/2 (Campbell, Viswanathan, Son, Barrett, and Kroos, unpublished). Tn*5 lac* Ω4414 proved to be an insertion in the developmentally regulated *dev* operon, which not only depended in part on C-signalling for its expression, but the insertion mutation greatly impaired sporulation (Kroos *et al.*, 1986, 1990; Kroos and Kaiser, 1987; Thony-Meyer and Kaiser, 1993). The gene into which Tn*5 lac* Ω4414 had inserted was called *devR*, and in-frame deletions in *devT* and *devS*, which lie upstream and downstream, respectively, of *devR*, cause similar sporulation defects (Boysen *et al.*, 2002; Viswanathan *et al.*, 2007a). DevT feeds back positively on *fruA* expression by an unknown mechanism (Boysen *et al.*, 2002) (Fig. 6.1). Subsequent work showed that the *dev* operon includes eight genes and downstream clustered regularly interspaced short palindromic repeats (CRISPR) (Viswanathan *et al.*, 2007a). The other genes include four typically CRISPR-associated (Cas) genes and a short open reading frame that was shown to be translated (Fig. 6.5). Hence, the *dev* operon can be considered a CRISPR-Cas system. These systems have been shown to mediate bacterial defence against phage and plasmid invasion *via* RNA interference (Bhaya *et al.*, 2011; Wiedenheft *et al.*, 2012). Whether *devTRS* are involved in RNA interference and whether the other genes in the *dev* operon are important for development remains to be determined. Transcriptional regulation of the *dev* operon is complex, including strong negative autoregulation involving DevS, and *cis*-regulatory elements spanning more than 1 kb that appear to influence each other by unknown mechanisms (Viswanathan *et al.*, 2007a). FruA binds to a site centred at about −90 (Viswanathan *et al.*, 2007b) and MrpC2 binds cooperatively just downstream (Campbell, Viswanathan, Son, Barrett, and Kroos, unpublished) in an arrangement similar to that at the *fmgA* promoter (Figs. 6.4 and 6.5). A third transcription factor, LadA, was identified by its affinity for a downstream *cis*-regulatory DNA element centred at about +350 (Viswanathan *et al.*, 2007b). LadA is a LysR-type transcription factor. These factors typically bind upstream of the promoter and activate transcription in response to a coinducer, so the downstream location of the LadA binding site is unusual and the mechanism by which LadA positively regulates *dev* transcription remains to be elucidated.

Genes required for spore coat formation depend on *dev*

Null mutations in *devRS* relieve the strong negative autoregulation of the *dev* promoter (Thony-Meyer and Kaiser, 1993; Viswanathan *et al.*, 2007a), which may result in overproduction of other gene products from the *dev* operon. Whether it is overproduction of other gene products or loss of DevR and DevS that impair sporulation and expression of other late genes is an open question. Genes at the Ω7536 locus fail to be expressed in a *devRS* mutant (Licking *et al.*, 2000) (Fig. 6.1). This locus was found to encode the FruA-regulated gene *fdgA* mentioned above (Ueki and Inouye, 2005b) plus eight additional genes that are co-transcribed to form the *exoABCDEFGHI* operon in which *fdgA* is renamed *exoA* (Muller *et al.*, 2012). Products of the *exo* operon appear to export polysaccharide that forms the spore coat. Proper deposition of polysaccharide to form a compact spore coat also requires products of the *nfs* operon, whose transcription depends in part on *dev* since expression of an *nfsA* reporter fusion was reduced in a *devRS* mutant after 24 h into development when nascent fruiting bodies had formed (Muller *et al.*, 2010, 2012) (Fig. 6.1). Unexpectedly, expression of the *nfsA* reporter fusion was induced earlier than normal in a *fruA* mutant, indicating that FruA negatively regulates *nfsA* directly or indirectly (Muller *et al.*, 2010). The different effects of *fruA* and *devRS* mutations on *nfsA* expression are intriguing. The *fruA* mutation is expected to eliminate expression of the entire *dev* operon (including *devRS*) (Viswanathan *et al.*, 2007b), whereas the *devRS* mutation may result in overproduction of other *dev* gene products (due to relief from negative

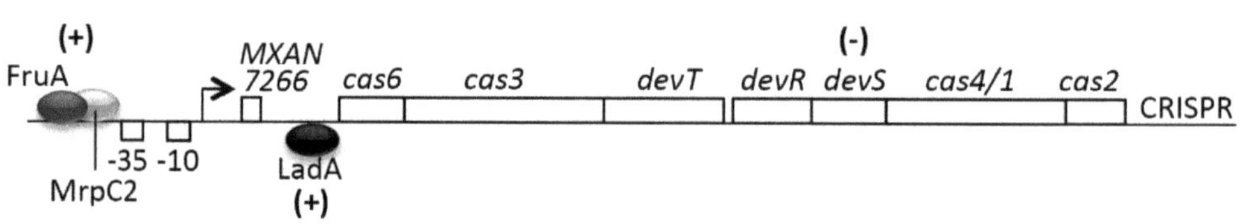

Figure 6.5 Regulation of the *dev* operon. FruA and MrpC2 appear to directly activate *dev* transcription. *MXAN7266* is a short open reading frame. LadA and DevS positively and negatively regulate *dev* transcription by unknown mechanisms. The *dev* operon includes *cas* and *dev* genes, as well clustered regularly interspaced short palindromic repeats (CRISPR). See the text for references.

autoregulation as noted above) (Thony-Meyer and Kaiser, 1993; Viswanathan *et al.*, 2007a), so the different effects on *nfsA* expression suggest that *dev* gene products other than DevR and DevS exert regulatory effects. DevT positively regulates *fruA* expression (Boysen *et al.*, 2002) (Fig. 6.1), so it is possible that DevT overproduction in the *devRS* mutant elevates the FruA concentration. Since FruA negatively regulates *nfsA* (Muller *et al.*, 2010), elevated FruA in the *devRS* mutant might account for its reduced *nfsA* expression late in development. While this scenario is plausible, it is also possible that other *dev* gene products exert regulatory effects. The mechanisms by which DevT positively regulates *fruA* and by which FruA negatively regulates *nfsA* remain to be elucidated. Given the negative regulation of *nfsA* by FruA, and given that C-signalling somehow activates FruA (Sogaard-Andersen *et al.*, 1996; Ellehauge *et al.*, 1998; Mittal and Kroos, 2009a), one would predict early and/or overexpression of *nfsA* in a *csgA* mutant, but *nfsA* expression was strongly impaired in a *csgA* mutant after 6 h into development (Muller *et al.*, 2010). This provides the first example of late gene expression regulated positively by CsgA and negatively by FruA. Determining the epistatic relationship of *csgA* and *fruA* mutations with respect to expression of the *nfs* operon, as well as whether the operon depends on extracellular C-signalling (as opposed to cell autonomous dependence on CsgA), may shed light on its unusual regulation.

A plethora of potential late genes revealed by studying glycerol-induced sporulation

The *nfs* operon was selected for characterization in part because it was found to be highly up-regulated during glycerol-induced spore formation (Muller *et al.*, 2010). Other selection criteria were that *nfs* genes code for proteins of unknown function that are predicted to localize to the cell envelope. Follow-up studies support the notion that Nfs proteins form a complex in the cell envelope involved in converting secreted polysaccharide into a compact spore coat (Muller *et al.*, 2012). In the microarray analysis that led to identification of the *nfs* genes, a total of 1486 genes were found to be up- or down-regulated during glycerol-induced sporulation (Muller *et al.*, 2010). Many of the 843 up-regulated genes are likely up-regulated during starvation-induced sporulation as well, since 9 out of 12 genes studied previously that appear to be part of the 'core' starvation-induced sporulation programme (i.e. when the gene is mutated it impairs sporulation but allows appreciable aggregation) were glycerol induced. One of the nine genes was *fdgA* (*exoA*) which, as noted above, is part of the FruA- and DevRS-regulated *exo* operon whose products appear to export polysaccharide that forms the spore coat (Licking *et al.*, 2000; Ueki and Inouye, 2005b; Muller *et al.*, 2012). Three others were *mspA*, *mspC*, and *cbgA* (Muller *et al.*, 2010), which are required for normal spore cortex formation (Tengra *et al.*, 2006; Dahl *et al.*, 2007). The *mspA* and *mspC* genes begin to be expressed late during starvation-induced development (Dahl *et al.*, 2007); late expression of *cbgA* remains to be examined. The remaining 5 were *actA*, *actB*, *mrpA*, *sigC*, and *sigB* (Muller *et al.*, 2010). ActA and ActB are predicted to be response regulators that function at the end of the EBP cascade to control C-signal production (Gronewold and Kaiser, 2001, 2002) and activate transcription of MXAN4899 (Giglio *et al.*, 2011), both of which govern FruA activation (Sogaard-Andersen *et al.*, 1996; Ellehauge *et al.*, 1998; Jelsbak *et al.*, 2005); however, a *fruA* mutant forms glycerol-induced spores (Ogawa *et al.*, 1996) but a mutation in *actA* or *actB* prevents glycerol-induced sporulation (Muller *et al.*, 2010), suggesting that ActA and ActB control other processes required for sporulation, perhaps not only in response to glycerol but in response to starvation as well. Several other proteins in the EBP cascade (Nla4, Nla6, and Nla18) control processes required for sporulation in response to both stimuli (Caberoy *et al.*, 2003; Diodati *et al.*, 2006; Ossa *et al.*, 2007; Giglio *et al.*, 2011). MrpA is a predicted HPK that may serve as a phosphatase of MrpB, which may activate transcription of *mrpC* (Sun and Shi, 2001b). Although effects of mutations in *mrpA* or *mrpB* on glycerol-induced sporulation have not been reported, a mutation in *mrpC* prevents it (Muller *et al.*, 2010), suggesting that MrpC is a key transcription factor for sporulation in response to both glycerol and starvation. SigC and SigB are predicted sigma factors that function early and late, respectively, during starvation-induced development, but mutations in *sigB* or *sigC* do not block glycerol-induced sporulation (Apelian and Inouye, 1990, 1993). The *sigB* gene is induced late during starvation-induced development, at the onset of sporulation (Apelian and Inouye, 1990)(Fig. 6.1), but the mechanism of regulation is unknown.

The microarray analysis identified two glycerol-induced genes (MXAN1101 and MXAN3026) that were shown to be important for starvation-induced development in a subsequent study (Muller *et al.*, 2010, 2012). Based on mutant phenotypes, MXAN3026 appears to function early, prior to aggregation, but MXAN1101 appears to function after aggregation (Muller *et al.*, 2012), so it might be induced late during starvation-induced development, although this remains to be tested. Both genes are predicted to code for proteins that affect polysaccharide synthesis. Although *mreB* was not regulated in response to glycerol induction (Muller *et al.*, 2010), experiments with an inhibitor of MreB polymerization suggest that this cytoskeletal protein is crucial for rod-to-spore conversion as well as for spore outgrowth to form a rod (Muller *et al.*, 2012). Since MreB appears to guide cell wall synthesis in other bacteria (Dominguez-Escobar *et al.*, 2011; Garner *et al.*, 2011), it was proposed that MreB spatially coordinates and/or controls the activity of peptidoglycan synthesis enzymes during rod-to-spore morphogenesis (Muller *et al.*, 2012). Candidates for such enzymes under MreB control include three predicted peptidoglycan binding proteins, PBP1A (MXAN5911), PBP1C (MXAN2419), and DacB or PBP4 (MXAN1070), since the corresponding genes were up-regulated during glycerol-induced sporulation (Muller *et al.*, 2010, 2012); however, they were not reported to be up-regulated during starvation-induced development (Huntley *et al.*, 2010).

Evolution of the GRN

The availability of genome sequences makes it possible to ask questions about the origin and conservation of genes in the GRN controlling *M. xanthus* development. An analysis of 78 genes required for fruiting body formation revealed that 22% have an altered codon bias and/or phylogeny suggestive of lateral gene transfer (Goldman *et al.*, 2007). Interestingly, AsgA involved in A-signal production, MrpC at the heart of the MrpC module (Fig. 6.1), and the transcription factor LadA as well as the DevTRS proteins it regulates, might have been acquired by lateral gene transfer based on phylogenetic analysis (Goldman *et al.*, 2007). Hence, several key components of the *M. xanthus* GRN might have come from 'alien' genomes rather than from a common ancestor of the δ-proteobacteria.

A comparative genomic analysis of δ-proteobacteria with completely sequenced genomes revealed diversity in the genetic programmes for fruiting body formation in the order Myxococcales (Huntley *et al.*, 2010). Analysis of 95 genes important for *M. xanthus* development (i.e. a mutation in the gene affects aggregation and/or sporulation but does not affect vegetative growth or motility) and 410 genes up-regulated during the first 24 h after starvation on a solid surface indicated that the programmes for development are highly similar in *M. xanthus* and *Stigmatella aurantiaca*, which belong to the same suborder, but significantly different in *Sorangium cellulosum* and *Haliangium ochraceum*, which belong to two other suborders. Extending this analysis to *Myxococcus fulvus* and *Corallococcus coralloides*, which are in the same suborder as *M. xanthus* and *S. aurantiaca*, Huntley *et al.* (Chapter 2) reached the same conclusion. Notably with respect to the three modules of the *M. xanthus* GRN (Fig. 6.1), *S. aurantiaca* has orthologues of all proteins in the EBP cascade (Huntley *et al.*, 2010; Kroos, unpublished) and all proteins in the MrpC module except Pkn8 (Huntley *et al.*, 2010), and it has orthologues of FruA and its transcriptional regulator HthA as well as proteins not shown in Fig. 6.1 that seem to affect FruA activation in response to C-signal (Huntley *et al.*, 2010), including the HPKs SdeK (Pollack and Singer, 2001), TodK (Rasmussen and Sogaard-Andersen, 2003) and RodK (Rasmussen *et al.*, 2005), and the response regulator RokA (Wegener-Feldbrugge and Sogaard-Andersen, 2009). In contrast, *S. cellulosum* and *H. ochraceum* lack orthologues of most of these proteins (except they have Nla4, ActCD,

MrpC, LonD, and Pkn14) (Huntley *et al.*, 2010). The differences in the genetic programmes comparing species in different suborders could be due to divergent evolution from a common ancestor that was a fruiting bacterium, or it could be that their last common ancestor was not a fruiting bacterium and they independently evolved the ability to form fruiting bodies (convergent evolution) (Huntley *et al.*, 2010) (see Chapter 2). In any case, it is clear that while key components of the *M. xanthus* GRN are conserved in its close relatives, they are not conserved in more distant relatives, begging the question of how fruiting body formation is controlled in these relatives.

Conclusions

Tremendous progress has been made towards understanding developmental gene regulation in *M. xanthus*. The GRN can be viewed as three interconnected modules of key transcription factors, all of which are required for aggregation and sporulation (Fig. 6.1). Nla4 and Nla18 are EBPs that regulate the transition from growth to development by initiating a cascade of EBPs including one, Nla28, that regulates the Mrp module. Positive autoregulatory and feedback loops abound within the EBP cascade and the Mrp module, presumably reinforcing developmental progression, yet accumulation of MrpC/2 is governed by phosphorylation (*via* Pkn8 and Pkn14) and proteolysis (*via* the Esp system and LonD), which so far seem to provide additional links to starvation apart from the EBP cascade. MrpC/2 activates transcription of *fruA*, and activity of FruA is believed to be regulated post-translationally. Many connections and feedback loops involving Act and MXAN4899 in the EBP cascade, the FruA module, C-signalling, and aggregation, are believed to activate FruA, forming FruA*, as shown in Fig. 6.1, and there are likely other inputs as well (see below). FruA and MrpC2 bind cooperatively to promoter regions of several late genes, subjecting them to control by the Mrp and FruA modules. Among these late genes is the *dev* operon, whose products somehow control expression of genes required for spore formation. Many of the key components of the starvation-induced GRN (Fig.6.1) are induced and/or required during sporulation in response to glycerol, with a few notable exceptions; *nla28*, *MXAN4899*, and *devRS* are neither induced (Muller *et al.*, 2010) nor required (Thony-Meyer and Kaiser, 1993; Caberoy *et al.*, 2003; Jelsbak *et al.*, 2005), and *fruA*, *sigB*, and *sigC* are induced (Muller *et al.*, 2010) but not required (Apelian and Inouye, 1990; Apelian and Inouye, 1993; Ogawa *et al.*, 1996). Phylogenetic analysis suggests that several important regulators in the starvation-induced GRN (MrpC, LadA, and DevTRS) were acquired by lateral gene transfer and comparative genomic analysis indicates that the programmes for development are highly similar in *M. xanthus* and its close relatives, but significantly different in *S. cellulosum* and *H. ochraceum*.

Questions and future directions

Much remains to be learned about the *M. xanthus* GRN and its evolution in other myxobacteria. Is A-signal sensed by Nla28S? How does Nla28 regulate A-signal production and what is the purpose of the apparent reciprocal regulation? What other signals are sensed by the EBP cascade? Why are many target genes of the EBP cascade activated early in development, since their products seem to function only much later during sporulation? While EBPs activate transcription by σ^{54} RNAP, which σ factors direct transcription of genes under MrpC/2 and/or FruA control? How early developmental gene regulation is coordinated with chromosomal replication, segregation, and cell division is largely unknown, but these events have been speculated to influence the eventual fate of individual cells in the developing population (Rosario and Singer, 2010) (see Chapter 4). Interestingly, the Mrp module controls expression of the Ω4408 locus (Sun and Shi, 2001a), which codes for the HPK SdeK that regulates a transient rise in *dnaA* expression during development (Rosario and Singer, 2010), and SdeK as well as two other HPKs (TodK and RodK) and the response regulator RokA appear to affect FruA* formation (Pollack and Singer, 2001; Rasmussen and Sogaard-Andersen, 2003; Rasmussen *et al.*, 2005; Wegener-Feldbrugge and Sogaard-Andersen, 2009), but this additional complexity is not shown in Fig. 6.1. Indeed, we do not even understand how FruA is activated in response to

C-signalling or how this controls cell movement so that aggregates form. Many other important questions remain to be answered about how MrpC/2 and FruA* operate in the GRN. What is the significance of MrpC2? Does MrpC2 accumulate preferentially in cells destined for a particular fate? The discovery that MrpC can act as an antitoxin to inhibit programmed cell death by the MazF toxin seemed to shed light on the question of cell fate determination (Nariya and Inouye, 2008), and MrpC and MrpC2 both differentially accumulate in aggregating versus nonaggregating cells (Lee *et al.*, 2012), but the importance of MazF in programmed cell death is strain dependent (Lee *et al.*, 2012). Better methods are needed to measure transcript and protein levels in individual cells undergoing development. The long-standing question of how genes depend to different extents on C-signalling for their expression seems to be answered by cooperative binding of MrpC/2 and FruA* in different arrangements (Mittal and Kroos, 2009a,b; Lee *et al.*, 2011; Son *et al.*, 2011; Campbell, Viswanathan, Son, Barrett, and Kroos, unpublished). If FruA* is formed in response to C-signalling, does MrpC/2 monitor persistent starvation, preventing commitment to cell shape change during the aggregation phase of development as has been proposed (Mittal and Kroos, 2009a)? Which proteins initiate cell shape change? Are they regulated by products of the *dev* operon, as are the Exo and Nfs proteins that appear to maintain the spherical shape of the spore by forming a compact spore coat (Muller *et al.*, 2012), and if so how? How can we account for the observations that key components of the starvation-induced GRN (Fig. 6.1) are not required for glycerol-induced sporulation, appear to have been acquired by lateral gene transfer into *M. xanthus*, or are absent from distantly related myxobacteria that nevertheless form fruiting bodies? These and countless other interesting questions await future generations of myxobacterial researchers with ever more powerful molecular and microscopic techniques at their disposal.

Acknowledgements

ZS was funded in part by an International Fellowship from the American Association of University Women (AAUW). Work in the laboratory of AG was supported by National Science Foundation Grant IOS-0950976. Work in the laboratory of LK was supported by National Science Foundation Grant MCB-0744343 and by Michigan AgBioResearch.

References

Anantharaman, V., Koonin, E.V., and Aravind, L. (2001). Regulatory potential, phyletic distribution and evolution of ancient, intracellular small-molecule-binding domains. J. Mol. Biol. *307*, 1271–1292.

Apelian, D., and Inouye, S. (1990). Development-specific σ-factor essential for late-stage differentiation of *Myxococcus xanthus*. Genes Dev. *4*, 1396–1403.

Apelian, D., and Inouye, S. (1993). A new putative sigma factor of *Myxococcus xanthus*. J Bacteriol *175*, 3335–3342.

Barrios, H., Valderrama, B., and Morett, E. (1999). Compilation and analysis of σ^{54}-dependent promoter sequences. Nucleic Acids Res. *27*, 4305–4313.

Bhaya, D., Davison, M., and Barrangou, R. (2011). CRISPR-Cas systems in bacteria and archaea: versatile small RNAs for adaptive defense and regulation. Annu. Rev. Genet. *45*, 273–297.

Blattner, F.R., Plunkett, G., 3rd, Bloch, C.A., Perna, N.T., Burland, V., Riley, M., Collado-Vides, J., Glasner, J.D., Rode, C.K., Mayhew, G.F., *et al.* (1997). The complete genome sequence of *Escherichia coli* K-12. Science *277*, 1453–1462.

Boysen, A., Ellehauge, E., Julien, B., and Sogaard-Andersen, L. (2002). The DevT protein stimulates synthesis of FruA, a signal transduction protein required for fruiting body morphogenesis in *Myxococcus xanthus*. J. Bacteriol. *184*, 1540–1546.

Brahmachary, P., Dashti, M.G., Olson, J.W., and Hoover, T.R. (2004). *Helicobacter pylori* FlgR is an enhancer-independent activator of σ^{54}-RNA polymerase holoenzyme. J. Bacteriol. *186*, 4535–4542.

Brandner, J.P., and Kroos, L. (1998). Identification of the Ω4400 regulatory region, a developmental promoter of *Myxococcus xanthus*. J. Bacteriol. *180*, 1995–2004.

Caberoy, N.B., Welch, R.D., Jakobsen, J.S., Slater, S.C., and Garza, A.G. (2003). Global mutational analysis of NtrC-like activators in *Myxococcus xanthus*: identifying activator mutants defective for motility and fruiting body development. J. Bacteriol. *185*, 6083–6094.

Cho, K., and Zusman, D.R. (1999). Sporulation timing in *Myxococcus xanthus* is contolled by the *espAB* locus. Mol. Microbiol. *34*, 714–725.

Curtis, P.D., Taylor, R.G., Welch, R.D., and Shimkets, L.J. (2007). Spatial organization of *Myxococcus xanthus* during fruiting body formation. J. Bacteriol. *189*, 9126–9130.

Dahl, J.L., Tengra, F.K., Dutton, D., Yan, J., Andacht, T.M., Coyne, L., Windell, V., and Garza, A.G. (2007). Identification of major sporulation proteins of *Myxococcus xanthus* using a proteomic approach. J. Bacteriol. *189*, 3187–3197.

Diodati, M.E., Ossa, F., Caberoy, N.B., Jose, I.R., Hiraiwa, W., Igo, M.M., Singer, M., and Garza, A.G. (2006).

Nla18, a key regulatory protein required for normal growth and development of *Myxococcus xanthus*. J. Bacteriol. *188*, 1733–1743.

Diodati, M.E., Gill, R.E., Plamann, L., and Singer, M. (2008). Initiation and early developmental events. In Myxobacteria: Multicellularity and Differentiation, D.E. Whitworth, ed. (Washington, DC: ASM Press), pp. 43–76.

Dominguez-Escobar, J., Chastanet, A., Crevenna, A.H., Fromion, V., Wedlich-Soldner, R., and Carballido-Lopez, R. (2011). Processive movement of MreB-associated cell wall biosynthetic complexes in bacteria. Science *333*, 225–228.

Dubnau, D., and Losick, R. (2006). Bistability in bacteria. Mol. Microbiol. *61*, 564–572.

Durocher, D., Taylor, I.A., Sarbassova, D., Haire, L.F., Westcott, S.L., Jackson, S.P., Smerdon, S.J., and Yaffe, M.B. (2000). The molecular basis of FHA domain:phosphopeptide binding specificity and implications for phospho-dependent signaling mechanisms. Mol. Cell *6*, 1169–1182.

Elias-Arnanz, M., Fontes, M., and Padmanabhan, S. (2008). Carotenogenesis in *Myxococcus xanthus*: a complex regulatory network. In Myxobacteria: Multicellularity and Differentiation, D.E. Whitworth, ed. (Washington, DC: ASM Press), pp. 211–226.

Ellehauge, E., Norregaard-Madsen, M., and Sogaard-Andersen, L. (1998). The FruA signal transduction protein provides a checkpoint for the temporal co-ordination of intercellular signals in *Myxococcus xanthus* development. Mol. Microbiol. *30*, 807–817.

Fisseha, M., Gloudemans, M., Gill, R., and Kroos, L. (1996). Characterization of the regulatory region of a cell interaction-dependent gene in *Myxococcus xanthus*. J Bacteriol *178*, 2539–2550.

Fisseha, M., Biran, D., and Kroos, L. (1999). Identification of the Ω4499 regulatory region controlling developmental expression of a *Myxococcus xanthus* cytochrome P-450 system. J. Bacteriol. *181*, 5467–5475.

Garner, E.C., Bernard, R., Wang, W., Zhuang, X., Rudner, D.Z., and Mitchison, T. (2011). Coupled, circumferential motions of the cell wall synthesis machinery and MreB filaments in *B. subtilis*. Science *333*, 222–225.

Garza, A.G., Pollack, J.S., Harris, B.Z., Lee, A., Keseler, I.M., Licking, E.F., and Singer, M. (1998). SdeK is required for early fruiting body development in *Myxococcus xanthus*. J. Bacteriol. *180*, 4628–4637.

Garza, A.G., Harris, B.Z., Greenberg, B.M., and Singer, M. (2000). Control of *asgE* expression during growth and development of *Myxococcus xanthus*. J. Bacteriol. *182*, 6622–6629.

Genschik, P., Drabikowski, K., and Filipowicz, W. (1998). Characterization of the *Escherichia coli* RNA 3′-terminal phosphate cyclase and its σ^{54}-regulated operon. J. Biol. Chem. *273*, 25516–25526.

Giglio, K.M., Eisenstatt, J., and Garza, A.G. (2010). Identification of enhancer binding proteins important for *Myxococcus xanthus* development. J. Bacteriol. *192*, 360–364.

Giglio, K.M., Caberoy, N., Suen, G., Kaiser, D., and Garza, A.G. (2011). A cascade of coregulating enhancer binding proteins initiates and propagates a multicellular developmental program. Proc. Natl. Acad. Sci. U.S.A. *108*, E431-E439.

Giglio, K.M., Zhu, C., Kummer, S., and Garza, A.G. The enhancer binding protein Nla6 regulates development genes that are important for quorum signal production and sporulation in *Myxococcus xanthus*. Submitted for publication.

Goldman, B.S., Nierman, W.C., Kaiser, D., Slater, S.C., Durkin, A.S., Eisen, J., Ronning, C.M., Barbazuk, W.B., Blanchard, M., Field, C., *et al.* (2006). Evolution of sensory complexity recorded in a myxobacterial genome. Proc. Natl. Acad. Sci. U.S.A. *103*, 15200–15205.

Goldman, B., Bhat, S., and Shimkets, L.J. (2007). Genome evolution and the emergence of fruiting body development in *Myxococcus xanthus*. PLoS ONE *2*, e1329.

Gorski, L., and Kaiser, D. (1998). Targeted mutagenesis of σ^{54} activator proteins in *Myxococcus xanthus*. J. Bacteriol. *180*, 5896–5905.

Gorski, L., Gronewold, T., and Kaiser, D. (2000). A σ^{54} activator protein necessary for spore differentiation within the fruiting body of *Myxococcus xanthus*. J. Bacteriol. *182*, 2438–2444.

Gronewold, T.M., and Kaiser, D. (2001). The *act* operon controls the level and time of C-signal production for *Myxococcus xanthus* development. Mol. Microbiol. *40*, 744–756.

Gronewold, T.M., and Kaiser, D. (2002). *act* operon control of developmental gene expression in *Myxococcus xanthus*. J. Bacteriol. *184*, 1172–1179.

Gronewold, T.M., and Kaiser, D. (2007). Mutations of the *act* promoter in *Myxococcus xanthus*. J. Bacteriol. *189*, 1836–1844.

Guo, D., Wu, Y., and Kaplan, H.B. (2000). Identification and characterization of genes required for early *Myxococcus xanthus* developmental gene expression. J. Bacteriol. *182*, 4564–4571.

Hager, E., Tse, H., and Gill, R.E. (2001). Identification and characterization of *spdR* mutations that bypass the BsgA protease-dependent regulation of developmental gene expression in *Myxococcus xanthus*. Mol. Microbiol. *39*, 765–780.

Harris, B.Z., Kaiser, D., and Singer, M. (1998). The guanosine nucleotide (p)ppGpp initiates development and A-factor production in *Myxococcus xanthus*. Genes Dev. *12*, 1022–1035.

Higgs, P.I., Jagadeesan, S., Mann, P., and Zusman, D.R. (2008). EspA, an orphan hybrid histidine protein kinase, regulates the timing of expression of key developmental proteins of *Myxococcus xanthus*. J. Bacteriol. *190*, 4416–4426.

Hodgkin, J., and Kaiser, D. (1979a). Genetics of gliding motility in *Myxococcus xanthus* (Myxobacterales): genes controlling movement of single cells. Mol. Gen. Genet. *171*, 167–176.

Hodgkin, J., and Kaiser, D. (1979b). Genetics of gliding motility in *Myxococcus xanthus* (Myxobacterales): two gene systems control movement. Mol. Gen. Genet. *171*, 177–191.

Horiuchi, T., Akiyama, T., Inouye, S., and Komano, T. (2002a). Analysis of *dofA*, a *fruA*-dependent developmental gene, and its homologue, *dofB*, in *Myxococcus xanthus*. J. Bacteriol. *184*, 6803–6810.

Horiuchi, T., Taoka, M., Isobe, T., Komano, T., and Inouye, S. (2002b). Role of *fruA* and *csgA* genes in gene expression during development of *Myxococcus xanthus*: analysis by two-dimensional gel electrophoresis. J. Biol. Chem. *277*, 26753–26760.

Huntley, S., Hamann, N., Wegener-Feldbrugge, S., Treuner-Lange, A., Kube, M., Reinhardt, R., Klages, S., Muller, R., Ronning, C.M., Nierman, W.C., *et al.* (2010). Comparative genomic analysis of fruiting body formation in Myxococcales. Mol. Biol. Evol. *28*, 1083–1097.

Jakobsen, J.S., Jelsbak, L., Welch, R.D., Cummings, C., Goldman, B., Stark, E., Slater, S., and Kaiser, D. (2004). σ^{54} enhancer binding proteins and *Myxococcus xanthus* fruiting body development. J. Bacteriol. *186*, 4361–4368.

Jelsbak, L., Givskov, M., and Kaiser, D. (2005). Enhancer-binding proteins with an FHA domain and the σ^{54} regulon in *Myxococcus xanthus* fruiting body development. Proc. Natl. Acad. Sci. U.S.A. *102*, 3010–3015.

Julien, B., Kaiser, A.D., and Garza, A. (2000). Spatial control of cell differentiation in *Myxococcus xanthus*. Proc. Natl. Acad. Sci. U.S.A. *97*, 9098–9103.

Kaiser, D., Robinson, M., and Kroos, L. (2010). Myxobacteria, polarity, and multicellular morphogenesis. Cold Spring Harb. Perspect. Biol. *2*, a000380.

Keseler, I., and Kaiser, D. (1995). An early A-signal-dependent gene in *Myxococcus xanthus* has a σ^{54}-like promoter. J. Bacteriol. *177*, 4638–4644.

Keseler, I., and Kaiser, D. (1997). σ^{54}, a vital protein for *Myxococcus xanthus*. Proc. Natl. Acad. Sci. U.S.A. *94*, 1979–1984.

Kim, S.K., and Kaiser, D. (1990a). C-factor: a cell–cell signaling protein required for fruiting body morphogenesis of *M. xanthus*. Cell *61*, 19–26.

Kim, S.K., and Kaiser, D. (1990b). Cell alignment required in differentiation of *Myxococcus xanthus*. Science *249*, 926–928.

Kim, S.K., and Kaiser, D. (1990c). Cell motility is required for the transmission of C-factor, an intercellular signal that coordinates fruiting body morphogenesis of *Myxococcus xanthus*. Genes Dev. *4*, 896–905.

Kim, S.K., and Kaiser, D. (1991). C-factor has distinct aggregation and sporulation thresholds during *Myxococcus* development. J. Bacteriol. *173*, 1722–1728.

Kirby, J.R., and Zusman, D.R. (2003). Chemosensory regulation of developmental gene expression in *Myxococcus xanthus*. Proc. Natl. Acad. Sci. U.S.A. *100*, 2008–2013.

Konovalova, A., Lobach, S., and Sogaard-Andersen, L. (2012). A RelA-dependent two-tiered regulated proteolysis cascade controls synthesis of a contact-dependent intercellular signal in *Myxococcus xanthus*. Mol. Microbiol. *84*, 260–275.

Kroos, L. (2007). The *Bacillus* and *Myxococcus* developmental networks and their transcriptional regulators. Ann. Rev. Genet. *41*, 13–39.

Kroos, L., and Inouye, S. (2008). Transcriptional regulatory mechanisms during *Myxococcus xanthus* development. In Myxobacteria: Multicellularity and Differentiation, D.E. Whitworth, ed. (Washington, DC: ASM Press), pp. 149–168.

Kroos, L., and Kaiser, D. (1987). Expression of many developmentally regulated genes in *Myxococcus* depends on a sequence of cell interactions. Genes Dev. *1*, 840–854.

Kroos, L., Kuspa, A., and Kaiser, D. (1986). A global analysis of developmentally regulated genes in *Myxococcus xanthus*. Dev. Biol. *117*, 252–266.

Kroos, L., Hartzell, P., Stephens, K., and Kaiser, D. (1988). A link between cell movement and gene expression argues that motility is required for cell–cell signaling during fruiting body development. Genes Dev. *2*, 1677–1685.

Kroos, L., Kuspa, A., and Kaiser, D. (1990). Defects in fruiting body development caused by Tn*5 lac* insertions in *Myxococcus xanthus*. J. Bacteriol. *172*, 484–487.

Kruse, T., Lobedanz, S., Berthelsen, N.M., and Sogaard-Andersen, L. (2001). C-signal: a cell surface-associated morphogen that induces and co-ordinates multicellular fruiting body morphogenesis and sporulation in *Myxococcus xanthus*. Mol. Microbiol. *40*, 156–168.

Kuspa, A., Kroos, L., and Kaiser, D. (1986). Intercellular signaling is required for developmental gene expression in *Myxococcus xanthus*. Dev. Biol. *117*, 267–276.

Kuspa, A., Plamann, L., and Kaiser, D. (1992). A-signalling and the cell density requirement for *Myxococcus xanthus* development. J. Bacteriol. *174*, 7360–7369.

Lancero, H., Caberoy, N.B., Castaneda, S., Li, Y., Lu, A., Dutton, D., Duan, X.Y., Kaplan, H.B., Shi, W., and Garza, A.G. (2004). Characterization of a *Myxococcus xanthus* mutant that is defective for adventurous motility and social motility. Microbiology *150*, 4085–4093.

Laub, M.T., and Goulian, M. (2007). Specificity in two-component signal transduction pathways. Ann. Rev. Genet. *41*, 121–145.

Lee, B., Holkenbrink, C., Treuner-Lange, A., and Higgs, P.I. (2012). *Myxococcus xanthus* developmental cell fate production: heterogeneous accumulation of developmental regulatory proteins and reexamination of the role of MazF in developmental lysis. J. Bacteriol. *194*, 3058–3068.

Lee, J., Son, B., Viswanathan, P., Luethy, P., and Kroos, L. (2011). Combinatorial regulation of *fmgD* by MrpC2 and FruA during *Myxococcus xanthus* development. J. Bacteriol. *193*, 1681–1689.

Li, S.-F., Lee, B., and Shimkets, L.J. (1992). *csgA* expression entrains *Myxococcus xanthus* development. Genes Dev. *6*, 401–410.

Licking, E., Gorski, L., and Kaiser, D. (2000). A common step for changing cell shape in fruiting body and starvation-independent sporulation of *Myxococcus xanthus*. J. Bacteriol. *182*, 3553–3558.

Lobedanz, S., and Sogaard-Andersen, L. (2003). Identification of the C-signal, a contact-dependent morphogen coordinating multiple developmental responses in *Myxococcus xanthus*. Genes Dev. *17*, 2151–2161.

Loconto, J., Viswanathan, P., Nowak, S.J., Gloudemans, M., and Kroos, L. (2005). Identification of the Ω4406 regulatory region, a developmental promoter of *Myxococcus xanthus*, and a DNA segment responsible for chromosomal position-dependent inhibition of gene expression. J. Bacteriol. *187*, 4149–4162.

McBride, M.J., and Zusman, D.R. (1993). FrzCD, a methyl-accepting taxis protein from *Myxococcus xanthus*, shows modulated methylation during fruiting body formation. J. Bacteriol. *175*, 4936–4940.

Mittal, S., and Kroos, L. (2009a). A combination of unusual transcription factors binds cooperatively to control *Myxococcus xanthus* developmental gene expression. Proc. Natl. Acad. Sci. U.S.A. *106*, 1965–1970.

Mittal, S., and Kroos, L. (2009b). Combinatorial regulation by a novel arrangement of FruA and MrpC2 transcription factors during *Myxococcus xanthus* development. J. Bacteriol. *191*, 2753–2763.

Muller, F.D., Treuner-Lange, A., Heider, J., Huntley, S.M., and Higgs, P.I. (2010). Global transcriptome analysis of spore formation in *Myxococcus xanthus* reveals a locus necessary for cell differentiation. BMC Genomics *11*, 264.

Muller, F.D., Schink, C.W., Hoiczyk, E., Cserti, E., and Higgs, P.I. (2012). Spore formation in *Myxococcus xanthus* is tied to cytoskeleton functions and polysaccharide spore coat deposition. Mol. Microbiol. *83*, 486–505.

Nariya, H., and Inouye, S. (2005). Identification of a protein Ser/Thr kinase cascade that regulates essential transcriptional activators in *Myxococcus xanthus* development. Mol. Microbiol. *58*, 367–379.

Nariya, H., and Inouye, S. (2006). A protein Ser/Thr kinase cascade negatively regulates the DNA-binding activity of MrpC, a smaller form of which may be necessary for the *Myxococcus xanthus* development. Mol. Microbiol. *60*, 1205–1217.

Nariya, H., and Inouye, M. (2008). MazF, an mRNA interferase, mediates programmed cell death during multicellular *Myxococcus* development. Cell *132*, 55–66.

Nielsen, M., Rasmussen, A.A., Ellehauge, E., Treuner-Lange, A., and Sogaard-Andersen, L. (2004). HthA, a putative DNA-binding protein, and HthB are important for fruiting body morphogenesis in *Myxococcus xanthus*. Microbiology *150*, 2171–2183.

O'Connor, K.A., and Zusman, D.R. (1988). Reexamination of the role of autolysis in the development of *Myxococcus xanthus*. J. Bacteriol. *170*, 4103–4112.

O'Connor, K.A., and Zusman, D.R. (1991). Development in *Myxococcus xanthus* involves differentiation into two cell types, peripheral rods and spores. J. Bacteriol. *173*, 3318–3333.

Ogawa, M., Fujitani, S., Mao, X., Inouye, S., and Komano, T. (1996). FruA, a putative transcription factor essential for the development of *Myxococcus xanthus*. Mol. Microbiol. *22*, 757–767.

Ossa, F., Diodati, M.E., Caberoy, N.B., Giglio, K.M., Edmonds, M., Singer, M., and Garza, A.G. (2007). The *Myxococcus xanthus* Nla4 protein is important for expression of stringent response-associated genes, ppGpp accumulation, and fruiting body development. J. Bacteriol. *189*, 8474–8483.

Pollack, J.S., and Singer, M. (2001). SdeK, a histidine kinase required for *Myxococcus xanthus* development. J. Bacteriol. *183*, 3589–3596.

Popham, D.L., Szeto, D., Keener, J., and Kustu, S. (1989). Function of a bacterial activator protein that binds to transcriptional enhancers. Science *243*, 629–635.

Rasmussen, A.A., and Sogaard-Andersen, L. (2003). TodK, a putative histidine protein kinase, regulates timing of fruiting body morphogenesis in *Myxococcus xanthus*. J. Bacteriol. *185*, 5452–5464.

Rasmussen, A.A., Porter, S.L., Armitage, J.P., and Sogaard-Andersen, L. (2005). Coupling of multicellular morphogenesis and cellular differentiation by an unusual hybrid histidine protein kinase in *Myxococcus xanthus*. Mol. Microbiol. *56*, 1358–1372.

Ray, J.C., Tabor, J.J., and Igoshin, O.A. (2011). Non-transcriptional regulatory processes shape transcriptional network dynamics. Nat. Rev. Microbiol. *9*, 817–828.

Rolbetzki, A., Ammon, M., Jakovljevic, V., Konovalova, A., and Sogaard-Andersen, L. (2008). Regulated secretion of a protease activates intercellular signaling during fruiting body formation in *M. xanthus*. Dev. Cell *15*, 627–634.

Romeo, J.M., and Zusman, D.R. (1991). Transcription of the myxobacterial hemagglutinin gene is mediated by a σ^{54}-like promoter and a *cis*-acting upstream regulatory region of DNA. J. Bacteriol. *173*, 2969–2976.

Rosario, C.J., and Singer, M. (2010). Developmental expression of *dnaA* is required for sporulation and timing of fruiting body formation in *Myxococcus xanthus*. Mol. Microbiol. *76*, 1322–1333.

Sager, B., and Kaiser, D. (1993). Spatial restriction of cellular differentiation. Genes Dev. *7*, 1645–1653.

Sarwar, Z., and Garza, A.G. (2012). The Nla28S/Nla28 two component signal transduction system regulates sporulation in *Mxyococcus xanthus*. J. Bacteriol. *194*, 4698–4708.

Sasse-Dwight, S., and Gralla, J.D. (1990). Role of eukaryotic-type functional domains found in the prokaryotic enhancer receptor factor σ^{54}. Cell *62*, 945–954.

Schramm, A., Lee, B., and Higgs, P.I. (2012). Intra- and inter-protein phosphorylation between two hybrid histidine kinases controls *Myxococcus xanthus* developmental progression. J. Biol. Chem. *287*, 25060–25072.

Shimkets, L.J., Gill, R.E., and Kaiser, D. (1983). Developmental cell interactions in *Myxococcus xanthus* and the *spoC* locus. Proc. Natl. Acad. Sci. U.S.A. *80*, 1406–1410.

Shimkets, L.J., and Rafiee, H. (1990). CsgA, an extracellular protein essential for *Myxococcus xanthus* development. J. Bacteriol. *172*, 5299–5306.

Shingler, V. (2011). Signal sensory systems that impact σ^{54}-dependent transcription. FEMS Microbiol. Rev. *35*, 425–440.

Singer, M., and Kaiser, D. (1995). Ectopic production of guanosine penta-and tetraphosphate can initiate early developmental gene expression in *Myxococcus xanthus*. Genes and Dev. *9*, 1633–1644.

Smits, W.K., Kuipers, O.P., and Veening, J.W. (2006). Phenotypic variation in bacteria: the role of feedback regulation. Nat. Rev. Microbiol. *4*, 259–271.

Sogaard-Andersen, L., and Kaiser, D. (1996). C factor, a cell-surface-associated intercellular signaling protein, stimulates the cytoplasmic Frz signal transduction system in *Myxococcus xanthus*. Proc. Natl. Acad. Sci. U.S.A. *93*, 2675–2679.

Sogaard-Andersen, L., Slack, F., Kimsey, H., and Kaiser, D. (1996). Intercellular C-signaling in *Myxococcus xanthus* involves a branched signal transduction pathway. Genes Dev. *10*, 740–754.

Son, B., Liu, Y., and Kroos, L. (2011). Combinatorial regulation by MrpC2 and FruA involves three sites in the *fmgE* promoter region during *Myxococcus xanthus* development. J. Bacteriol. *193*, 2756–2766.

Stein, E.A., Cho, K., Higgs, P.I., and Zusman, D.R. (2006). Two Ser/Thr protein kinases essential for efficient aggregation and spore morphogenesis in *Myxococcus xanthus*. Mol. Microbiol. *60*, 1414–1431.

Studholme, D.J., and Dixon, R. (2003). Domain architectures of σ^{54}-dependent transcriptional activators. J. Bacteriol. *185*, 1757–1767.

Sun, H., and Shi, W. (2001a). Analyses of *mrp* genes during *Myxococcus xanthus* development. J. Bacteriol. *183*, 6733–6739.

Sun, H., and Shi, W. (2001b). Genetic studies of *mrp*, a locus essential for cellular aggregation and sporulation of *Myxococcus xanthus*. J. Bacteriol. *183*, 4786–4795.

Tengra, F.K., Dahl, J.L., Dutton, D., Caberoy, N.B., Coyne, L., and Garza, A.G. (2006). CbgA, a protein involved in cortex formation and stress resistance in *Myxococcus xanthus* spores. J. Bacteriol. *188*, 8299–8302.

Thony-Meyer, L., and Kaiser, D. (1993). *devRS*, an autoregulated and essential genetic locus for fruiting body development in *Myxococcus xanthus*. J. Bacteriol. *175*, 7450–7462.

Ueki, T., and Inouye, S. (2003). Identification of an activator protein required for the induction of *fruA*, a gene essential for fruiting body development in *Myxococcus xanthus*. Proc. Natl. Acad. Sci. U.S.A. *100*, 8782–8787.

Ueki, T., and Inouye, S. (2005a). Activation of a development-specific gene, *dofA*, by FruA, an essential transcription factor for development of *Myxococcus xanthus*. J. Bacteriol. *187*, 8504–8506.

Ueki, T., and Inouye, S. (2005b). Identification of a gene involved in polysaccharide export as a transcription target of FruA, an essential factor for *Myxococcus xanthus* development. J. Biol. Chem. *280*, 32279–32284.

Viswanathan, P., and Kroos, L. (2003). *cis* elements necessary for developmental expression of a *Myxococcus xanthus* gene that depends on C signaling. J. Bacteriol. *185*, 1405–1414.

Viswanathan, K., Viswanathan, P., and Kroos, L. (2006). Mutational analysis of the *Myxococcus xanthus* Ω4406 promoter region reveals an upstream negative regulatory element that mediates C-signal dependence. J. Bacteriol. *188*, 515–524.

Viswanathan, P., Murphy, K., Julien, B., Garza, A.G., and Kroos, L. (2007a). Regulation of *dev*, an operon that includes genes essential for *Myxococcus xanthus* development and CRISPR-associated genes and repeats. J. Bacteriol. *189*, 3738–3750.

Viswanathan, P., Ueki, T., Inouye, S., and Kroos, L. (2007b). Combinatorial regulation of genes essential for *Myxococcus xanthus* development involves a response regulator and a LysR-type regulator. Proc. Natl. Acad. Sci. U.S.A. *104*, 7969–7974.

Wedel, A., and Kustu, S. (1995). The bacterial enhancer-binding protein NTRC is a molecular machine: ATP hydrolysis is coupled to transcriptional activation. Genes Dev. *9*, 2042–2052.

Wegener-Feldbrugge, S., and Sogaard-Andersen, L. (2009). The atypical hybrid histidine protein kinase RodK in *Myxococcus xanthus*: spatial proximity supersedes kinetic preference in phosphotransfer reactions. J. Bacteriol. *191*, 1765–1776.

Wiedenheft, B., Sternberg, S.H., and Doudna, J.A. (2012). RNA-guided genetic silencing systems in bacteria and archaea. Nature *482*, 331–338.

Wösten, M.M. (1998). Eubacterial sigma factors FEMS Microbiol. Rev. *22*, 127–150.

Wu, S.S., and Kaiser, D. (1997). Regulation of expression of the *pilA* gene in *Myxococcus xanthus*. J. Bacteriol. *179*, 7748–7758.

Xu, H., and Hoover, T.R. (2001). Transcriptional regulation at a distance in bacteria. Curr. Opin. Microbiol. *4*, 138–144.

Yoder-Himes, D., and Kroos, L. (2006). Regulation of the *Myxococcus xanthus* C-signal-dependent Ω4400 promoter by the essential developmental protein FruA. J. Bacteriol. *188*, 5167–5176.

Yoder, D., and Kroos, L. (2004a). Mutational analysis of the *Myxococcus xanthus* Ω4400 promoter region provides insight into developmental gene regulation by C signaling. J. Bacteriol. *186*, 661–671.

Yoder, D., and Kroos, L. (2004b). Mutational analysis of the *Myxococcus xanthus* Ω4499 promoter region reveals shared and unique properties in comparison with other C-signal-dependent promoters. J. Bacteriol. *186*, 3766–3776.

Abundance and Complexity of Signalling Mechanisms in Myxobacteria

7

José Muñoz-Dorado, Penelope I. Higgs and Montserrat Elías-Arnanz

Abstract

The complex lifestyle of myxobacteria depends on a vast array of signalling mechanisms. Cells must not only monitor their environment and induce adaptive responses, but they must also communicate in order to coordinate and successfully complete their multicellular life cycle. Recent genomics data have revealed that the signalling potential of myxobacteria is far greater than predicted, since the number of two-component systems, extracytoplasmic function σ factors, and eukaryotic-like protein kinases and phosphatases encoded in their genomes is very high. Characterization of some of these signal-transduction systems has revealed unusual complexity, featuring multiple components and interconnections which suggests that many myxobacterial signalling mechanisms function as intricate networks, rather than as linear pathways. Undoubtedly, a comprehensive elucidation of these signalling mechanisms will be one of the main future goals in order to understand the life cycle of these bacteria.

Introduction

Myxobacteria, like all other organisms, must cope with a great variety of external and internal fluctuations. To ensure that they survive all these changes, bacterial genomes are equipped with a large number of genes that encode proteins capable of sensing and transducing different signals in order to trigger a precise response to the corresponding stimuli. However, in addition to these general adaptive mechanisms, myxobacteria are social bacteria which exhibit a complex lifestyle in which the coordinated behaviour of a large number of cells is essential for the successful completion of their life cycle (Whitworth, 2008). This coordination is required during development, where cells organize themselves into one of at least three possible cell fates: formation of multicellular spore-filled fruiting bodies, programmed cell death, or differentiation into a persister-like state termed peripheral rods (see Chapter 3 for descriptions of the cell fates and Chapte 6 for details on regulation of the developmental programme). It is known that developing cells communicate with each other, sequentially producing and responding to five different signals (Kaiser, 2004). Furthermore, as many myxobacteria are predators, even during the growth (vegetative) phase of their life cycle, cells must also be capable of coordinated behaviour and movement. In order to attack their prey, myxobacterial cells organize themselves and act as a 'wolf pack'. This behaviour implies that they must be capable of sensing the presence of the prey and producing what they need to destroy it. Simultaneously, they have to defend themselves against the prey response (Berleman and Kirby, 2009). Given this complex and unique behaviour, it is reasonable to predict that the number of adaptive mechanisms encoded in a myxobacterial genome will be greater than in other bacteria.

Four different types of signal-transduction mechanisms have been reported in bacteria (Muñoz-Dorado *et al.*, 2012): one-component systems, two-component systems (TCSs), extracytoplasmic function (ECF) σ-factors, and eukaryotic-like Ser/Thr protein kinases (STPKs)/protein phosphatases (PPs). Although not all of these components are universal among bacteria, they are all present in myxobacteria. The simplest of these four mechanisms are one-component systems, which are, interestingly, underrepresented in myxobacteria (Goldman *et al.*, 2006; Schneiker *et al.*, 2007). In general, there is a linear correlation between the number of one-component systems encoded in a genome and the genome size (Ulrich *et al.*, 2005). However, the number of one-component systems encoded in the *M. xanthus* genome is roughly half of that expected (Goldman *et al.*, 2006). Some of these regulators are discussed in Chapter 6. In contrast, the other three types of are overrepresented in most myxobacterial genomes (see Chapter 2 for detailed information). Owing to the abundance of TCS, ECF and STPK/PP signalling systems in myxobacterial genomes, this chapter will focus on current knowledge of the abundance and complexity of these three types of signalling mechanisms and on their significance in the life cycle of myxobacteria.

Two-component systems

His-Asp phosphorelay signalling proteins, often termed two-component signal transduction (TCS) proteins, are widely employed by bacteria to mediate appropriate responses to specific stimuli (Stock *et al.*, 2000; West and Stock, 2001). Proteins involved in these systems can be identified by the presence of one or more highly conserved signalling modules: a histidine kinase (HK) module [comprising a dimerization and histidine phosphorylation (DHp) domain preceding a catalytic ATP hydrolysis (CA) domain], and a receiver (REC) module. Occasionally, histidine phosphotransferase modules (HPT) are also involved in signal transmission, although these modules are less well conserved at the sequence level. HK modules autophosphorylate by CA domain-mediated transfer of the γ-phosphoryl group of ATP to an invariant His residue in the DHp domain (Stock *et al.*, 2000; West and Stock, 2001). REC modules catalyse transfer of a phosphoryl group from an HK, HPT, or sometimes small molecule phosphodonor, onto an invariant Asp residue in the REC module. Phosphorylation promotes a global conformational change within the REC module leading to the active conformation. HPT modules, which structurally resemble a DHp domain, function solely to transmit phosphoryl groups between two REC modules, whereby a phosphoryl group from a donor REC module is accepted onto an invariant HPT His residue and subsequently donated to a second REC module.

The paradigm TCS contains a HK and a REC module which are organized into two separate proteins consisting of a sensor histidine protein kinase (HPK) and a response regulator (RR), respectively (Fig. 7.1A) (Stock *et al.*, 2000; West and Stock, 2001). Typically, these two proteins (termed cognate pairs) are encoded together in an operon. The HPK contains a variable sensing (input) domain, which upon stimulus detection, modulates the autophosphorylation activity of the associated HK module (Stewart, 2010). The RR consists of the REC module followed by a variable effector (output) domain. Signal transmission is mediated by transfer of the phosphoryl group from the HPK HK to the RR REC and the resulting active conformation of the REC module modulates the activity of an associated output domain (Bourret, 2010). Signal termination can be controlled by HPK-dependent phosphatase activity, or by the autophosphatase activity inherent in REC modules. Evolutionary analyses suggest that TCSs evolved from one-component systems because of the need to couple signals from outside of the cell to modulation of gene expression in the cytoplasm (Ulrich *et al.*, 2005). Indeed, most TCSs consist of a membrane-associated HPK and a cytoplasmic RR containing a DNA-binding output domain. While these simple TCSs are the predominant systems in many bacteria, it has become clear that the HK and REC signal transmission modules are highly flexible and can be readily arranged into more complex, multicomponent systems. Such systems are often associated with regulation of complex behaviour where multiple stimuli, which vary in spatial location and time, must be integrated and/or coupled to coordinated responses.

The myxobacteria are excellent examples of microorganisms which engage in complex behaviour and exhibit TCS signalling complexity. From a bioinformatics perspective, the large number of TCS genes, the preponderance of TCS genes encoded as orphans (without an

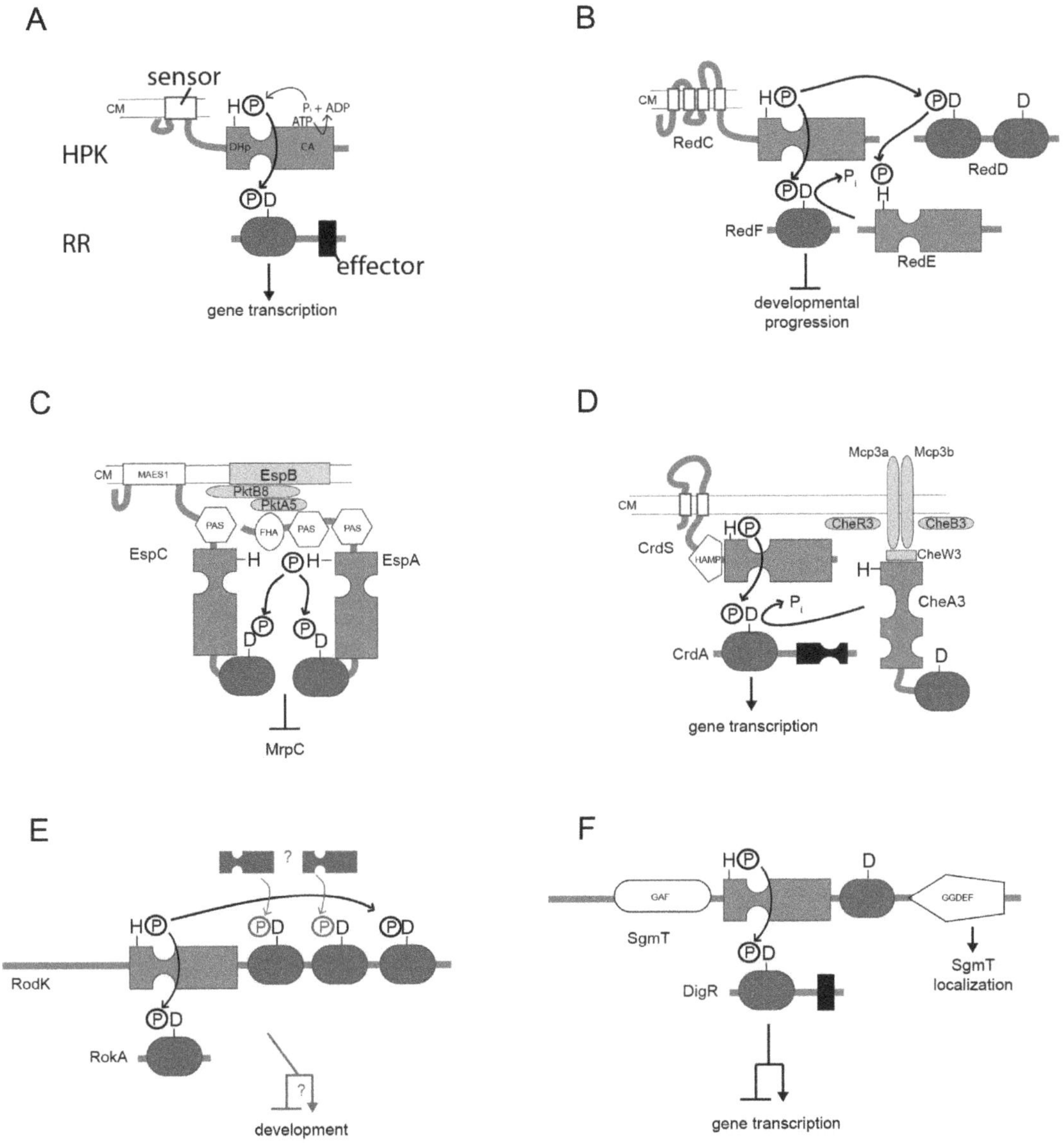

Figure 7.1 Schematic of His-Asp phosphorelay signalling mechanisms. (A) A paradigm linear two-component signalling system. Extracellular signals sensed by a HPK are transmitted to a DNA-binding RR in order to modulate gene transcription. Signal transmission occurs via His-Asp phosphorelay between conserved HK (green rectangle) and REC (blue oval) modules. The HK module consists of a DHp and a CA domain and, in response to activation by the sensor domain, autophosphorylates on an invariant His (H) residue at the expense of ATP. The phosphoryl group (P) is transferred to an invariant Asp (D) in the REC module which activates the associated DNA binding effector domain. (B) The Red signalling pathway. An HPK (RedC) and HPK-like dedicated phosphatase (RedE) and two RR lacking output domains (RedD and RedF) form a 'four-component' signalling system which controls progression through the developmental programme. (C) The Esp signalling pathway. Two HyHK (EspA and EspC), two STPKs (PktA5 and PktB8) and a transmembrane protein (EspB) form signalling complex that regulates turnover of the important transcriptional regulator of development, MrpC. (D) The Crd/Che3 multicomponent signalling pathway. The HPK CrdS and the Che3 chemosensory pathway converge on the terminal RR CrdA to regulate the developmental programme. (E) The Rod signalling pathway. The HyHK RodK, together with unidentified HPK(s) (grey rectangles) regulates the developmental programme via the single domain RR RokA and unknown downstream components. (F) The SgmT/DigR signalling pathway. The HyHK SgmT and RR DigR control gene expression of metabolic and cell surface components. SgmT localization is controlled by GGDEF dependent ci-d-GMP reception. See text for details of all signalling systems. A colour version of this figure is located in the plate section at the back of the book.

adjacent gene encoding a cognate HK or RR) or in complex clusters (more than one HK or RR), the prevalence of atypical TCS protein domain architectures, and the apparent rapid expansion of TCS genes in the myxobacteria all predict the capacity for complex and novel signalling mechanisms (Chapter 2) (Schneiker *et al.*, 2007; Shi *et al.*, 2008; Whitworth and Cock, 2008a,b; Huntley *et al.*, 2011). Specifically, in the genome of the model myxobacterium, *Myxococcus xanthus*, only 29% of the 251 (non-chemosensory) TCS genes are encoded as HPK–RR pairs; the majority are encoded as orphans (55%) or in complex gene clusters (16%) (Shi *et al.*, 2008). Transcriptome analysis indicates that orphan and complex TCS genes are over-represented in the category of genes which were transcriptionally up-regulated during the developmental programme; these data support the suggestion that atypical systems may indeed be favoured in the regulation of complex behaviour (Shi *et al.*, 2008). In the last few years, several complex TCSs have been analysed biochemically and these analyses have confirmed (and sometimes exceeded) the expectation of novel signalling arrangements. This section will focus on some recently described complex signalling systems.

Recently characterized *M. xanthus* TCSs

The Red (regulation of early development) signal transduction system is an example of a TCS encoded as a complex gene cluster; it contains a HPK (RedC), a kinase-like protein (RedE), and two RRs: a dual REC module protein (RedD) and a stand-alone REC module (RedF) (Fig. 7.1B) (Higgs *et al.*, 2005). These four genes are encoded together from the seven gene *redA-G* operon which is expressed during vegetative conditions and down-regulated during development. The Red system is thought to control progression through the developmental programme because a Δ*redCDEF* mutant enters the developmental programme appropriately but begins to form fruiting bodies approximately 12 h more rapidly than the wild type (premature developmental progression *aka* 'early development phenotype'). Interestingly, in-frame deletion of *redA*, *redB*, or *redG*, (encoding putative ribosome modulation, hypothetical, and aminopeptidase proteins, respectively) does not significantly affect developmental progression. The signalling flow between RedC, D, E and F was elucidated by a combination of *in vitro* autophosphorylation/phosphotransfer analyses and systematic genetic mutation and epistasis analyses (Jagadeesan *et al.*, 2009). Together, these analyses indicate that the Red system is a 'four-component' TCS involving a novel signalling mechanism. *redC* or *redF* in-frame deletion mutants, or mutants encoding a RedC or RedF protein with an Ala substitution in its phospho-accepting residue ($RedC_{H253}$ or $RedF_{D61}$) each produce an early development phenotype. In contrast, equivalent mutations in *redE* and *redD* produce a delayed developmental phenotype, suggesting the Red system has a switch-like output control. Genetic epistasis analysis together with *in vitro* biochemical activity assays suggest that the single domain RR (sdRR) RedF is the system output which represses developmental progression when phosphorylated. RedF is phosphorylated by RedC and has an unusually long-lived phosphorylated state ($T_{1/2} \geq 48$ h). To relieve the block on developmental progression, RedC instead phosphorylates the first receiver domain of RedD. Surprisingly, the phosphoryl group is passed from RedD to RedE. RedE acts as a nucleotide (ATP or ADP)-dependent phosphatase on RedF. RedE itself is apparently incapable of autophosphorylation, consistent with a poorly conserved Dhp domain. The invariant His residue, which accepts the phosphoryl group from RedD, is required for the protein function *in vivo*, but is not required for its phosphatase activity on RedF~P *in vitro*. These results were interpreted to mean that *in vivo* phosphorylation of RedE (RedE~P) mediates accessibility to RedF~P, perhaps by releasing RedE from RedD, but this hypothesis has not yet been directly tested. Although the signal flow has been elucidated, several questions remain:

1 What are the signals which regulate the system? It is presumed that RedC senses an unknown stimulus via its membrane-associated putative sensing domain which likely regulates RedC kinase/phosphatase activity on RedF or RedD. Additionally, however, the second REC module of RedD, for which no

function is known, may be a connection to an additional TCS; perhaps RedD-REC2~P influences interaction with RedC or RedE and thereby modulates the levels of RedF~P.
2 What is the exact output of the Red system? RedF~P is the output of the signalling pathway, but since RedF is an sdRR which lacks an associated output domain, exactly how development progression is controlled is not understood.
3 Why is such a complex system necessary to control developmental progression? While answering the first two questions will no doubt clarify the latter, a simple hypothesis is that the modularity of the four-component Red system can be exploited to allow integration of information from diverse sources (i.e. through the RedC sensing domain, RedD-REC2, and/or unknown factors that could influence localization/accessibility of the four Red TCS components) which may allow *M. xanthus* to adjust the developmental programme in response to specific (combinations of) environmental and intercellular stimuli.

An additional example of a complex system controlling developmental progression is the Esp (early sporulation) signalling system. This system consists of two orphan hybrid histidine kinases (HyHK; HPK–REC fusions; see Fig. 7.1C), EspA (Cho and Zusman, 1999) and EspC (Lee *et al.*, 2005), a putative transport protein (EspB) (Cho and Zusman, 1999), and two STPKs (PktA5 and PktB8) (Stein *et al.*, 2006) (Fig. 7.1C). Like the Red system, the Esp system controls developmental progression using a switch-like mechanism: deletion or substitution of phospho-accepting residues in either *espA* or *espC* causes an early development phenotype (Cho and Zusman, 1999; Lee *et al.*, 2005; Higgs *et al.*, 2008; Schramm *et al.*, 2012), while deletion of *espB*, *pktA5* or *pktB8* causes a delayed development phenotype (Cho and Zusman, 1999; Stein *et al.*, 2006). Genetic epistasis experiments strongly suggest that EspB, PktA5, and PktB8 form a signalling unit that modulates the activity of EspAC because: (1) A triple *espB pktA5 pktB8* mutant produces a delayed developmental phenotype indistinguishable from each of the single mutants; and (2) The early aggregation phenotype of *espA* or *espC* is epistatic to that of *espB*, *pktA5*, or *pktB8* mutations (Cho and Zusman, 1999; Stein *et al.*, 2006). These data suggest a model in which EspB/PktA5/PktB8 act together to modulate the activity of EspAC, although this remains to be demonstrated biochemically. EspA appears to interact with PktA5 and or PktB8 (Stein *et al.*, 2006). Interestingly, the EspA-sensing region contains a forkhead associated (FHA) domain which are generally thought to bind phospho-Thr containing substrates (Durocher *et al.*, 2000) and PktA5 autophosphorylates on Thr residues (Stein *et al.*, 2006). Thus, it appears that the Esp signalling system consists of both TCS and STPK proteins.

A novel mechanism of HyHK signalling was recently revealed by demonstrating that the EspA and EspC function together intimately and it was shown that the HyHK EspA is a phosphodonor not only to its own REC module, but also to the REC module of the orphan HyHK EspC (Schramm *et al.*, 2012). *espA*, *espC*, and *espAC* strains produce an identical early developmental phenotype. Ala substitution of the phospho-accepting His residue in the EspA HK module ($espA_{H407}$) and/or the phospho-accepting Asp in the EspA or EspC REC modules ($espA_{D696}$ and $espC_{D749}$, respectively), all mimic the null early development phenotype (Higgs *et al.*, 2008; Schramm *et al.*, 2012). Surprisingly however, Ala substitution of the phospho-accepting His residue in the EspC HK module ($espC_{H461}$), or even deletion of the entire CA domain ($espC_{\Delta CA}$) produces a developmental phenotype indistinguishable from wild type, suggesting EspC kinase activity is not necessary *in vivo* (Schramm *et al.*, 2012). The EspC HK module is capable of efficient autophosphorylation on His_{461} *in vitro*. These discrepancies were resolved by demonstrating that the EspA HK, but not the EspC HK, could phosphorylate the EspC REC module. Together with the genetic analyses, these data strongly suggest a model in which the simultaneous phosphorylation of the EspA and EspC REC modules are required for the system output (repression of the developmental programme) and that EspA HK is the phosphodonor for both REC modules. Thus, this system defines a novel signalling mechanism involving both inter- and intra-hybrid histidine kinase phosphotransfer (Schramm *et al.*, 2012). Both EspA and EspC contain multiple sensing domains; in addition to the FHA domain, EspA contains two Per-Arnt-Sim (PAS) domains, and EspC contains a PAS domain and a MASE1 domain. PAS domains are typically thought to be involved in responding to redox potential, alone or with a diverse range of cofactors (Henry and Crosson, 2011). The exact stimuli are difficult to identify from the sequence alone and the signal(s) sensed by these domains in EspA and EspC are unknown. MASE1 domains are comprised of 8 predicted transmembrane segments and are commonly found associated with signalling proteins, but the exact function is unknown (Nikolskaya *et al.*, 2003).

What does the Esp system regulate? Recent evidence suggests that this complex signalling system regulates proteolytic turnover of the important developmental transcriptional regulator, MrpC (Sun and Shi, 2001a; Sun and Shi, 2001b; Ueki and Inouye, 2003). The observation that MrpC protein, but not *mrpC* mRNA levels are up-regulated earlier in the Δ*esp* mutant relative to wild type suggested Esp may control MrpC accumulation either via translational regulation or protein turnover (Higgs *et al.*, 2008). It was subsequently demonstrated that MrpC is rapidly turnover in wild type cells ($T_{1/2}$ ~30 min), but not in Δ*espAC* mutant (turnover not detected) (Schramm *et al.*, 2012). Since MrpC is essential for both aggregation and sporulation (Sun and Shi, 2001b; Ueki and Inouye, 2003; Nariya and Inouye, 2005), inappropriate accumulation of MrpC in the *esp* mutants likely explains the associated early development phenotype. Thus, it is currently assumed that the combined phosphorylation of EspAC REC modules somehow stimulates an unknown protease to degrade MrpC. It is unknown whether this stimulation is via direct interaction or occurs via continued signal transmission.

Yet another example demonstrating TCS plasticity involves a complex system in which a TCS consisting of an orphan RR and orphan HPK (CrdA and CrdS, respectively) appears to function intimately with a chemosensory system (Che3) to regulate the developmental programme (Kirby and Zusman, 2003; Willett and Kirby, 2011) (Fig. 7.1D). Using a comparative genomics and co-occurrence bioinformatics approach, the HPK CrdS was identified as a candidate cognate kinase for CrdA. Like a *crdA* insertion mutant, a Δ*crdS* mutant displays a delayed developmental phenotype, and the CrdS HK module specifically phosphorylates, and dephosphorylates, CrdA *in vitro* (Willett and Kirby, 2011).

Interestingly, CrdA appears to be additionally regulated by Che3 system, which is encoded adjacent to *crdA* (Kirby and Zusman, 2003). Prototypical chemosensory systems, which can be considered specialized TCSs, regulate flagella motility, but some have been described to be associated with alternative cellular functions (Wuichet and Zhulin, 2010). The protypical Che system consists of a class II HPK, CheA, which autophosphorylates on an invariant His residue and then donates the phosphoryl group to a cognate sdRR, CheY. CheY interacts directly with the switch complex of the flagellar motor to control direction or speed of flagellar rotation. CheA kinases lack a sensing domain; their kinase activity is instead modulated (via the coupling protein, CheW) by one or more methyl-accepting chemosensory proteins (MCPs). MCPs typically consist of an extracytoplasmic ligand-binding domain and a cytoplasmic highly conserved domain which contains signature residues that can be reversibly methylated by methyltransferase (CheR) and methylesterase (CheB) enzymes. Ligand binding by the MCP(s) alters the CheA kinase activity on the cognate RR CheY, and MCP methylation can quench this signal enabling an adaption response to the current signalling state (reviewed in Hazelbauer and Lai, 2010).

The *che3* system encodes homologues of two MCPs (MCP3A and 3B), a CheA (with an additional C-terminal REC module), a CheW, a CheR, and a CheB. CheA3 interacts specifically with CrdA by yeast two-hybrid analysis, the *crdA* insertion mutation (delayed development) is epistatic to a Δ*mcp3B* mutant (early development) (Kirby and Zusman, 2003). Consistent with the opposing mutant phenotypes of *cheA3* and *crdA* mutants, CheA3 acts as a specific phosphatase on CrdA~P *in vitro* (Willett and Kirby, 2011). Deletions of *mcp3B*, and apparently *mcp3A*, *crdB* (encoding a predicted periplasmic protein), and *cheBR*, display a similar early developmental phenotype to the *cheA3* insertion mutant suggesting the respective proteins function together (Kirby and Zusman, 2003). These *che3* mutants are reported to effect entry into the developmental programme because (unlike *red* or *esp* mutants) they can form some fruiting bodies on rich media or on nutrient-limited media at cell densities too low to induce development in wild-type cells. Together, these results suggest a model in which CrdS phosphorylates CrdA to induce the developmental programme, but CheA3, likely in complex with other components of the Che3 chemosensory system, could be necessary to remove the phosphoryl group under conditions close to the borderline for development (e.g. low nutrient levels which are not yet sufficient for starvation).

CrdS and CrdA belong to the TCS subfamily which regulate the transcription from sigma factor 54 (σ^{54})-dependent promoters. CrdA is a bacterial enhancer-binding protein (EBP) with an amino-terminal REC module, a central triple AAA+ (ATPases associated with various cellular activities) domain, and a carboxy-terminal DNA binding domain. When activated (in the case of CrdA, presumably by REC phosphorylation), EBPs stimulate transcription from σ^{54}-promoters by binding to specific regulatory regions upstream of σ^{54} promoters and energy from ATP-hydrolysis is used to convert the inherently stable RNAP-σ^{54} closed complex into the open complex necessary to initiate transcription (reviewed in Bush and Dixon, 2012). It has been shown that a number of developmental genes are mis-regulated in *che3* mutant backgrounds, but it is unknown which of the many σ^{54}-dependent *M. xanthus* genes are directly regulated by CrdA (Kirby and Zusman, 2003).

To start to understand the evolutionary advantage of co-regulation of CrdA by the Che3 system, it will be important to determine whether CrdA, CrdS, and at least CheA3 share the same protein accumulation patterns during vegetative and developmental time courses, and to analyse the effects of functional residue substitution at wild type expression levels. It will also be important to determine which signals regulate the system through CrdS and Che3. CrdS contains a putative sensing region consisting of two predicted transmembrane segments (suggesting a periplasmic loop) followed by a HAMP domain, and the signals which likely modulate its kinase/phosphatase activity on CrdA are unknown. The ligands which Mcp3A and B would be predicted to bind are not known, but it has been proposed that signal input may involve two other proteins (CrdB and CrdC) which are predicted to be located in the periplasm (Willett and Kirby, 2011).

An additional TCS composed of orphan components is that of the HPK SgmT and RR DigR (Petters *et al.*, 2012) (Fig. 7.1E). DigR is an orphan RR composed of a REC module and a helix–turn–helix DNA binding domain (Overgaard *et al.*, 2006). *digR* mutants are incapable of type four-pili mediated social (S)- motility and appear to misregulate extracellular polymeric substances (EPS) (Overgaard *et al.*, 2006) (see Chapter 10 for details of S-motility and EPS). A cognate HPK was identified by searching for orphan HPKs reported to display an S-motility phenotype, and SgmT was identified (Petters *et al.*, 2012). Δ*sgmT* and *sgmT digR* mutants display similar phenotypes as *digR* mutants: no S-motility, overproduction of EPS and reduced levels of O-antigen (Petters *et al.*, 2012). Microarray analysis of the individual *digR* and *sgmT* mutants indicated good overlap between significantly up- and down-regulated genes, consistent with the identical mutant phenotypes. SgmT has a complex domain architecture consisting of a GAF domain, HK module, and REC module followed by a C-terminal GGDEF domain. GAF domains are poorly defined sensing domains commonly associated with signalling systems (Aravind and Ponting, 1997), and GGDEF domains are involved c-di-GMP second messenger signalling (Hengge, 2009). It was demonstrated that the GAF domain is essential for SgmT activity and that SgmT autophosphorylates and donates a phosphoryl group to the DigR REC module in a process that is independent of the SgmT REC module (Petters *et al.*, 2012). Interestingly, deletion or mutation of the GGDEF domain does not affect kinase activity or phosphotransfer *in vitro* and does not produce a detectable motility or developmental phenotype. It is, however, required for localization of a SgmT-YFP fusion protein to 1–2 clusters centred in a position in the range of

40–70% of the vegetative cell length. The GGDEF domain, which contains lacks a conserved dcG synthesis motif (A-site), binds c-di-GMP at a conserved binding (I-site) (Petters *et al.*, 2012), and like other A– I+ GGDEF domains (Hengge, 2009), is proposed to function as a receptor for c-di-GMP. However, given the lack of detectable phenotype of SgmT GGDEF mutants, the functional significance of c-di-GMP-dependent localization is not yet understood.

Global transcriptome analysis of *sgmT* and *digR* mutants under vegetative conditions indicated that co-regulated genes (104) were either up- (86) or down- (18) regulated, and promoter binding assays suggested that DigR bound representative promoters in positions consistent with a repressor or an activator of target gene expression, respectively (Petters *et al.*, 2012). Based on the predicted function of significantly regulated genes, Petters *et al.*, (2012) propose that SgmT/DigR are necessary for appropriate transcriptional regulation of genes coding for proteins secreted to the extracellular matrix and in secondary metabolism (which then indirectly effects EPS production), rather than by directly effecting S-motility or EPS production. The authors' model is that the *sgmT digR* mutant S-motility defect is due to inappropriate accumulation of EPS to compensate for loss of proteins involved in maintaining the cell envelope. Clearly, an open question is the functional significance of the c-di-GMP dependent localization of SgmT. One intriguing possibility is that SgmT cluster localization has functional significance during the production of vegetative and developmental heterogeneity (see Chapter 3) which may be overlooked in standard motility/developmental phenotype assays. Consistently, at least two factors influenced by SgmT/DigR (i.e. EPS and FibA) are produced heterogeneously in both vegetative and developmental populations (Berleman *et al.*, 2011; Lee *et al.*, 2011; Lee *et al.*, 2012).

An additional characterization of a complex *M. xanthus* signalling pathway is that of RodK/RokA (Fig. 7.1F). RodK is an unusual HyHK containing a putative N-terminal sensing domain with no obvious motifs, a HK module, and three consecutive REC modules (Rasmussen *et al.*, 2005). *rodK* is transcribed in an operon with *rokA* (which encodes a sdRR) and two genes encoding hypothetical proteins (Wegener-Feldbrugge and Sogaard-Andersen, 2009). Δ*rodK* mutants produce disorganized fruiting bodies in which sporulation is uncoupled from aggregation (Rasmussen *et al.*, 2005). Δ*rokA* mutants are delayed in development, and the more severe Δ*rodK* phenotype is epistatic to that of Δ*rokA* (Wegener-Feldbrugge and Sogaard-Andersen, 2009). These genetic results suggest that RodK and RodA function together to regulate fruiting body formation, but that RokA is not the sole output of the system and may instead function to modulate signal flow. Investigation of the signalling flow in these proteins revealed that inactivation of the phospho-accepting residue in the RodK HK module mimicked the Δ*rodK* null phenotype (Rasmussen *et al.*, 2006). Interestingly, however, individual substitution of the phospho-accepting residue in the three REC modules, or combinatorial double mutants, each produced distinct developmental phenotypes which affect the timing and/or organization of fruiting bodies. *In vitro* biochemical analyses indicated that RodK autophosphorylates and donates a phosphoryl group only to REC-3 which is then rapidly hydrolysed. RodK will also efficiently donate a phosphoryl group to RokA with a kinetic preference for REC-3 only if REC-3 is present as the full-length protein (Wegener-Feldbrugge and Sogaard-Andersen, 2009). RodK and RokA share the same protein accumulation pattern and their respective protein stability depends on each other. Together, these results suggest a model in which RodK serves as a central signal integration protein. RodK autophosphorylation and phosphotransfer to REC-3 or RokA could be controlled by the unknown stimuli through the putative sensing domain. Phosphorylation of REC-1 and REC-2, presumably from an as yet unidentified additional HPK, are important for the specific system output, possibly by effecting phosphotransfer from RodK HK~P to RokA or REC-3 *in vivo*. Exactly how the system output is mediated is not clear. As with many of the *M. xanthus* HyHK or sdRRs which have so far been characterized, it is not clear whether these systems participate in a multistep phosphorelay, or whether the output is mediated by direct interaction between the REC module and a target protein. In the case of RodK, it is also possible that both REC-3 and RodK simply act as phosphate sinks, which influence phosphotransfer from RodK HK to an as yet unidentified cognate RR.

Perspectives of TCS organization in *M. xanthus*

Perhaps the most obvious common feature of the complex TCSs described here is the abundance of stimuli sensing domains wired into each system. In some cases, multiple sensing domains are present in a single HPK (e.g. EspA, EspC, and SgmT), and in other cases by combining more than one HPK into a single signalling system (e.g. EspA/C, CrdS/CheA3 and, likely via 'extra' REC modules in RodK REC-2 and -3 and RedD REC-2) (Fig. 7.1). The presence of multiple sensor domains presumably allows different signals to be integrated into a coordinated output(s), but it is not obvious how signal integration is achieved. In the multi-component systems where phenotypic and biochemical analyses indicate that one of the components functions primarily in removing phosphoryl groups from the output module (e.g. Che3A or RedE), different stimuli likely control these alternate states. It is also possible, however, that these systems are designed such that multiple stimuli must be simultaneously present (co-incidence detection) or that full system activation can be achieved by one of several different stimuli. Alternatively, the example from SgmT (Fig. 7.1F) indicates that sensor domains (i.e. GGDEF dependent c-di-GMP binding) may control localization of the protein. As with many simple TCSs, identification of the stimuli which control the system remains a challenge for the future. In these complex systems, however, an additional challenge is understanding how each of the sensing domains contributes to activation of the system, and a full understanding of why these complex systems have evolved so frequently in the myxobacteria awaits these analyses.

Another common feature of these complex signalling systems described here are the predominance of sdREC modules (i.e. REC modules which are not directly linked to effector domains). Non-effector linked REC modules have been shown to function by mediating signal output by directly interacting with target proteins. For example, the *Escherichia coli* CheY sdRR in its phosphorylated state directly interacts with the flagellar switch protein to control the direction of flagellar rotation (Stock and Surette, 1994). Alternatively, REC modules, both as sdRRs and as HyHK, can feed into a multistep (His-Asp-His-Asp) phosphorelay system in which the phosphoryl group is first passed to an HPT module and then to a terminal RR (Burbulys *et al.*, 1991; Freeman and Bassler, 1999). Finally, non-effector REC modules may instead modulate signal flow by acting as phosphate sinks or by sterically hindering HK activity (Chang *et al.*, 1996; Sourjik and Schmitt, 1998; Inclan *et al.*, 2008; Paul *et al.*, 2008). In the Red, and Esp signalling systems (Fig. 7.1B and C, respectively), genetic analyses strongly suggest that the RedF and EspA/C REC module(s) are the major output of the respective systems, but it is not clear whether these REC modules mediate output by direct interaction with target proteins or whether they participate in a multistep phosphorelay to a terminal RR. There are only two (non-CheA) recognizable HPT-module containing proteins in the *M. xanthus* genome (MIST2 database; Ulrich and Zhulin, 2010). However, domains with His phosphotransferase activity can sometimes be difficult to identify by sequence analysis alone (Biondi *et al.*, 2006), and it is possible that unrecognized domains serve this purpose in these systems. In the case of the RodK/RokA system, it is not yet clear whether upon phosphorylation [by either RodK (REC-3, RokA) or additional HK modules (REC-1,2)] each REC module mediates a distinct output, or whether they influence RodK HK kinase/phosphatase activity on an unidentified terminal RR. Thus, identification of the exact means by which non-effector linked REC modules function in each system is also an important priority in understanding the functional implications of complex systems. Consistent with the systems described here, the *M. xanthus* genome encodes a total REC/transmitter (HK or HPT) module factor of 1.35. This trend, however, is not necessarily conserved in the myxobacteria (*Sorangium cellulosum*: 1.1) nor in organisms which are known to exhibit complex behaviour such as differentiation (i.e. *Bacillus subtilis*: 0.9; *Streptomyces coelicolor*: 0.59; or *Anabaena variabilis*: 1.09) (Table 7.1).

Table 7.1 Comparison of His-Asp signal transduction module counts in representative bacteria

Organism	Number of His-Asp signalling modules[a] REC	Transmitter (HK, HPT)	Total	REC/(HK+HPT)
M. xanthus	198	147	345	1.35
S. cellulosum	187	168	355	1.11
A. variabilis ATCC 29413	152	140	292	1.09
A. dehalogenans 2CP-1	124	90	214	1.38
Pseudomonas aeruginosa PAO1	95	78	173	1.22
S. coelicolor A3(2)	90	152	242	0.59
Caulobacter crescentus NA1000	75	63	138	1.19
E. coli W3110	37	35	72	1.06
B. subtilis	35	39	74	0.90

[a]Module counts from MiST2 (http://mistdb.com/) (Ulrich and Zhulin, 2010).

In summary, the TCSs described herein illustrate that *M. xanthus* (and likely most bacteria with complex behaviour) have exploited the plasticity of the basic His-Asp signal transmission modules to produce multicomponent systems which approach the signalling complexity in higher eukaryotes. These systems illustrate that proteins containing His-Asp signalling modules should no longer be, by default, considered 'two-component' signalling proteins, and each system requires careful dissection using a combination of bioinformatic, genetic, biochemical, and cell biology analyses.

ECF σ factors

The σ factor is a crucial, dissociable component of the multisubunit (two α and one each of the β, β′, and ω subunits) bacterial RNA polymerase (RNAP) that determines promoter recognition, selection and DNA melting for transcription initiation. Promoter recognition by the essential primary or major 'housekeeping' σ factor, such as *E. coli* σ^{70} or *M. xanthus* σ^{A}, initiates transcription of genes required for normal vegetative growth, while alternative σ factors are often employed to recognize other distinct promoters and activate normally silent genes in response to a specific signal (Helmann, 2002; Gruber and Gross, 2003; Österberg *et al.*, 2011). With the exception of the σ^{54} family of σ factors, alternative σ factors are related in sequence and structure to σ^{70}. The largest and most diverse group of such alternative σ^{70} family factors is that constituted by the ECF σ factors (henceforth σ^{ecf}s), which are employed to activate specific genes in response to extracytoplasmic stimuli (light, temperature, osmolarity, nutrient availability, etc.) and during morphological development (Helmann, 2002; Gruber and Gross, 2003; Österberg *et al.*, 2011). Some bacterial species have no σ^{ecf}s, but others, especially those with large genomes and complex lifestyles, can have many σ^{ecf}s. Thus, compared with seven σ^{ecf}s in *B. subtilis*, as many as 51 have been identified in *S. coelicolor*. In *M. xanthus*, whose 9.14 Mb genome is among the largest known in bacteria and ~0.4 Mb larger than that of *S. coelicolor*, 41 candidate σ^{ecf}s distributed throughout the genome have been included in the MiST2 database, a comprehensive genomic resource on microbial signal transduction (Ulrich and Zhulin, 2010). Details about the number of σ^{ecf}s encoded in other myxobacterial genomes can be found in Table 2.5.

Most σ^{ecf}s, including those in *M. xanthus*, have been identified from genome data using sequence comparisons. The essential primary σ factor and its non-essential paralogues contain four conserved regions (designated σ_1, σ_2, σ_3 and σ_4, each with its own subregions) that occur contiguously along the sequence as distinct structural modules connected by flexible linkers, and a non-conserved region (NCR) next to σ_2 (Fig. 7.2A) (Gruber and Gross, 2003; Österberg *et al.*, 2011). Only the σ_2 and σ_4 modules occur in σ^{ecf}, which are therefore usually of a smaller size (~20–25 kDa) than other σs. The σ_4 and σ_2 regions, respectively, mediate recognition of the two 6-bp DNA segments centred around the −35 and −10 positions relative to the transcription start site that define a typical σ^{70} family promoter (Fig. 7.2A). In addition, the σ_2 region is involved in binding to the core RNAP and in DNA melting, and the σ_3 region, which is absent in σ^{ecf}s, in the recognition of the −10 extended motif found in a number of σ^{70} family promoters. Since the σ_1 region prevents promoter-binding by the major σ when it is not associated with RNAP, its absence in σ^{ecf}s suggests that these σ factors can bind promoter DNA on their own, as is observed for some, but not all, σ^{ecf}s (Brooks and Buchanan, 2008). Most σ^A–dependent promoters studied in the GC-rich *M. xanthus* have a −35 region resembling the TTGACA consensus for the major σ-dependent promoters in other bacteria, but the −10 region is more divergent from the *E. coli* TATAAT consensus (Fig.7.2B). The σ^{ecf}-dependent promoters are defined by −35 and −10 regions with sequences that typically contain conserved AAC and CGT motifs, respectively, as observed in *M. xanthus* (Fig. 7.2B) (Helmann, 2002; Lane and Darst, 2006; Staroń *et al.* 2009).

Signalling of σ^{ecf} activity and its regulation

Because most σ^{ecf}s have been identified based on genome analysis, the functions and modes of action for the vast majority remain unknown. Besides confirming the prediction that it is a σ^{ecf}, it is also necessary to define the specific extra-cytoplasmic stimulus that governs the action of the given σ^{ecf}, determine whether more than one signal is involved and, if so, examine whether they act independently or coordinately. Furthermore, it is also necessary to address the identity of target genes and regulons, the specificity or redundancy with other σ^{ecf}s, the molecular mechanisms controlling σ^{ecf} activity, and the competition among the different σs for RNAP.

One level of control is, as noted earlier, the promoter design and the determinants in its sequence. Other cellular strategies also enable the action of a specific σ^{ecf} for precise gene activation in response to a given signal. Thus, most known σ^{ecf}s autoregulate their own expression, and the activity is often, though not always, controlled by inhibition via a specific cognate anti-σ^{ecf}. This usually involves direct physical interaction between the σ^{ecf} and anti-σ^{ecf}, the stoichiometry of which is frequently maintained by coexpression of the two factors from adjacent genes in the same operon. The anti-σ^{ecf}s are mostly membrane

A

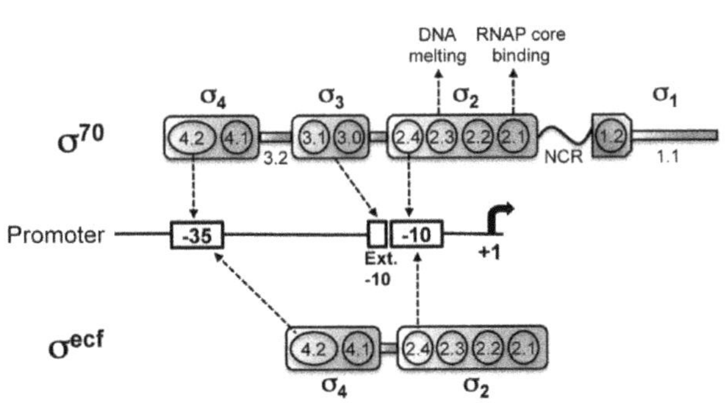

B

Some *M. xanthus* σ^A and σ^{ecf} promoter sequences

σ factor	-35	-10	Gene
σ^A	TAGACA	TAAGGG	*vegA*
	TTGAGT	TTCGTT	*lonV*
	TTGCCA	TACGTT	*lonD*
	TTGCCA	TACGTT	Ω4514
	TTGACA	TAGAAG	*rrnD*
	TTGACA	TACCTC	*carB*
σ^{ecf} (CarQ)	TGTAAC	GCGTTC	*crtIb*
	CGGAAAC	CCGTAG	*carQRS*
σ^{ecf} (RpoE1)	GGGAAT	GTCGTG	*rpoE1*

Figure 7.2 σ^{ecfs} in *M. xanthus*. (A) Schematic representation of σ^{70} (top) and σ^{ecfs} (bottom) showing the conserved regions (σ_1 to σ_4) and subregions (1.1 to 4.2), as well as a nonconserved region (NCR) present in some σ factors. The functional assignment of these regions is also indicated. The principal elements of a typical σ^{70} family promoter is shown in the middle with conserved −35, −10, and extended −10 DNA regions from the start (+1) site. (B) Promoter regions recognized by *M. xanthus* RNAP bound to σ^A or to the σ^{ecf} CarQ and RpoE1.

proteins (Helmann, 2002; Staroń *et al.*, 2009), and their inactivation (by directed proteolysis, phosphorylation, and/or conformational change in response to an extracellular signal) releases the σ^{ecf} and enables its activity (Brooks and Buchanan, 2008; Ho and Ellermeier, 2012). Although a diverse group of proteins, ~40% of anti-σ^{ecf}s may contain a conserved zinc-binding $HisX_3CysX_2Cys$ motif termed ZAS (Zn^{2+} anti-sigma), located in an N-terminal module that is fused to different classes of soluble or membrane-spanning C-terminal signalling domains (Campbell *et al.*, 2007). Finally, factors such as anti-anti-σ^{ecf}s, which are antagonists of anti-σ^{ecf} activity and other specific or global transcriptional regulators, expand the repertoire of available mechanisms for controlling σ^{ecf} activity.

σ^{ecf}s in *M. xanthus*

In *M. xanthus*, 28 out of the 41 putative σ^{ecf} genes lie adjacent to a gene which expresses a putative membrane protein that may be the cognate anti-σ^{ecf}. In 24 of these 28, the σ^{ecf} gene is located upstream of the putative anti-σ^{ecf} gene, while in the remaining 4 cases the σ^{ecf} gene is downstream. In most of them, the pair appears to be translationally coupled. Also, 20 of the 28 putative membrane anti-σ^{ecf} proteins contain the ZAS module mentioned above. The remaining 13 predicted σ^{ecf} genes either have a flanking gene that encodes a protein lacking predicted membrane segments (two of which contain the ZAS motif) or occur as orphan genes. The latter may either lack a cognate anti-σ^{ecf}, may have one expressed at a different genome locus, or may have their activity controlled intramolecularly (for instance, one of these presumably orphan σ^{ecf}s has a ZAS motif, and as described below, a C-terminal Cys-rich domain appears to control the activity of CorE in response to copper).

Thus far, experimental studies have been reported for only three of the *M. xanthus* σ^{ecf}s. Interestingly, the best characterized of these, CarQ, was among the first examples classified as a σ^{ecf} (Lonetto *et al.*, 1994). The specific extracytoplasmic signal that activates CarQ is blue light, the end-result of which is the biosynthesis of carotenoids: lipophilic, isoprenoid derivatives that protect cells against photo-oxidative damage (reviewed in Elías-Arnanz *et al.*, 2008; Elías-Arnanz *et al.*, 2011). Three translationally coupled genes that constitute the *carQRS* operon produce CarQ, its membrane-associated anti-σ^{ecf} CarR, and CarS. Outside of the myxobacteria orthologues, CarR (with six predicted transmembrane regions) does not share significant sequence similarity with any other protein in the databases, and hence is unlike any known anti-σ^{ecf} (Galbis-Martínez *et al.*, 2012). CarS is an antirepressor that controls the activation of other genes involved in this light-induced response (see León *et al.*, 2010). In the dark, CarQ is inactive because it is physically sequestered to the membrane by CarR. Light neutralizes CarR (apparently via its degradation), freeing CarQ to associate with RNAP and activate two promoters: its own promoter (P_{QRS}), and that of *crtIb* (P_I), which encodes a carotenogenic enzyme (Gorham *et al.*, 1996; Martínez-Argudo *et al.*, 1998; Browning *et al.*, 2003). Both P_{QRS} and P_I contain the AAC and CGT motifs at the −35 and −10 regions, respectively, which characterize σ^{ecf}-dependent promoters, as mentioned earlier (Fig. 7.2B).

In addition to the promoter determinants and an anti-σ^{ecf} (CarR), light-induced activity of CarQ is subject to regulation by transcriptional factors and an anti-anti-σ^{ecf}, CarF. Thus, light-inactivation of CarR absolutely requires CarF, with which it physically interacts (Fontes *et al.*, 2003; Galbis-Martínez *et al.*, 2008). CarF, which is also membrane-associated, shares sequence similarity with proteins of unknown function called Kua that constitute the protein superfamily Kua-UEV1_localn (PF10520 in the protein family database Pfam: http://pfam.sanger.ac.uk; Punta *et al.*, 2012). Kua proteins are widely distributed in eukaryotes, but CarF-like proteins are virtually unknown in other bacteria, except for *Bradyrhizobium japonicum* and *Leptospira borgpetersenii* and its orthologues in closely related myxobacteria (Galbis-Martínez *et al.*, 2012). One striking feature of CarF (and Kua) is the presence of several conserved His residues in the predicted cytoplasmic segments. Several of these residues are essential for function and are distributed in a manner similar to those that form a di-iron cluster in some fatty acid hydroxylases and desaturases (Fontes *et al.*, 2003; Galbis-Martínez *et al.*, 2008).

Recent work has shown that $^1O_2$ is the signal that activates the CarQ-dependent P_{QRS} promoter, and that transmission of the $^1O_2$ signal absolutely requires CarF (Galbis-Martínez *et al.*, 2012). $^1O_2$ itself is produced when protoporphyrin IX (that can accumulate in the membrane) is excited by blue light and reacts with molecular oxygen. The molecular details of how CarF mediates the transmission of the $^1O_2$ signal and CarR inactivation remain to be resolved.

Activation of its promoters by the CarQ σ^{ecf} is known to require at least three other factors. Two, CarD and CarG, form a global regulatory complex that also regulates other processes such as fruiting body development (reviewed in Elías-Arnanz *et al.*, 2010). CarD and its myxobacterial orthologues are the only known examples of bacterial proteins with a C-terminal, DNA-binding domain that resembles eukaryotic high-mobility group A proteins (non-histone components of chromatin involved in numerous cellular processes). CarG is a zinc-associated transcriptional adaptor that does not bind DNA but acts via its interaction with the CarD N-terminal domain. The latter also interacts with RNAP and is the defining member of the large CarD_CdnL_TRCF bacterial protein family (PF02559 in http://pfam.sanger.ac.uk). A specific CarD-binding site upstream of P_{QRS} which consists of two 3-bp long AT-rich tracts separated by a 7-bp segment is essential for activating this promoter *in vivo*. Thus, the CarD-CarG complex appears to act by simultaneously binding to a region of the promoter DNA as well as to RNAP. However, the other CarQ-dependent promoter, $P_{I'}$, does not contain a similar CarD-binding site, and the effect of the CarD-CarG complex on CarQ-activation of this promoter may be indirect. The third factor that affects CarQ activity at P_{QRS} is the integration host factor (IHF), a crucial nucleoid-associated bacterial protein involved in various DNA transactions (Elías-Arnanz *et al.*, 2008, 2010, 2011). Several questions remain, and are being currently addressed, about the molecular mechanisms by which the CarD-CarG complex and IHF regulate CarQ-mediated activation of its promoters.

An intriguing finding has been that copper can trigger the CarQ-dependent carotenogenesis cascade without the need for light or CarF (Moraleda-Muñoz *et al.*, 2005). Thus, copper-activation of P_{QRS} requires CarQ, CarD and IHF but not CarF, even though CarF is physically associated with CarR, and CarR with CarQ. It is unknown how this copper effect is achieved and whether additional factors are involved. It does, however, illustrate that two distinct signals can modulate a given σ^{ecf} activity.

Interestingly, in a recent study, copper was shown to be required for the action of CorE, a σ^{ecf} distinct from CarQ (Gómez-Santos *et al.*, 2011). Unlike CarQ and many other typical σ^{ecf}, CorE appears to lack a cognate anti-σ^{ecf}. The most downstream in a gene cluster involved in copper homeostasis, *corE* is not translationally coupled to the preceding gene in the cluster. When copper is present as Cu(II), but not as Cu(I), CorE alone (without RNAP) can bind to a DNA fragment containing a target promoter. The promoter sequences to which CorE binds have not yet been mapped, but segments with an AAC motif at a putative −35 region of candidate promoters have been identified (Gómez-Santos *et al.*, 2011). Although direct copper-binding by CorE remains to be demonstrated, four Cys residues in a 38-residue C-terminal extension of the σ_4 region are important for CorE function and have been suggested to provide a coordination environment for copper. Moreover, 21 putative σ^{ecf}s with similar C-terminal extensions and conserved Cys residues were identified in some other bacteria, leading to the proposal that CorE and its homologues form a distinct σ^{ecf} subclass (Gómez-Santos *et al.*, 2011).

A third *M. xanthus* σ^{ecf} for which experimental data have been reported is RpoE1 (Ward *et al.*, 1998). Immediately downstream of *rpoE1*, but not translationally coupled to it, is a gene encoding a putative transmembrane regulator with a ZAS motif which may be the anti-σ^{ecf} for RpoE1, but this remains to be demonstrated. A σ^{ecf}-dependent promoter upstream of *rpoE1* has been proposed. This, as well as the genes transcriptionally regulated by RpoE1 (its regulon), awaits identification.

In conclusion, the large number of σ^{ecf}s in *M. xanthus* suggests these proteins are a substantial and critical part of the cellular tools used by this free-living soil bacterium with a complex lifestyle to deal with the innumerable extracytoplasmic signals. The few cases studied in detail have revealed the elaborate mechanisms employed to regulate the specific response. The immediate challenges will be to assign the specific signals and regulons, and the molecular mechanisms of action, for the large number of remaining σ^{ecf}s.

Eukaryotic-like Ser/Thr protein kinases and phosphatases

The fourth type of signal-transduction mechanism widely used by myxobacteria is mediated by eukaryotic-like protein kinases and PPs. The term 'eukaryotic-like' refers to the fact that this protein superfamily was historically thought to be present only in eukaryotes. However, it has been known since 1991 that they are also encoded by prokaryotes (Muñoz-Dorado *et al.*, 1991). Eukaryotic protein kinases, in contrast to the proteins in TCSs, phosphorylate protein substrates on Ser, Thr and/or Tyr residues to form phosphoesters (Hanks and Hunter, 1995). This phosphorylation is reversible and dephosphorylation is mediated by PPs. Together, these kinases and phosphatases function as biological switches that turn the signal-transduction pathways in which they participate on and off. The superfamily of eukaryotic protein kinases is divided into two families, Ser/Thr protein kinases and Tyr protein kinase, based on the residue which is phosphorylated and the mode of regulation (Hanks and Hunter, 1995). Bacterial eukaryotic-like protein kinases are related to the Ser/Thr protein kinase family (Pereira *et al.*, 2011). Accordingly, this type of proteins will be referred in this chapter to as STPKs.

Unlike the other three signal-transduction mechanisms, STPKs are not universal among the prokaryotes. Mining 626 prokaryotic genomes revealed that 35% of these genomes did not encode any STPKs. Interestingly, only 25 strains encoded more than 25 STPKs, and most of these exhibited a complex lifestyle, in which cell cooperation and coordination is necessary. This observation points to multicellular behaviour as the main evolutionary drive towards an extensive kinome (Pérez *et al.*, 2008). Nevertheless, STPKs have been reported to be involved in the control of a wide variety of physiological processes in bacteria, such as aggregation, differentiation, spore germination, cell division, motility, central and secondary metabolism, and virulence. Some have even been shown to be essential for cell viability (for an extensive and recent review, see Pereira *et al.*, 2011).

Although phosphorylation at Ser/Thr/Tyr residues can also be carried out by other kinases in bacteria (LaPorte *et al*, 1985; Najafi *et al.*, 1995; Kravanja *et al.*, 1999; Grangeasse *et al.*, 2007; Kimura *et al.*, 2011b), STPKs clearly differ from these other proteins in sequence. Despite the great diversity of STPK sequences, all of these proteins share a catalytic domain with 12 clearly differentiated subdomains. These subdomains are defined by a number of conserved residues (Hanks and Hunter, 1995). Furthermore, the catalytic domains also share a similar two-lobed structure, in which ATP is sandwiched in the deep cleft between the two lobes (Madhusudan *et al.*, 2002; Ortiz-Lombardia *et al.*, 2003; Young *et al.*, 2003). In order to function as biological switches, STPKs alternate between an active and an inactive state. This transition must be tightly controlled, and may be modulated by a variety of mechanisms, such as binding to specific effectors which cause significant conformational changes in the protein to allow the phosphotransfer reaction to take place (Johnson *et al.*, 1996). In many cases, activation of the protein is mediated by phosphorylation of a region of the catalytic domain known as the activation segment, which is located between subdomains VII and VIII. This region contains one or more residues whose phosphorylation promotes interaction with the substrate and catalysis. Proteins regulated by these mechanisms exhibit a conserved Arg and Asp in subdomain VIb, and hence they are termed RD kinases. In contrast, STPKs that are not regulated by this mechanism lack this Arg residue in this subdomain and are termed non-RD kinases (Johnson and Lewis, 2001).

The highly complexity of some characterized TCS pathways (see above), the presumably high number of components that participate in a signal-transduction pathway mediated by STPKs (as it happens in eukaryotes), and the interconnection of pathways of different nature (see Fig. 7.1C and D, where TCSs and STPKs or Che systems are components of the same pathway), ensure a more finely tuned adjustment of the cell response to the corresponding stimulus. This is especially necessary when the response involves coordinating thousands of cells to enable them to successfully complete their life cycle.

The myxobacterial kinome

Myxobacteria exhibit by far the largest kinomes among the prokaryotes (Pérez *et al.*, 2008). Bearing in mind that *M. xanthus* encodes 99 STPKs, it is not surprising that the first prokaryotic STPK was identified in this organisms and that a large number of paralogous genes were reported even before we entered the genomics era and the complete sequence of the first myxobacterial genome became available (Muñoz-Dorado *et al.*, 1991; Zhang *et al.*, 1992; Udo *et al.*, 1995; Inouye *et al.*, 2000). At the time of this writing, the sequences of 11 myxobacterial genomes are available in the NCBI database (http://www.ncbi.nlm.nih.gov/genome/browse/). In addition, we have obtained information about the kinome of *Myxococcus stipitatus* DSM 14675, which has been kindly provided by the Max Planck Institute at Marburg. These 12 strains represent the three suborders of the *Myxococcales* (see the Chapter 2 for details of myxobacterial genomics). Ten of these genomes are complete, whilst two are split into a small number of contigs (*P. pacifica* SIR-1 and *M. stipitatus* DSM 14675). One interesting feature of the myxobacterial genomes is that, with the exception of the strains of the genus *Anaeromyxobacter* (genome size around 5 Mb), their sizes are among the largest reported in prokaryotes. In fact, *S. cellulosum* (13.03 Mb) harbours the largest prokaryotic genome, which is even larger than that of the yeast *Saccharomyces cerevisiae* (12.45 Mb).

When the myxobacterial kinome was first analysed in 2008 (Pérez *et al.*, 2008), only seven genomes were available, three of which were incomplete. The results obtained after mining the 12 myxobacterial genomes currently analysed confirm that the conclusions drawn in that report are still valid. One of these conclusions is that the number of STPKs in a myxobacterial genome seems to increase exponentially with size (Fig. 7.3). Furthermore, the per cent of the genome encoding STPK genes also increases with the genome size (Pérez *et al.*, 2008). For instance, while 35,643 bp (0.71% of the genome) are devoted to encode STPKs in *A. dehalogenans* 2CP-C, this amount raises to 207,555 bp in *M. xanthus* (2.28% of the genome), and to

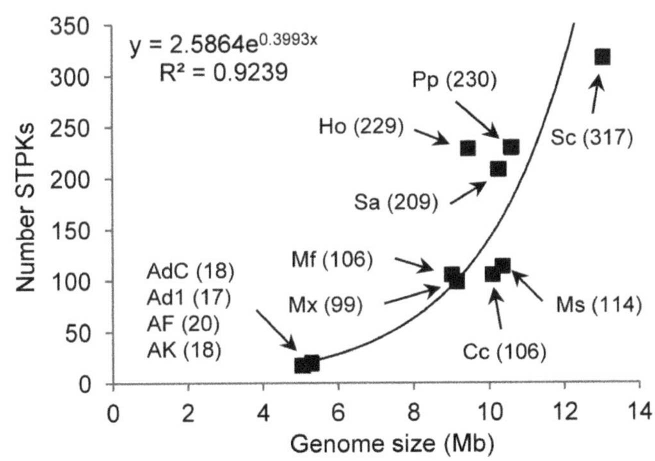

Figure 7.3 Number of STPKs encoded in the genomes of the sequenced myxobacterial strains versus genome size. Abbreviations used for the strains are as follows: AdC, *Anaeromyxobacter dehalogenans* 2CP-C; Ad1, *A. dehalogenans* 2CP-1; AF, *Anaeromyxobacter* sp. Fw109–5; AK, *Anaeromyxobacter* sp. K; Mf, *Myxococcus fulvus* HW-1; Mx, *M. xanthus* DK1622; Ho, *Haliangium ochraceum* DSM 14365; Cc, *Corallococcus coralloides* DSM 2259; Sa, *S. aurantiaca* DW4/3–1; Ms, *M. stipitatus* DSM 14675; Pp, *Plesiocystis pacifica* SIR-1; and Sc, *S. cellulosum* So ce56. Digits in parentheses indicate the number of STPKs encoded by each strain. Minor discrepancies between numbers provided here and those in Chapter 2 are presumably due to differences in the methods of identification. Numbers shown here have been manually curated.

808,467 bp (6.21% of the genome) in *S. cellulosum*. These data indicate that duplication of genes encoding STPKs can be considered as a significant contribution to the genome expansion of this bacterial group. Moreover, the fact that all of the myxobacteria, and especially those with more complex lifestyles, encode a large number of STPKs reinforces the link between a large kinome and multicellular behaviour. Although it appears plausible that the number of STPKs encoded in a myxobacterial genome is related to the complexity of the fruiting bodies, this speculation should be made with caution. In the case of the suborder *Cystobacterinae*, to which the majority of the myxobacterial genomes sequenced belong, this correlation seems to hold. Hence, cells of the genus *Anaeromyxobacter* (around 20 STPKs) hardly aggregate, although they form spore-like structures (Sanford *et al.*, 2002). In contrast, the fruiting bodies of the genera *Myxococcus* and *Corallococcus* (approximately 100 STPKs per genome) are mounds with or without a stalk or branched coral-like structures, while those of *Stigmatella aurantiaca* (with more than 200 STPKs) are very complex, consisting of a stalk with several sporangioles (Stamm *et al.*, 2005). However, although the myxobacteria of the other two suborders have larger kinomes than *S. aurantiaca* (Fig. 7.3), their fruiting bodies do not seem to exhibit a correspondingly higher degree of complexity to account for the enrichment in STPKs in their genomes (Fudou *et al.*, 2002; Iizuka *et al.*, 2003; Schneiker *et al.*, 2007) which suggests that the STPKs may be implicated in other unknown functions. However, since neither the structures of the fruiting bodies of these other myxobacteria nor their lifestyles are well known, it is hard to predict which processes could be regulated by STPKs.

Myxococcus xanthus STPKs

M. xanthus is the model myxobacterium studied in most laboratories. As several strains are used by different groups (DK1622, DZ2, DZF1, etc.), STPKs have been characterized in different strains. However, the only publicly available genome sequence for this species is that of the strain DK1622 (Goldman *et al.*, 2006). Consequently, genomic analyses of STPKs can only be carried out with this strain. However, it should be noted that the STPK sequences examined in other strains are so far identical to those of DK1622, suggesting that it may also be possible to extrapolate the observations made with the sequenced genome to the genomes of the other common laboratory strains.

The genome size of *M. xanthus* DK1622 is 9.14 Mb, and it encodes 99 STPKs. This number is quite similar to that found in the other species of *Myxococcus* (Fig. 7.3). Of the 99 *M. xanthus* STPKs, 88 contained all the conserved residues known to be essential for catalysis, while the other 11 lack one or several of these residues, suggesting that they may be inactive (Pérez *et al.*, 2008). Accordingly, they must be considered pseudokinases, which are known to function as regulatory factors in several cellular processes in both eukaryotes and prokaryotes (Boudeau *et al.*, 2006; Gee *et al.*, 2012).

Most of the *M. xanthus* STPKs (81 of the 88 complete kinases) are classified as RD kinases, and therefore, are predicted to be activated by phosphorylation of the activation segment. Regulation of kinase activity will also undoubtedly be modulated by additional regions found in most of the STPKs. These extra portions may also function as ouput/input domains of the signal-transduction pathways, and they are mainly located at the C-terminus of the proteins. When these regions are analysed in the database of Pfam (Punta *et al.*, 2012), the extra portions of 62 STPKs do not exhibit significant similarities to any of the 13,672 protein families defined in the database (Fig. 7.4), indicating that they are uncharacterized and undefined domains, many of which are unique among myxobacteria. In contrast, the other 37 STPKs exhibit a modular architecture, with one or several domains that are found in the database (Fig. 7.4). The variety of these domains, together with the high number of domains whose function is unknown, is good evidence of the complexity of the signal-transduction pathways that operate in *M. xanthus* (Pérez *et al.*, 2008). Furthermore, some of these architectures reveal that different kinds of cyclic nucleotides (cAMP, cGMP or c-di-GMP) modulate several pathways, and that direct connections between STPKs and other signal-transduction mechanisms (such as TCSs or σ^{ecf}s) are also crucial in some adaptive processes (Fig.

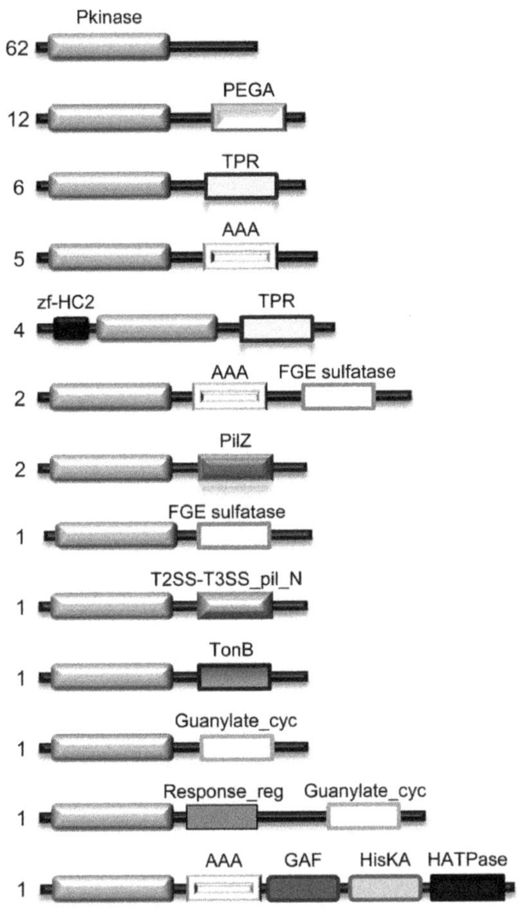

Figure 7.4 Domain organization of the *M. xanthus* STPKs. The size of each domain has not been drawn to scale. The names above each domain correspond to the abbreviations given in the Pfam database, and their corresponding definitions and numbers are as follows: Pkinase, protein kinase domain (PF00069); PEGA, PEGA domain (PF08308); TPR, tetratricopeptide repeat (PF13424); AAA, AAA ATPase domain (PF13192); zf-HC2, putative zinc-finger (PF13490); FGE-sulphatase, formylglycine-generating sulphatase enzyme (PF03781); PilZ, PilZ domain (PF07238); T2SS-T3SS_pil_N, pilus formation protein N terminal region (PF13629); TonB, Gram-negative bacterial TonB protein (PF03544); Guanylate_cyc, adenylate and guanylate cyclase catalytic domain (PF00211); Response_Reg, response regulator receiver domain (PF00072); GAF, GAF domain (PF01590); HisKA, His Kinase A (phospho-acceptor) domain (PF00512); HATPase, His kinase-, DNA gyrase B-, and HSP90-like ATPase (PF02518). The number of STPKs exhibiting the different domain architectures is shown on the left of each drawing.

7.4 shows that several modular STPKs comprise a receiver or a histidine kinase domain, or a Zn-binding motif similar to that found in many anti-σ factors).

Most likely, the enrichment, the diversity, and the complexity of the signal-transduction pathways mediated by STPKs explains why not many of these proteins have been characterized as to their exact mechanism and physiological role, despite of the fact that more than 20 years have passed since the first STPK was found in this bacterium. Nevertheless, a total of 13 STPKs have been studied (Inouye *et al.*, 2008), which have been found to be involved in various cellular processes, such as negative (Pkn2, Pkn5, Pkn8 and Pkn14) or positive (Pkn1, Pkn4, Pkn6, Pkn9, PktA5 and PktB8) regulation of development and motility (MasK). The role of the other two STPKs (SkpA and SkpB) is not well elucidated (Kimura *et al.*, 2009). The best characterized of all these pathways is that of Pkn8 and Pkn14. In this pathway, Pkn8 is able to phosphorylate Pkn14 when nutrients are abundant in the medium. Phosphorylated Pkn14 is then activated to phosphorylate the regulator MrpC, which belongs to the group of one-component systems (and is also regulated by the Esp TCS system as described above). This posttranslational modification is thought to prevent the cleavage of MrpC to MrpC2, the active form of the regulator. As MrpC2 regulates the expression of *fruA*, a key response regulator (one of the components of the TCS pathways) for the progression of the developmental programme, activation of the pathway negatively regulates development (see Chapter 6 for details). Upon starvation, the phosphorylation cascade Pkn8/Pkn14 is not activated. Therefore, MrpC is not phosphorylated and it can be processed to form MrpC2 which positively activates the expression of *fruA* and development can proceed (Inouye *et al.*, 2008).

The remaining characterized STPKs were extensively reviewed in the previous book about myxobacteria (see Inouye *et al.*, 2008) and not many novel data have been published since then. Therefore, specific details about the other STPKs will not be dealt with in this chapter.

The myxobacterial phosphatome

Since phosphorylation at Ser/Thr involves a stable phosphoester bond, PPs are required to dephosphorylate the substrates of the STPKs. Accordingly, regulation of the activity of PPs will also be essential for fine-tune control of the cell response to the stimulus that triggers the signal-transduction pathway where STPKs participate.

PPs are more diverse in sequence and substrate specificity than STPKs. Hence, three major families have been defined (Shi *et al.*, 1998; Shi, 2009): phosphoprotein phosphatases (PPPs), metal-dependent protein phosphatases (PPMs), and phosphotyrosine phosphatases (PTPs). Moreover, some proteins with motifs of members of the PPP family may not function as PPs, but rather as phosphodiesterases, nucleotidases, or nucleases. Hence, the determination of the phosphatome of a genome is more complex and less accurate than the determination of the kinome. To date, the phosphatome has only been analysed in six myxobacterial genomes: the four sequenced *Anaeromyxobacter* strains, *M. xanthus* DK1622, and *S. cellulosum* So ce56 (Inouye *et al.*, 2008; Treuner-Lange, 2010). These analyses have revealed that each *Anaeromyxobacter* strain encodes three or four PPMs, while *M. xanthus* and *S. cellulosum* encode 5 and 39, respectively. As with STPKs, the number of PPMs increases with the genome size. This enrichment is not observed when we search for PTPs in the genomes, because *Anaeromyxobacter* strains encode three or four PTPs, but this number only rises to five and six in *M. xanthus* and *S. cellulosum*, respectively. The *M. xanthus* phosphatome has been analysed by two different groups (Inouye *et al.*, 2008; Treuner-Lange, 2010), and the numbers of PPMs and PTPs obtained for this bacterium are quite similar. However, the numbers of PPPs obtained in both studies differed significantly: 26 (Inouye *et al.*, 2008) versus 4 (Treuner-Lange, 2010). Nevertheless, and as mentioned above, this large discrepancy can be explained by the fact that proteins similar to PPPs have been demonstrated to exhibit different activities. The bioinformatics analyses are certainly not robust enough to determine the phosphatome in a genome, and biochemical characterizations of the proteins will be required to unambiguously classify these proteins as PPs. In the study by Treuner-Lange (2010), the number of PPPs in *Anaeromyxobacter* ranges between four and five, while *S. cellulosum* encodes seven. These data also indicate that there is no relationship between the number of this type of proteins and the genome size. One interesting conclusion that can be drawn is that the number of PPs encoded in a given myxobacterial genome is always much lower than the respective number of STPKs, indicating that each phosphatase must operate in the signal-transduction pathways of several STPKs.

Characterized *Myxococcus xanthus* PPs

PPs have only been characterized in *M. xanthus*. Two of them, Pph1 and Pph3, belong to the PPM family (Treuner-Lange *et al.*, 2001; Kimura *et al.*, 2011a), while Pph2 is a member of the PPP family (García-Hernández *et al.*, 2009). Although the three proteins have been reported to be required for normal aggregation and sporulation, the mechanism by which they modulate development clearly differs. The Δ*pph1* mutant exhibits only a few pili, which are necessary for type 4 pili mediated S-motility (Chapter 10), a low level of methylation of the chemotactic protein FrzCD, and a low reversal frequency (Treuner-Lange *et al.*, 2001). It is therefore thought that these defects in oriented motility prevent cells from developing.

In the case of the Δpph2 mutant, aggregation is considerably delayed; the yield of myxospores is low, and their morphology is aberrant. Furthermore, germination efficiency is reduced. However, these anomalies are related to the contribution of Pph2 to the optimization of the scarce energy resources available during development (García-Hernández *et al.*, 2009).

Regarding Pph3, the protein has not been characterized in detail, and the reasons why development is impaired in the *pph3* mutant remain unknown (Kimura *et al.*, 2011a). However, this protein exhibits a modular architecture, with a cyclic nucleotide monophosphate-binding domain located at the C-terminal region. Moreover, it has been demonstrated that the activity of Pph3 increases in the presence of cAMP (Mori *et al.*, 2012), indicating that the pathway where this

PP operates is modulated by the pool of cyclic nucleotides in the cytoplasm.

Finally, it should be mentioned that no relationship has been so far established between any of the *M. xanthus* STPKs and PPs characterized.

Concluding remarks and future perspectives

Although some signal-transduction mechanisms that have been reported in *M. xanthus* are involved in the adaptation to environmental fluctuations (light or copper concentration, for instance), the abundance and diversity of these types of mechanisms in myxobacteria are closely related to their complex and unique life cycle among the bacteria. To date, most of the pathways characterized have been shown to be involved in motility, which is essential for both predation and development, or directly involved in regulation of aggregation and/or sporulation. Although many factors involved in signal transduction have been studied, our knowledge of all the factors that participate in each pathway is far from complete. Moreover, the genomics data unambiguously show that the potential number of signalling mechanisms encoded in a myxobacterial genome is much larger than the number of characterized pathways. For instance, less than 10% of the proteins involved in signal transduction have been studied in the model myxobacterium *M. xanthus*. Finally, it should be kept in mind that interconnections between signal-transduction mechanisms of different types have been reported in several *M. xanthus* pathways, indicating that signalling mechanisms in myxobacteria function as intricate networks, where many elements are involved. This complexity favours the possibility of regulation at many distinct levels and, consequently, a more precise response. The exact elucidation of the myxobacterial signalling mechanisms will be one of the key future challenges that must be overcome to move forward in our comprehension of the myxobacterial life cycle.

Acknowledgements

This work has been supported by MICINN (BFU2009–07565 and CSD2009–00006 to JMD, and BFU2009-12445-C02-01 to MEA) and the Max Planck Society (to PIH). We thank Stuart Huntley (Max Planck Institute, Marburg) for providing information about STPKs in *M. stipitatus*.

References

Aravind, L., and Ponting, C.P. (1997). The GAF domain: an evolutionary link between diverse phototransducing proteins. Trends Biochem. Sci. *22*, 458–459.

Berleman, J.E., and Kirby, J.R. (2009). Deciphering the hunting strategy of a bacterial wolfpack. FEMS Microbiol. Rev. *33*, 942–957.

Berleman, J.E., Vicente, J.J., Davis, A.E., Jiang, S.Y., Seo, Y.E., and Zusman, D.R. (2011). FrzS regulates social motility in *Myxococcus xanthus* by controlling exopolysaccharide production. PLoS ONE *6*, e23920.

Biondi, E.G., Reisinger, S.J., Skerker, J.M., Arif, M., Perchuk, B.S., Ryan, K.R., and Laub, M.T. (2006). Regulation of the bacterial cell cycle by an integrated genetic circuit. Nature *444*, 899–904.

Boudeau, J., Miranda-Saavedra, D., Barton, G.J., and Alessi, D.R. (2006). Emerging roles of pseudokinases. Trends Cell Biol. *16*, 443–452.

Brooks, B.E., and Buchanan, S.K. (2008). Signaling mechanisms for activation of extracytoplasmic function (ECF) sigma factors. Biochim. Biophys. Acta *1778*, 1930–1945.

Browning, D.F., Whitworth, D.E., and Hodgson, D.A. (2003). Light-induced carotenogenesis in *Myxococcus xanthus*: functional characterization of the ECF sigma factor CarQ and antisigma factor CarR. Mol. Microbiol. *48*, 237–251.

Bourret, R.B. (2010). Receiver domain structure and function in response regulator proteins. Curr. Opin. Microbiol. *13*, 142–149.

Burbulys, D., Trach, K.A., and Hoch, J.A. (1991). Initiation of sporulation in *B. subtilis* is controlled by a multicomponent phosphorelay. Cell *64*, 545–552.

Bush, M., and Dixon, R. (2012). The role of bacterial enhancer binding proteins as specialized activators of sigma54-dependent transcription. Microbiol. Mol. Biol. Rev. *76*, 497–529.

Campbell, E.A., Greenwell, R., Anthony, J.R., Wang, S., Lim, L., Das, K., Sofia, H.J., Donohue, T.J., and Darst, S.A. (2007). A conserved structural module regulates transcriptional responses to diverse stress signals in bacteria. Mol. Cell *27*, 793–805.

Chang, C.H., Zhu, J., and Winans, S.C. (1996). Pleiotropic phenotypes caused by genetic ablation of the receiver module of the *Agrobacterium tumefaciens* VirA protein. J. Bacteriol. *178*, 4710–4716.

Cho, K., and Zusman, D.R. (1999). Sporulation timing in *Myxococcus xanthus* is controlled by the *espAB* locus. Mol. Microbiol. *34*, 714–725.

Durocher, D., Taylor, I.A., Sarbassova, D., Haire, L.F., Westcott, S.L., Jackson, S.P., Smerdon, S.J., and Yaffe, M.B. (2000). The molecular basis of FHA domain:phosphopeptide binding specificity and implications for phospho-dependent signaling mechanisms. Mol. Cell *6*, 1169–1182.

Elías-Arnanz, M., Fontes, M., and Padmanabhan, S. (2008). Carotenogenesis in *Myxococcus xanthus*: a complex regulatory network. In Myxobacteria, Multicellularity and Differentiation, D.E. Whitworth, ed. (Washington, DC: ASM Press), pp. 211–225.

Elías-Arnanz, M., Padmanabhan, S., and Murillo, F.J. (2010). The regulatory action of the myxobacterial CarD/CarG complex: a bacterial enhanceosome? FEMS Microbiol. Rev. *34*, 764–778.

Elías-Arnanz, M., Padmanabhan, S., and Murillo, F.J. (2011). Light-dependent gene regulation in nonphototrophic bacteria. Curr. Opin. Microbiol. *14*, 128–135.

Fontes, M., Galbis-Martínez, L., and Murillo, F.J. (2003). A novel regulatory gene for light-induced carotenoid synthesis in the bacterium *Myxococcus xanthus*. Mol. Microbiol. *47*, 561–571.

Freeman, J.A., and Bassler, B.L. (1999). Sequence and function of LuxU: a two-component phosphorelay protein that regulates quorum sensing in *Vibrio harveyi*. J. Bacteriol. *181*, 899–906.

Fudou, R., Jojima, Y., Iizuka, T., and Yamanaka, S. (2002). *Haliangium ochraceum* gen. nov., sp. nov. and *Haliangium tepidum* sp. nov.: novel moderately halophilic myxobacteria isolated from coastal saline environments. J. Gen. Appl. Microbiol. *48*, 109–116.

Galbis-Martínez, L., Galbis-Martínez, M., Murillo, F.J., and Fontes, M. (2008). An anti-antisigma factor in the response of the bacterium *Myxococcus xanthus* to blue light. Microbiology *154*, 895–904.

Galbis-Martínez, M., Padmanabhan, S., Murillo, F.J., and Elías-Arnanz, M. (2012). CarF mediates signaling by singlet oxygen, generated via photoexcited protoporphyrin IX, in *Myxococcus xanthus* light-induced carotenogenesis. J. Bacteriol. *194*, 1427–1436.

García-Hernández, R., Moraleda-Muñoz, A., Castañeda-García, A., Pérez, J., and Muñoz-Dorado, J. (2009). *Myxococcus xanthus* Pph2 is a manganese-dependent protein phosphatase involved in energy metabolism. J. Biol. Chem. *284*, 28720–28728.

Gee, C.L., Papavinasasundaram, K.G., Blair, S.R., Baer, C.E., Falick, A.M., King, D.S., Griffin, J.E., Venghatakrishnan, H., Zukauskas, A., Wei, J.R., *et al.* (2012). A phosphorylated pseudokinase complex controls cell wall synthesis in mycobacteria. Sci. Signal. *5*, ra7.

Goldman, B.S., Nierman, W.C., Kaiser, D., Slater, S.C., Durkin, A.S., Eisen, J.A., Ronning, C.M., Barbazuk, W.B., Blanchard, M., Field, *et al.* (2006). Evolution of sensory complexity recorded in a myxobacterial genome. Proc. Natl. Acad. Sci. U S A. *103*, 15200–15205.

Gómez-Santos, N., Pérez, J., Sánchez-Sutil, M.C., Moraleda-Muñoz, A., and Muñoz-Dorado, J. (2011). CorE from *Myxococcus xanthus* is a copper-dependent RNA polymerase sigma factor. PLoS Genet. *7*, e1002106.

Gorham, H.C., McGowan, S.J., Robson, P.R., and Hodgson, D.A. (1996). Light-induced carotenogenesis in *Myxococcus xanthus*: light-dependent membrane sequestration of ECF sigma factor CarQ by anti-sigma factor CarR. Mol. Microbiol. *19*, 171–186.

Grangeasse, C., Cozzone, A.J., Deutscher, J., and Mijakovic, I. (2007). Tyrosine phosphorylation: an emerging regulatory device of bacterial physiology. Trends Biochem. Sci. *32*, 86–94.

Gruber, T.M., and Gross, C.A. (2003). Multiple sigma subunits and the partitioning of bacterial transcription space. Annu. Rev. Microbiol. *57*, 441–466.

Hanks, S.K., and Hunter, T. (1995). The eukaryotic protein kinase superfamily: Kinase (catalytic) domain structure and classification. FASEB J. *9*, 576–596.

Hazelbauer, G.L., and Lai, W.C. (2010). Bacterial chemoreceptors: providing enhanced features to two-component signaling. Curr. Opin. Microbiol. *13*, 124–132.

Helmann, J.D. (2002). The extracytoplasmic function (ECF) sigma factors. Adv. Microb. Physiol. *46*, 47–110.

Hengge, R. (2009). Principles of c-di-GMP signalling in bacteria. Nat. Rev. Microbiol. *7*, 263–273.

Henry, J.T., and Crosson, S. (2011). Ligand-binding PAS domains in a genomic, cellular, and structural context. Annu. Rev. Microbiol. *65*, 261–286.

Higgs, P.I., Cho, K., Whitworth, D.E., Evans, L.S., and Zusman, D.R. (2005). Four unusual two-component signal transduction homologs, RedC to RedF, are necessary for timely development in *Myxococcus xanthus*. J. Bacteriol. *187*, 8191–8195.

Higgs, P.I., Jagadeesan, S., Mann, P., and Zusman, D.R. (2008). EspA, an orphan hybrid histidine protein kinase, regulates the timing of expression of key developmental proteins of *Myxococcus xanthus*. J. Bacteriol. *190*, 4416–4426.

Ho, T.D., and Ellermeier, C.D. (2012). Extra cytoplasmic function σ factor activation. Curr. Opin. Microbiol. *15*, 182–188.

Huntley, S., Hamann, N., Wegener-Feldbrugge, S., Treuner-Lange, A., Kube, M., Reinhardt, R., Klages, S., Müller, R., Ronning, C.M., Nierman, W.C., and Sogaard-Andersen, L. (2011). Comparative genomic analysis of fruiting body formation in *Myxococcales*. Mol. Biol. Evol. *28*, 1083–1097.

Iizuka, T., Jojima, Y., Fudou, R., Hiraishi, A., Ahn, J.W., and Yamanaka, S. (2003). *Plesiocystis pacifica* gen. nov., sp. nov., a marine myxobacterium that contains dihydrogenated menaquinone, isolated from the Pacific coasts of Japan. Int. J. Syst. Evol. Microbiol. *53*, 189–195.

Inclan, Y.F., Laurent, S., and Zusman, D.R. (2008). The receiver domain of FrzE, a CheA-CheY fusion protein, regulates the CheA histidine kinase activity and downstream signalling to the A- and S-motility systems of *Myxococcus xanthus*. Mol. Microbiol. *68*, 1328–1339.

Inouye, S., Jain, R., Ueki, T., Nariya, H., Xu, C.Y., Hsu, M.Y., Fernandez-Luque, B.A., Muñoz-Dorado, J., Fárez-Vidal, E., and Inouye, M. (2000). A large family of eukaryotic-like protein Ser/Thr kinases of *Myxococcus xanthus*, a developmental bacterium. Microb. Comp. Genomics 5, 103–120.

Inouye, S., Nariya, H., and Muñoz-Dorado, J. (2008). Protein Ser/Thr kinases and phosphatases in *Myxcoccus xanthus*. In Myxobacteria, Multicellularity and Differentiation, D.E. Whitworth, ed. (Washington, DC: ASM Press), pp. 181–210.

Jagadeesan, S., Mann, P., Schink, C.W., and Higgs, P.I. (2009). A novel 'four-component' two-component signal transduction mechanism regulates developmental progression in *Myxococcus xanthus*. J. Biol. Chem. *284*, 21435–21445.

Johnson, L.N., and Lewis, R.J. (2001). Structural basis for control by phosphorylation. Chem. Rev. *101*, 2209–2242.

Johnson, L.N., Noble, M.E.M., and Owen, D.J. (1996). Active and inactive protein kinases: Structural basis for regulation. Cell 85,149–158.

Kaiser, D. (2004). Signaling in myxobacteria. Annu. Rev. Microbiol. 58, 75–98.

Kimura, Y., Kakernizu, A., Matsubara, Y., and Takegawa, K. (2009). Enzymatic characteristics of a Ser/Thr protein kinase, SpkA, from *Myxococcus xanthus*. J. Biosci. Bioeng. *107*, 10–15.

Kimura, Y., Mori, Y., Ina, Y., and Takegawa, K. (2011a). Enzymatic and functional analysis of a protein phosphatase, Pph3, from *Myxococcus xanthus*. J. Bacteriol. *193*, 2657–2661.

Kimura, Y., Yamashita, S., Mori, Y., Kitajima, Y., and Takegawa, K. (2011b). A *Myxococcus xanthus* bacterial tyrosine kinase, BtkA, is required for the formation of mature spores. J. Bacteriol. *193*, 5853–5857.

Kirby, J.R., and Zusman, D.R. (2003). Chemosensory regulation of developmental gene expression in *Myxococcus xanthus*. Proc. Natl. Acad. Sci. U.S.A. *100*, 2008–2013.

Kravanja, M., Engelmann, R., Dossonnet, V., Blüggel, M., Meyer, H.E., Frank, R., Galinier, A., Deutscher, J., Schnell, N., and Hengstenberg, W. (1999). The *hprK* gene of *Enterococcus faecalis* encodes a novel bifunctional enzyme: the HPr kinase/phosphatase. Mol. Microbiol. *31*, 59–66.

Lane, W.J., and Darst, S.A. (2006). The structural basis for promoter −35 element recognition by the group IV sigma factors. PLoS Biol. 4, e269.

LaPorte, D.C., and Chung, T. (1985). A single gene codes for the kinase and phosphatase which regulate isocitrate dehydrogenase. J. Biol. Chem. *260*, 15291–15297.

Lee, B., Higgs, P.I., Zusman, D.R., and Cho, K. (2005). EspC is involved in controlling the timing of development in *Myxococcus xanthus*. J. Bacteriol. *187*, 5029–5031.

Lee, B., Mann, P., Grover, V., Treuner-Lange, A., Kahnt, J., and Higgs, P.I. (2011). The *Myxococcus xanthus* spore cuticula protein C is a fragment of FibA, an extracellular metalloprotease produced exclusively in aggregated cells. PLoS ONE *6*, e28968.

Lee, B., Holkenbrink, C., Treuner-Lange, A., and Higgs, P.I. (2012). *Myxococcus xanthus* developmental cell fate production: heterogeneous accumulation of developmental regulatory proteins and reexamination of the role of MazF in developmental lysis. J. Bacteriol. *194*, 3058–3068.

León, E., Navarro-Avilés, G., Santiveri, C.M., Flores-Flores, C., Rico, M., González, C., Murillo, F.J., Elías-Arnanz, M., Jiménez, M.A., and Padmanabhan, S. (2010). A bacterial antirepressor with SH3 domain topology mimics operator DNA in sequestering the repressor DNA recognition helix. Nucleic Acids Res. *38*, 5226–5241.

Lonetto, M.A., Brown, K.L., Rudd, K.E., and Buttner, M.J. (1994). Analysis of the *Streptomyces coelicolor sigE* gene reveals the existence of a subfamily of eubacterial RNA polymerase sigma factors involved in the regulation of extracytoplasmic functions. Proc. Natl. Acad. Sci. U.S.A. *91*, 7573–7577.

Madhusudan, P., Akamine, N., Xuong, H., and Taylor, S.S. (2002). Crystal structure of a transition state mimic of the catalytic subunit of cAMP-dependent protein kinase. Nat. Struct. Biol. *9*, 273–277.

Martínez-Argudo, I., Ruiz-Vázquez, R.M., and Murillo, F.J. (1998). The structure of an ECF-sigma-dependent, light-inducible promoter from the bacterium *Myxococcus xanthus*. Mol. Microbiol. *30*, 883–893.

Moraleda-Muñoz, A., Pérez, J., Fontes, M., Murillo, F.J., and Muñoz-Dorado, J. (2005). Copper induction of carotenoid synthesis in the bacterium *Myxococcus xanthus*. Mol. Microbiol. *56*, 1159–1168.

Mori, Y., Takegawa, K., and Kimura, Y. (2012). Function analysis of conserved amino acid residues in a Mn^{2+}-dependent protein phosphatase, Pph3, from *Myxococcus xanthus*. J. Biochem. *15*, 269–274.

Muñoz-Dorado, J., Inouye, S., and Inouye, M. (1991). A gene encoding a protein serine/threonine kinase is required for normal development of *M. xanthus*, a Gram-negative bacterium. Cell *67*, 995–1006.

Muñoz-Dorado, J., Gómez-Santos, N., and Pérez, J. (2012). A novel mechanism of bacterial adaptation mediated by copper-dependent RNA polymerase σ factors. Transcription 3, 63–67.

Najafi, S.M., Willis, A.C., and Yudkin, M.D. (1995). Site of phosphorylation of SpoIIAA, the anti-anti-sigma factor for sporulation-specific sigma F of *Bacillus subtilis*. J. Bacteriol. *177*, 2912–2913.

Nariya, H., and Inouye, S. (2005). Identification of a protein Ser/Thr kinase cascade that regulates essential transcriptional activators in *Myxococcus xanthus* development. Mol. Microbiol. *58*, 367–379.

Nikolskaya, A.N., Mulkidjanian, A.Y., Beech, I.B., and Galperin, M.Y. (2003). MASE1 and MASE2: two novel integral membrane sensory domains. J. Mol. Microbiol. Biotechnol. *5*, 11–16.

Ortiz-Lombardia, M., Pompeo, F., Boitel, B., and Alzari, P.M. (2003). Crystal structure of the catalytic domain of the PknB serine/threonine kinase from *Mycobacterium tuberculosis*. J. Biol. Chem. *278*, 13094–13100.

Österberg, S., del Peso-Santos, T., and Shingler, V. (2011). Regulation of alternative sigma factor use. Annu. Rev. Microbiol. *65*, 37–55.

Overgaard, M., Wegener-Feldbrugge, S., and Sogaard-Andersen, L. (2006). The orphan response regulator DigR is required for synthesis of extracellular matrix fibrils in *Myxococcus xanthus*. J. Bacteriol. *188*, 4384–4394.

Paul, R., Jaeger, T., Abel, S., Wiederkehr, I., Folcher, M., Biondi, E.G., Laub, M.T., and Jenal, U. (2008). Allosteric regulation of histidine kinases by their cognate response regulator determines cell fate. Cell *133*, 452–461.

Pereira, S.F., Goss, L., and Dworkin, J. (2011). Eukaryote-like serine/threonine kinases and phosphatases in bacteria. Microbiol. Mol. Biol. Rev. *75*, 192–212.

Pérez, J., Castañeda-García, A., Jenke-Kodama, H., Müller, R., and Muñoz-Dorado, J. (2008). Eukaryotic-like protein kinases in the prokaryotes and the myxobacterial kinome. Proc. Natl. Acad. Sci. U.S.A. *105*, 15950–15955.

Petters, T., Zhang, X., Nesper, J., Treuner-Lange, A., Gómez-Santos, N., Hoppert, M., Jenal, U., and Sogaard-Andersen, L. (2012). The orphan histidine protein kinase SgmT is a c-di-GMP receptor and regulates composition of the extracellular matrix together with the orphan DNA binding response regulator DigR in *Myxococcus xanthus*. Mol. Microbiol. *84*, 147–165.

Punta, M., Coggill, P.C., Eberhardt, R.Y., Mistry, Tate, J., J., Boursnell, C., Pang, N., Forslund, K., Ceric, G., Clements, J., Heger, A., Holm, L., Sonnhammer, E.L.L., Eddy, S.R., Bateman, A., and Finn, R.D. (2012). The Pfam protein families database. Nucleic Acids Res. Database Issue 40, D290-D301.

Rasmussen, A.A., Porter, S.L., Armitage, J.P., and Sogaard-Andersen, L. (2005). Coupling of multicellular morphogenesis and cellular differentiation by an unusual hybrid histidine protein kinase in *Myxococcus xanthus*. Mol. Microbiol. *56*, 1358–1372.

Rasmussen, A.A., Wegener-Feldbrugge, S., Porter, S.L., Armitage, J.P., and Sogaard-Andersen, L. (2006). Four signalling domains in the hybrid histidine protein kinase RodK of *Myxococcus xanthus* are required for activity. Mol. Microbiol. *60*, 525–534.

Sanford, R.A., Cole, J.R., and Tiedje, J.M. (2002). Characterization and description of *Anaeromyxobacter dehalogenans* gen. nov., sp. nov., an aryl-halorespiring facultative anaerobic myxobacterium. Appl. Environ. Microbiol. *68*, 893–900.

Schneiker, S., Perlova, O., Kaiser, O., Gerth, K., Alici, A., Altmeyer, M.O., Bartels, D., Bekel, T., Beyer, S., Bode, *et al.* (2007). Complete genome sequence of the myxobacterium *Sorangium cellulosum*. Nat. Biotechnol. *25*, 1281–1289.

Schramm, A., Lee, B., and Higgs, P.I. (2012). Intra- and inter-protein phosphorylation between two hybrid histidine kinases controls *Myxococcus xanthus* developmental progression. J. Biol. Chem. *287*, 25060–25072.

Shi, Y. (2009). Serine/threonine phosphatases: mechanism through structure. Cell *139*, 468–484.

Shi, L., Potts, M., and Kennelly, P.J. (1998). The serine, threonine, and/or tyrosine-specific protein kinases and protein phosphatases of prokaryotic organisms: a family portrait. FEMS Microbiol. Rev. *22*, 229–253.

Shi, X., Wegener-Feldbrugge, S., Huntley, S., Hamann, N., Hedderich, R., and Sogaard-Andersen, L. (2008). Bioinformatics and experimental analysis of proteins of two-component systems in *Myxococcus xanthus*. J. Bacteriol. *190*, 613–624.

Sourjik, V., and Schmitt, R. (1998). Phosphotransfer between CheA, CheY1, and CheY2 in the chemotaxis signal transduction chain of *Rhizobium meliloti*. Biochemistry *37*, 2327–2335.

Stamm, I., Lottspeich, F., and Plaga, W. (2005). The pyruvate kinase of *Stigmatella aurantiaca* is an indole binding protein and essential for development. Mol. Microbiol. *56*, 1386–1395.

Staroń, A., Sofia, H.J., Dietrich, S., Ulrich, L.E., Liesegang, H., and Mascher, T. (2009). The third pillar of bacterial signal transduction: classification of the extracytoplasmic function (ECF) sigma factor protein family. Mol. Microbiol. *74*, 557–581.

Stein, E.A., Cho, K., Higgs, P.I., and Zusman, D.R. (2006). Two Ser/Thr protein kinases essential for efficient aggregation and spore morphogenesis in *Myxococcus xanthus*. Mol. Microbiol. *60*, 1414–1431.

Stewart, R.C. (2010). Protein histidine kinases: assembly of active sites and their regulation in signaling pathways. Curr. Opin. Microbiol. *13*, 133–141.

Stock, A.M., Robinson, V.L., and Goudreau, P.N. (2000). Two-component signal transduction. Annu. Rev. Biochem. *69*, 183–215.

Stock, J., and Surette, M. (1994). Bacterial chemotaxis. The motor connection. Curr. Biol. *4*, 143–144.

Sun, H., and Shi, W. (2001a). Analyses of *mrp* genes during *Myxococcus xanthus* development. J. Bacteriol. *183*, 6733–6739.

Sun, H., and Shi, W. (2001b). Genetic studies of *mrp*, a locus essential for cellular aggregation and sporulation of *Myxococcus xanthus*. J. Bacteriol. *183*, 4786–4795.

Treuner-Lange, A. (2010). The phosphatomes of the multicellular myxobacteria *Myxococcus xanthus* and *Sorangium cellulosum* in comparison with other prokaryotic genomes. PLoS ONE *5*, e11164.

Treuner-Lange, A., Ward, M.J., and Zusman, D.R. (2001). Pph1 from *Myxococcus xanthus* is a protein phosphatase involved in vegetative growth and development. Mol. Microbiol. *40*, 126–140.

Udo, H., Muñoz-Dorado, J., Inouye, M., and Inouye, S. (1995). *Myxococcus xanthus*, a Gram-negative bacterium, contains a transmembrane protein serine/threonine kinase that blocks the secretion of beta-lactamase by phosphorylation. Genes Dev. *15*, 972–983.

Ueki, T., and Inouye, S. (2003). Identification of an activator protein required for the induction of *fruA*, a gene essential for fruiting body development in *Myxococcus xanthus*. Proc. Natl. Acad. Sci. U.S.A. *100*, 8782–8787.

Ulrich, L.E., Koonin, E.V., and Zhulin, I.B. (2005). One-component systems dominate signal transduction in prokaryotes. Trends Microbiol. *13*, 52–56.

Ulrich, L.E., and Zhulin, I.B. (2010). The MiST2 database: a comprehensive genomics resource on microbial signal transduction. Nucleic Acids Res. *38*, D401–407.

Ward, M.J., Lew, H., Treuner-Lange, A., and Zusman, D.R. (1998). Regulation of motility behavior in *Myxococcus xanthus* may require an extracytoplasmic-function sigma factor. J. Bacteriol. *180*, 5668–5675.

Wegener-Feldbrugge, S., and Sogaard-Andersen, L. (2009). The atypical hybrid histidine protein kinase RodK in *Myxococcus xanthus*: spatial proximity supersedes kinetic preference in phosphotransfer reactions. J. Bacteriol. *191*, 1765–1776.

West, A.H., and Stock, A.M. (2001). Histidine kinases and response regulator proteins in two-component signaling systems. Trends Biochem. Sci. *26*, 369–376.

Whitworth, D.E. (2008). Myxobacteria: Multicellularity and Differentiation (Washington, DC: ASM Press).

Whitworth, D.E., and Cock, P.J.A. (2008a). Two-component systems of the myxobacteria: structure, diversity and evolutionary relationships. Microbiology *154*, 360–372.

Whitworth, D.E., and Cock, P.J.A. (2008b). Two-component signal transduction systems of the myxobacteria. In Myxobacteria, Multicellularity and Differentiation, D.E. Whitworth, ed. (Washington, DC: ASM Press), pp. 169–189.

Willett, J.W., and Kirby, J.R. (2011). CrdS and CrdA comprise a two-component system that is cooperatively regulated by the Che3 chemosensory system in *Myxococcus xanthus*. mBio 2, e00110–11.

Wuichet, K., and Zhulin, I.B. (2010). Origins and diversification of a complex signal transduction system in prokaryotes. Sci. Signal. *3*, ra50.

Young, T.A., Delagoutte, B., Endrizzi, J.A., Falick, A.M., and Alber, T. (2003). Structure of *Mycobacterium tuberculosis* PknB supports a universal activation mechanism for Ser/Thr protein kinases. Nat. Struct. Biol. *10*, 168–174.

Zhang, W., Muñoz-Dorado, J., Inouye, M., and Inouye, S. (1992). Identification of a putative eukaryotic-like protein kinase family in the developmental bacterium *Myxococcus xanthus*. J. Bacteriol. *174*, 5450–5453.

Computational Biology: From Observation to Statistical Image Analysis to Modelling and Back to Biology

8

Cameron W. Harvey, Oleg A. Igoshin, Roy D. Welch, Mark Alber and Lawrence J. Shimkets

Abstract

Emergence refers to the manner in which complex behaviours self-organize from a multitude of relatively simple events. *Myxococcus xanthus* is one of the premier organisms for studying emergence as genetic and biochemical approaches can be wielded in equal measure with computational approaches. Computational modelling has been successfully used to study a range of topics related to myxobacterial self-organization. By reviewing a number of modelling studies from the myxobacteria literature, this chapter aims to demonstrate the cycle of computational modelling. The path from experimental observations to modelling and predictions is traced for three emergent behaviours: swarming, formation of ripples, and fruiting body development. A number of important questions regarding emergent behaviour can potentially benefit from computational modelling including the motility mechanisms, signalling during development and biochemical pathways driving direction reversals. Computational models can provide critical insight and predictive power, and are becoming important tools in the study of bacterial behaviour.

Introduction

Emergent behaviour refers to the way complex patterns arise from relatively simple events that cannot be explained by the properties of individual components. Examples of emergence in biological systems include multicellular development or swarming behaviour by flocking birds. Emergent systems lack higher order thought processes that enable the complex choreography. Rather, the behaviour emerges in an iterative fashion from effects that become causes until they course through the population. For example, mathematical models showed that flocking behaviour may arise from three simple rules: avoid crowding (separation), steer towards the average heading of your neighbours (alignment), and steer towards the average position of your neighbours (centring). While separation, alignment and centring can explain some types of flocking behaviour, they cannot explain all types of emergence. Nor do they provide a genetic or biochemical basis for the behaviour. Model systems are being used to explore different types of emergence at a deeper and more physiological level.

The myxobacteria are one of the simplest organisms where genetic, biochemical and behavioural approaches can be used to examine emergent behaviour. Emergence is evident in swarming as well as two more complex behaviours, rippling and fruiting body development. A thorough study of *Myxococcus xanthus* self-organization requires an iterative approach, combining information about gene and biochemical pathways with computational modelling guided by image analysis. *Myxococcus xanthus* has an excellent genetic system and enough phylogenetic proximity to *Escherichia coli* to benefit from detailed biochemical and genetic studies of the world's most thoroughly studied organism.

Predictive computer simulations, or in silico experiments, are widely used for obtaining additional insight into different biological problems and are now being used to study emergent behaviour in *M. xanthus*. With increased computational power and introduction of new parallelization algorithms for running simulations on large computer clusters, the latest models can account for large numbers of biologically relevant details. Simulations can be run in a realistic time for a large range of parameters to account for known properties of the system while producing new hypotheses that can be experimentally tested.

In this chapter we demonstrate the path from observation to modelling and predictive simulations for three processes: swarming, rippling, and fruiting body development. We also review several models to demonstrate different modelling approaches and their importance in describing possible mechanisms of *M. xanthus* emergent behaviour. While modelling can never completely capture or explain a biological system, it can be extremely useful in guiding the traditional experimental approach. Computer models can explore a wide range of possible variations in the system and predict specific aspects that are most critical for laboratory experiments. Furthermore, a computational modeler does not typically have traditional biological training and often explores possible scenarios and explanations that biologists would not consider leading to fruitful collaborations and new insights.

Computational modelling has been successfully used to study a range of topics including travelling waves (Igoshin *et al.*, 2001, 2004a; Börner *et al.*, 2002; Zhang *et al.*, 2012), aggregation and fruiting body formation (Sozinova *et al.*, 2006; Starruß *et al.*, 2007; Holmes *et al.*, 2009; Harvey *et al.*, 2012), coordinated cell motility in swarms (Wu *et al.*, 2007, 2009; Janulevicius *et al.*, 2010), and chemical pathway networks regulating motility (Igoshin *et al.*, 2004b). Computational modelling of a biological system begins with an in-depth examination of the experimental data. Then several main assumptions and simplifications about the behaviour of a biological system are used to form a biological model. This model is implemented on computer clusters and a variety of simulations explore the emergent behaviour of the model. Ideally, this is followed by comparison with experimental data for model verification and calibration, which can suggest additional experiments to test model predictions.

An early step in modelling involves determining what scale the model will capture. The model can look at the molecular scale and try to capture the kinetics of different processes within a single cell or focus on the multicellular scale to describe the behaviour of interacting cells. It can also focus on a meso-scale that tries to capture population wide effects. Multi-scale models were developed to combine features at two or more of these scales (Newman *et al.*, 2008; Du *et al.*, 2011; and Xu *et al.*, 2012). After determining the scale, the modeler makes a series of assumptions that simplify the system so that a computational representation can be made. This representation can be the concentrations of different proteins, use of spherical or rod-shaped particles for cells moving on a surface, or density-fields for a population of cells. The model then uses a set of rules that prescribe how the system evolves during a simulation. These rules can be based on well-defined experimental data or on a best guess scenario; a protein becomes methylated or binds another protein at a given rate, a cell moves in the direction of its long axis and secretes a chemical at a defined rate, or the density of cells is updated by reaction–diffusion equations with prescribed diffusion and reaction rates. Finally, in silico experiments can vary different parameters within the model (kinetic rates, movement behaviours, etc.) to identify factors that impact the emerging phenomenon being studied. Ideally, the in silico experiments result in predictions that can be tested experimentally.

The next section provides a discussion on modelling methods. We also included a Glossary at the end of the chapter that provides definitions of technical terms which are underlined when they first appear in the text.

General introduction to mathematical and computational modelling

In order to introduce the concepts of computational modelling to readers who may be less familiar with the computational techniques, this section will present an overview of methods that are applied to specific biological questions.

A mathematical model is, according to Parker (2003) 'a mathematical representation of a process, device, or concept by means of a number of variables which are defined to represent the inputs, outputs, and internal states of the device or process, and a set of equations and inequalities describing the interaction of these variables'. Equations describing relationships between variables and their rates of change can be continuous or discrete. Continuous processes vary smoothly and can be modelled with calculus such as differentiation and integration. Discrete processes have distinct, separated values that do not vary smoothly and rely on completely different mathematical approaches (discrete mathematics). A process can be as simple as a chemical reaction or as complicated as the immune system's response to an infectious disease (Thakar *et al.*, 2010), blood clot formation (Riddel *et al.*, 2007) or an ecological system (Austin, 2007). The variables of the model can be the concentration of chemicals, number of species, size of a tumour, or any other dynamic feature that can be quantified. The equations that describe the dynamics of a system may have known solutions that can be studied to determine the behaviour of the system, for example arriving at equilibrium. Most complex systems will not have a known solution and must be solved using numerical methods with computers. In essence, computational modelling involves using computers to produce and analyse a mathematical model that describes the process of interest.

Computational models can involve continuous and discrete representations of components in the system, and be deterministic or stochastic in nature depending on the available experimental data and specific biological problem. The following will introduce the reader to these basic distinctions in computational modelling.

Continuous versus discrete models

One basic distinction in a computational model is whether the model is continuous (a continuum model) or discrete. A *continuum model* represents the dynamics of a process such as changes in the density of a population of cells or the local concentrations of chemicals using differential equations (DEs) (either ordinary (ODEs) or partial (PDEs). For example, the concentration ($C(t)$) may follow the reaction equation:

$$\frac{d}{dt}c(t) = kc(t)$$

and have the solution:

$$c(t) = c_0 e^{-kt}$$

when allowed to take on any value between zero and some maximum value. While solutions of ODEs represent homogeneous distributions of chemicals (or cells in models of population kinetics), non-homogenous conditions may be characteristic of the system and require a spatial component. This type of biological process would use PDEs for the model. One common approach in biological modelling is to use reaction–diffusion equations to model chemicals or cells diffusing in space (Holmes *et al.*, 1994).

Continuous models, particularly when applied to cell distribution patterns, can predict macroscopic properties of the cell population. However, they 'often overlook or oversimplify microscopic dynamics and interactions of cells' as mentioned in (Wu *et al.*, 2011). *Discrete models* have variables such as position in space and time that can take on non-continuous values, e.g. integers only, and can be used to represent individual cells rather than cell densities. Discrete systems can move from one state to another at fixed intervals. More specifically, in a continuous model of a cell population in some defined region, the equations would describe the dynamics of the cell concentration c(**x**,t), which averages behaviour of cells in some local neighbourhood of space position **x** at time t (**x** can be a vector in 1D, 2D or 3D space). In a discrete model, cells can be represented individually as points or clusters of points on a grid or *lattice*. The discrete size of a cell would be the number of grid locations (or lattice sites) occupied by a cell. Individual cell representation is the basic set-up of a cellular automata (CA) described by John Von Neumann (Neumann, 1966) and at the heart of Conway's 'game of life' (Gardner, 1970). A set of rules describes the behaviour of cells on the lattice, prescribing how they move, survive or die. Computational models that adopt this framework are known as lattice models [popular instances include the Ising model (Chandler, 1987) in physics or the predator–prey model in ecology (Pekalski, 2004)]. The space or domain of a lattice model can be one-dimensional (1D), two-dimensional (2D), or three-dimensional (3D).

One popular and biologically relevant example of the lattice-based cellular automata is the *cellular Potts model* (CPM) (Graner and Glazier, 1992). In this case, a cell is not confined to a single lattice site, but has a particular size and contains a collection of lattice sites that evolve through system energy manipulation. Multiple types of cells, e.g. normal and cancerous cells, can be modelled within the grid and each lattice site is given an array of integer numbers each corresponding to one of the types of cells. Lattice-sites can change from one state to another. During simulations, biologically justified local rules that describe the interactions between the different types of lattice sites result in the fluctuation of cell shape, cell movement, cell–cell adhesion and cell division.

An off-lattice model allows individual nodes connected with each other to represent a cell. The cell moves along continuous curves in 2D and 3D space instead of jumping from one node of a lattice to another. There is room for confusion since the discrete cells are allowed to move in a continuous space. Models that use discrete cells and have a set of rules that determine their shape and behaviour are often referred to as *cell-based models*. An equivalent term is *agent-based model* (ABM) or individual-based model (IBM) if the objects in the model are not cells, but rather, people, animals, insects, or particles. Macal and North (2010) provide a nice introduction to the growing field of ABM. Generally speaking, all cell-based models are discrete models, but not all discrete models are cell-based. For example, a population model that uses the logistic map (a discrete version of the logistic equation) to specify the change in population requires integer values, and is discrete but not cell-based.

Another type of discrete modelling approach that has emerged in recent years is known as subcellular elements (Newman, 2005; Harvey *et al.*, 2011). A *subcellular element model* uses the technique of partitioning a cell into interacting subregions (elements) whose positions are tracked during a simulation. This can be used to model localized processes such as adhesion or signalling within a cell in order to enhance the biological details incorporated into the model.

Deterministic versus stochastic models

An important distinction within computational models is whether the model is deterministic or stochastic. If the outcome of a model simulation is fully determined by the parameters and input values, then the model is a *deterministic model*. ODE and PDE models are deterministic since the equations always yield the same results. Studies using deterministic models might perform sensitivity analysis on specific parameters to better understand the behaviour of the system.

When a model attempts to incorporate unknown or partially understood aspects of a process through the use of probability, a different modelling approach is required. A *stochastic model* uses random selection from a set of possible outcomes to determine how the process evolves. One common approach in stochastic models is to use a Monte Carlo method that repeatedly samples from a distribution of outcomes in order to determine a particular outcome of an event. For a given set of parameters and inputs, the outcome of the simulation may vary with each iteration. Stochastic model simulations are run repeatedly and the results are averaged to provide information about the process to be compared with experimental data.

A complete review of all modelling approaches is beyond the scope of this chapter. Rather, several recent examples demonstrating the cycle of biological modelling will be described. Most computational researchers are trained in application of a particular modelling approach for a particular biological system. This is why providing a review of different approaches applied to the same biological system is important in training a new generation of computational biologists.

Biochemical control of reversals in *Myxococcus xanthus*

The three types of social behaviour examined in this chapter, swarming, rippling, and fruiting body development, involve different types of cell behaviour. Individual *M. xanthus* cells reverse their gliding direction through a coordinated reversal in the polarity of the two types of motility engines (Leonardy *et al.*, 2007; Kaiser, 2008). Cells simply stop forward motion and resume motion in the opposite direction along the long axis of the cell. The Frz signal transduction pathway controls the frequency of the reversals (Kaiser *et al.*, 2005; Y. Zhang *et al.*, 2012). Frz is a traditional chemosensory pathway except that the putative chemoreceptor, FrzCD, is cytoplasmic rather than an integral membrane protein (Mauriello *et al.*, 2009). The precise signal that FrzCD responds to, if it is a chemoreceptor, remains unknown. The hallmark of chemosensory pathways is regulation of protein function by covalent modification, in this case both phosphorylation and methylation. Wild type cells typically reverse their direction of movement on average every 7 min (Blackhardt and Zusman, 1985). A C-signal producing (*csgA*) mutant exhibited differences in gliding speed and reversal frequency that were corrected following addition of a MalE-CsgA fusion protein (Jelsback and Sogaard-Andersen, 1999). As C-signal stimulation of motility depended on the Frz signal transduction system, the two became linked in a modelling effort. While many of the models in the rest of the chapter simulate the spatial position or density of cells, the model presented in this section pertains to the protein pathway of a single cell and simulates the dynamic concentrations of proteins in either an inactive or active state.

Igoshin *et al.*, (2004b) introduced a model that is continuous and deterministic to study how the Frz pathway can regulate directional reversals in *M. xanthus*. The study showed that negative feedback in the signal transduction cascade containing the Frz proteins could cause the active form of the proteins to oscillate between high and low levels. Because the active (phosphorylated) form of FrzE induces reversals (Ward and Zusman, 1997), this biochemical oscillator could explain the periodic directional reversals exhibited by individual cells. The work demonstrates how the pathway can become an oscillator when feedback is introduced into the cascade. In order to model the frizilator, enzyme-controlled reaction equations of the Michaelis–Menten kinetics are used. Parameter values were chosen to account for the activation-deactivation of FrzF, the methylation–demethylation of FrzCD, and the phosphorylation-dephosphorylation of FrzE. Negative feedback was introduced by having FrzE-P deactivate FrzF. The resulting oscillations in FrzCD-Me and FrzE-P are shown in Fig. 8.1.

Alternative feedback loops were also discussed in the paper including FrzE acting to promote FrzG or FrzE acting on FruA. The authors elected to use the scheme with FrzF for the following reasons. The FruA scheme would not explain reversals by cells that were not undergoing development. The authors excluded FrzG since deletion mutants of FrzG did not create non-reversing phenotypes similar to those observed by deletion of FrzF, FrzE, or FrzCD. It is proposed that C-signal induced activation of FrzF would lead to a shift in reversal frequency. This can explain how contact between cells producing the C-signal induces cells to reverse ahead of the natural period established by the biochemical cycle ('clock'). Additionally, the model assumes the presence of a refractory period in the cycle when cells do not respond to C-signal. The model proposed that FrzE-P deactivates FrzF, which is the key event that makes the system oscillate. A comparable reaction has not been yet detected in any known chemosensory system nor has this reaction been verified in the Frz system. This is an important prediction to be examined experimentally since the frizilator has been invoked to explain aspects of swarming and rippling.

Swarm formation

Many bacteria are able to spread rapidly over the surface using a self-organizational strategy called swarming. *Myxococcus* has two surface motility systems known as A and S. A-motility is described in detail in Chapter 9. S-motility is described in Chapter 10. Sensory regulation and coordination of the two motility systems is described in Chapter 11. Swarm behaviour sometimes requires groups of cells to engage in synchronized directional changes and oriented movement. On a nutrient-rich surface, *M. xanthus* cells expand outward from the point of inoculation as a swarm with the

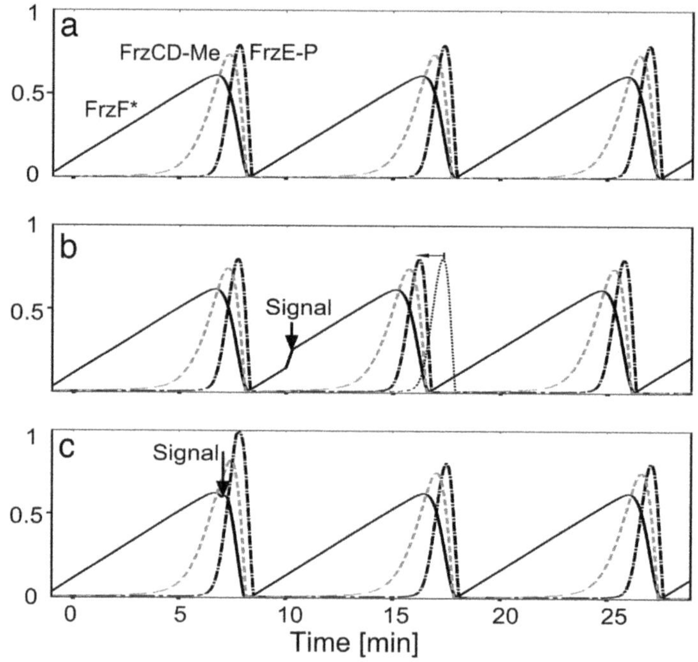

Figure 8.1 (Figure 2 from Igoshin *et al.*, 2004b. Proc. Natl. Acad. Sci. U.S.A. *101*, 15760–15765. Copyright (2004) National Academy of Sciences, U.S.A.). Simulations of protein activation in the Frz pathway. The plots show the concentration of activated FrzF (FrzF*), methylated FrzCD (FrzCD-Me), and phosphorylated FrzE (FrzE-P) over time. An initial increase in activated FrzF* is followed by increases in FrzCD-Me and FrzE-P. The negative feedback by FrzE-P increases the deactivation of FrzF* and concentrations reduce until the cycle repeats. (a) A 10-minute cycle without signalling. (b) A pulse in FrzF activation due to C-signalling can shift the cycle to a shorter reversal period. (c) The pulse applied at a different time in the phase of the cycle does not shift the period, consistent with the idea of a refractory period.

edge of a swarm containing patterns that depend on different motility systems. Smooth undulating edges are observed for A^-S^+ mutants and a mix of peninsular protrusions called flairs, clusters of cells, and isolated cells for A^+S^+ and A^+S^- strains (Spormann, 1999; Pelling *et al.*, 2006; Wu *et al.*, 2007; Kaiser and Warrick, 2011). Over a period of hours, these flares merge when they make contact, thereby increasing the area covered by a swarm. Single cells can move forward or backwards, and reverse their direction of movement. Directional reversals require reorientation of many proteins associated with the two motors to transform the lagging pole into the leading pole (Leonardy *et al.*, 2007; Leonardy *et al.*, 2010). The methods by which myxobacteria synchronize their behaviour are largely unknown and different hypotheses have been studied by modelling. The following sections contain examples demonstrating how a different focus leads to a specific modelling approach. While certain models can be similar in many ways, the simulations performed often explore very different questions and use different analyses to make sense of the simulations.

Myxococcus xanthus cell reversal is optimized for higher expansion rate

A two-dimensional cell-based model was developed by Wu *et al.* (2007) to study the impact of cell aspect ratio (the ratio of the length to the width) and the frequency of reversals on the collective behaviour. As a cell-based model, this is an example of a discrete model, but it is also a stochastic model since the set of rules that describes a cell's movement includes a random component. This model represents each cell as three connected nodes that move in a 2-dimensional domain. The movement of each cell is tracked through the simulation and the collective behaviour of spreading is connected to experimentally measured expansion rates. The algorithms for cell movement are designed based upon observed cell behaviour.

Cells in the model are self-propelled, follow slime trails, align with neighbouring cells, and regularly reverse their direction. Simulations with wild type and motility mutants emulated the expansion rates of colonies over a range of densities. This study also compared simulations using only self-propelled motility with simulations of self-propelled motility in conjunction with a model for S-motility in which cells attempt to align with their neighbours. Using a measurement of alignment between cells at a given distance – i.e. a global order parameter – the comparison showed that a population of cells using social interactions between neighbouring cells was better aligned than a population of cells using only self-propelled motility. The ratio of the order parameter between wild type and A^+S^- cells was strongly correlated with the swarming efficiency of the two strains.

A follow up simulation study (Wu *et al.*, 2009) made a prediction that the swarm expansion rate could be optimized by cells reversing on average every 8 min. The model predictions that both shorter and longer reversal periods would have slower expansion rates were confirmed with a number of Frz mutants that had reversal periods different from the wild type cells (Fig. 8.2). One of the conclusions drawn by the authors was that the reversal period evolved under selective pressure to maximize the swarming rate. Higher swarm expansion rates would mean less competition between individual cells within the swarm for nutrients. Ultimately a population with an 8-minute reversal period would acquire a selective foraging advantage.

Flexiblity impacts cell alignment during *Myxococcus xanthus* swarming

Janulevicius *et al.* (2010) developed what they refer to as a 'mass-spring model' to study the effects of flexibility on motion and cell alignment. This is a cell-based model that is also stochastic, but probability is introduced quite differently than the previous model. While the movement of cells is fully determined by initial cell position and the positions of neighbouring cells, the probability of reversal varies. Although cells reverse direction on average every 8 min, because the time between reversals is random for a cell, each simulation has a different outcome. Each cell is represented with ten points connected by springs. Unlike the model in the previous section

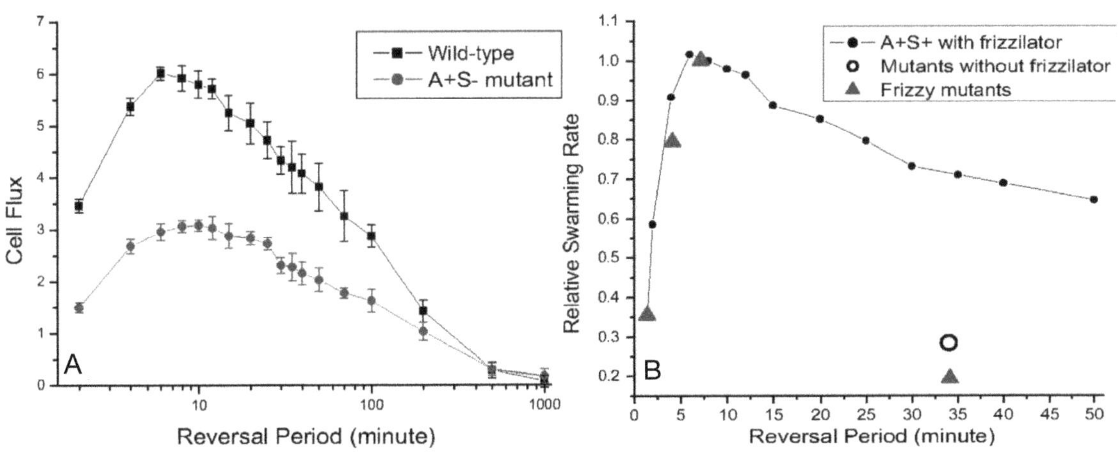

Figure 8.2 (Figs. 3 and 6 from Wu *et al.*, 2009). The impact of reversal period on swarming rate. Left: Measurements of the cell flux (equivalent to the swarmming rate) are plotted for different reversal periods for simulations of wild type and A^+S^- strains. The maximum cell flux or fastest swarming rate occurs when wild type cells reverse every six to eight minutes. (Right): Measurements of swarm rates are plotted against the reversal period from four experimental measurements (grey triangles) and simulations of A^+S^+ strains with frizzilator (black dots) and without frizzilator (open circles). Triangles are plots of two reversal mutants with shorter reversal periods, the wild type (~8-minute reversal period), and one mutant with a longer reversal period. Open circles correspond to a simulation using a long stall time during a 34-minute reversal period that better matches the experimental measurement. The simulations predicted a dependence of the swarm rate on reversal period that was tested and confirmed by experiments.

(Wu *et al.*, 2007) where cells could only bend in the middle, this model allows more realistic bending. In order to focus on the effects of flexibility and A-motility, the model assumptions exclude slime trail following and S-motility (i.e. the model describes A^+S^- strains). In addition, the model was used to test two opposing biological models of A-motility. In one model, the force is applied at the lagging pole consistent with the slime-engine model for A-motility. In the second model, the same force is distributed along the length of the cell consistent with the model that requires cells to push against the substrate at multiple points along the length of the cell. Simulations of single cells showed that lower bending stiffness resulted in a flailing motion for the polar engine (RE) but not for a distributed engine (DE). While flailing has been seen in experiments where the leading end of a cell was stuck (Sporemann and Kaiser, 1995), the flailing motion seen in simulations occurred while the cell was moving. For both models of A-motility, the alignment of two cells was measured in simulations of two cells colliding (Fig. 8.3). The alignment following a collision was comparable only above a certain bending stiffness threshold. The flailing motion of the RE model affected how well cells could align when the bending stiffness decreased, but the DE showed improved alignment in that range of flexibility. While the RE modelled aligned more poorly at low bending stiffness, estimates of the bending stiffness from experimental measurements of the Young's modulus (Pelling *et al.*, 2005) found the bending stiffness to be just above the threshold for flailing. The studies found that both models of A-motility behaved similarly during collisions in the range of the estimated bending stiffness, but that DE model cells would glide along the long axis independent of bending stiffness whereas RE model cells will flail if the stiffness is too small.

In addition to studies on cell collisions, a different set of simulations looked at how flexibility impacted large populations of cells. For large populations, simulations showed that flexibility could interfere with alignment. This was analysed by comparing the average orientation autocorrelation of cells during simulations of rigid vs. flexible cells. Rigid cells did not reorient well during collisions which allowed them to maintain their orientation (i.e. have higher autocorrelation for longer times) while flexible cells reoriented quickly. While experimental measurements of flexibility are difficult to perform, this paper was able to use previous and new theoretical estimates for bending stiffness to justify the range of values used in simulations. From the findings of the simulations, the authors suggested that bending stiffness may arise from a trade-off between being rigid enough to align effectively, but flexible enough to reorient during a high jamming environment such as in fruiting body aggregates. The authors speculated that a possible regulatory mechanism that could control stiffness at different stages of development might be desirable. For example, more rigid cells could align in streams or ripples while more flexible cells could resolve jamming during aggregation. The ability to make accurate measurements of bending stiffness throughout the life

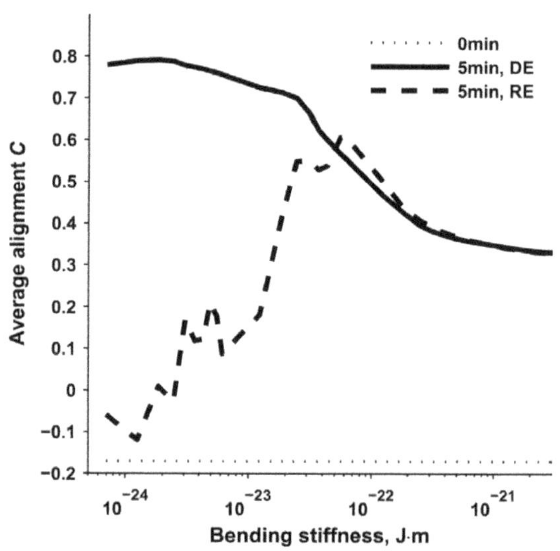

Figure 8.3 (Fig. 4 from Janulevicius *et al.*, 2010) Alignment of two cells during collision. The alignment (C) measures the orientation of two cells where a value of +1 means cells have identical orientation and −1 means cells are perpendicular to each other. The alignment (C) was measured for simulated collisions between cells with a distributed engine (DE) or rear-engine (RE) model over a range of bending stiffnesses. More than 150 simulations were run for each bending stiffness value. More flexible cells using the DE model align better. For the RE model, cells exhibit flailing motion at low bending stiffness, which reduces alignment. The two models show similar alignment values for less flexible cells.

cycle would allow these predictions to be tested and could provide new insight into the evolution and regulation of flexibility.

Patterns of cell clusters determined by physical properties

A cellular Potts model (CPM) was developed by Starruß *et al.* (2007) to study the collective motion of cells. The modelers chose this approach because it provides cell shape approximation, the possibility of simulating cell–cell and cell-substrate adhesion, and because it is relatively easy to implement and run. The CPM is a lattice-based model that represents cells as connected sets of nodes within the lattice. Cells interact with neighbouring cells or the surrounding medium by the interaction energy which is part of the Hamiltonian free energy used to determine the energetic cost of cell shape changes and cell–cell interactions. In this model, each cell is a cluster of nodes on a lattice. Interactions between cell segments determine the shape of a cell and control cell shape changes during simulations. The interaction energy for this model was calibrated by comparing the bending motion of simulated cells to images of a flailing cell (Sporemann and Kaiser, 1995). The calibrated model was used to observe the emergent behaviour of a collection of cells moving in a 2D domain. The authors used a newly defined metric – the intensity of collective motion – that measured the average of the maximum cluster size as a way to determine how readily clusters of aligned cells formed.

The authors point out that many other models use cooperative mechanisms such as directional changes based on the orientation of neighbouring cells or cell signalling in order to capture the swarming patterns of *M. xanthus.* The authors use the model to suggest that cooperative mechanisms such as adhesion and signalling are not essential for pattern formation. Instead, the physical properties of individual cells – the aspect ratio, stiffness, and active motion – might be sufficient for simulating clustering of aligned cells. By studying the impact of cell properties on the intensity of collective motion, the authors found that a minimal stiffness and minimal aspect ratio are essential for intense collective motion during *M. xanthus* swarming. As shown in Fig. 8.4, randomly arranged cells form well-aligned clusters as they migrate without any cooperative mechanism. This computational study suggested that mutant strains lacking adhesion could be examined experimentally for clustering intensity in order to test this model.

Swarm expansion depends on nutrient concentration but not on nutrient diffusion rate

Two different continuum models of swarming were introduced by Gallegos *et al.* (2006) to examine swarm expansion. Instead of using

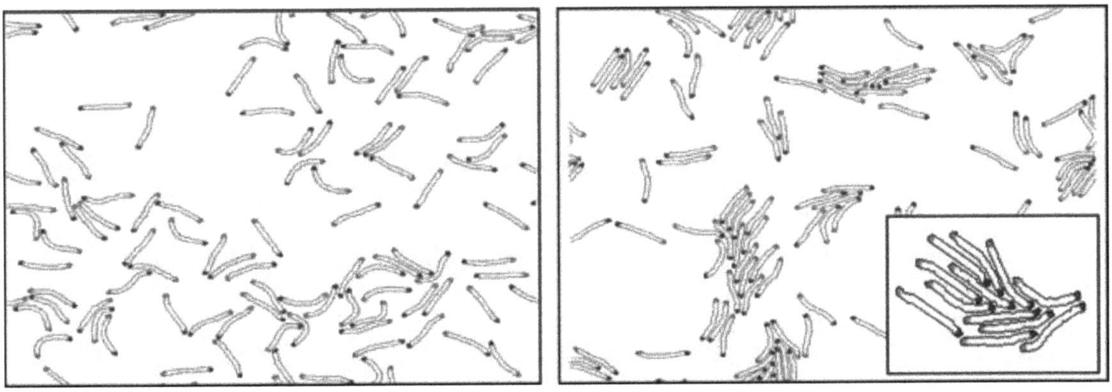

Figure 8.4 (Fig. 6 From Starruß *et al.*, 2007). Simulation showing that mechanical interactions of actively moving cells can form clusters. (Left) The spatial arrangement of a collection of cells is randomized at the beginning of a simulation. (Right) Simulation clusters form with time. (Inset) Cluster with an arrow-like arrangement showing that cell axes often point towards the centre of cluster. Simulations found that clusters can change direction as leading cells make random turns.

cell-based models like those described in previous sections, continuum models update fields that represent densities of cells and nutrient concentrations. Both models are deterministic models. One advantage of this approach is that population growth is easy to model whereas cell-based models can sometimes be limited by either computational memory or numerical performance. Their first model, which uses a 1D diffusion equation, examined short time-scale spreading, less than five hours, which allowed the model to neglect growth and cell division so that swarm expansion was driven solely by diffusion. In the second model, a reaction-diffusion equation was used to study long time scales where cell division and nutrient assimilation become important. The latter modelling approach captured the macroscopic behaviour of the spreading edge of a colony, which behaved as a travelling wave with a constant velocity – consistent with experiments. Although this swarm expansion rate (i.e. the velocity of the expanding edge) was constant in time, the rate was not independent of all parameters.

Simulations predicted that the swarm expansion rate is independent of nutrient diffusion. However, the model simulations also found that the spreading rate is proportional to the square root of the initial nutrient concentration, which agrees with experimental observations (Burchard, 1974). The expansion rate was also found to be proportional to the square roots of the growth rate and cell diffusion (motility). In the paper, the authors provide the details of how cell diffusion is estimated based on cell velocity and the frequency of reversals. They also provide a formula that introduces additional dependence of cell diffusion on the local cell density. The use of this more biologically realistic cell-density dependent diffusion rate sets this model apart from previous works in continuum modelling using constant diffusion rates. The authors cite good agreement between the simulated spreading rates and published results. The researchers suggested that experiments could test two predictions of their second model: (1) the expansion rate is independent from the nutrient diffusion and (2) the spreading colony density is dependent on the initial concentration of nutrients.

Ripple formation

One of the behaviours unique to myxobacteria is the formation of travelling waves, also called ripples because of their resemblance to ripples on the surface of water. Ripples are ridges of cells moving in opposing directions that give the appearance of passing through each other. In fact, collision of the two ridges causes synchronized direction reversal of the cells within them (Welch and Kaiser, 2001). Thus the cells do little more than traverse back and forth over the same spot. Initial attempts to understand the molecular mechanism of rippling have focused on C-signalling as a means of mediating the directional change and the Frz chemosensory system as a means of executing the reversals. Despite extensive modelling, there is precious little biochemical evidence that C-signalling and the Frz system perform the roles ascribed to them in the models. Recent reports of transient alignment of FrzCD protein clusters in cells making side-to-side contact that appear to stimulate reversals (Mauriello *et al.*, 2009) and growing evidence of rippling during predation (Berleman *et al.*, 2006, 2008), where C-signalling may not be active, call for a new rippling model.

Ripples may be explained by contact-based signaling during a reversal cycle

A *mean field model* that describes the density of cells in space was introduced by Igoshin *et al.* (2001) to explore how waves form due to contact-mediated signalling and cell motility (Fig. 8.5). The mean field model is another example of a continuous and deterministic model. The model includes C-signal exchange between colliding cells, reversal of movement, and the dependence of reversals on the level of C-signal. Direction reversals in the model are controlled by a reversal cycle with a *phase velocity* that is C-signal dependent. C-signal levels were in turn, dependent upon the density of cells travelling in the opposite direction. The model was able to demonstrate both unidirectional waves and counter migrating waves. The model required a refractory period where phase velocity of the reversal clock was insensitive to C-signal for some period following a reversal.

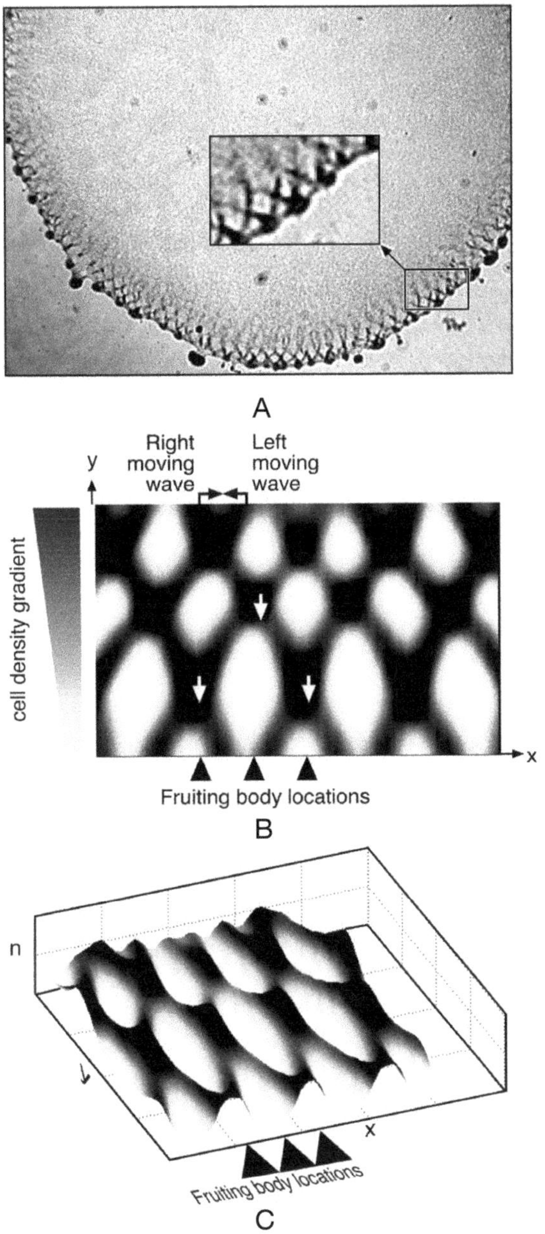

◁ **Figure 8.5** (Fig. 5 From Igoshin *et al.*, 2001.) Travelling waves in experiments and simulations. (a) Early stage of fruiting body formation showing travelling waves along the periphery of a spot containing *M. xanthus* cells. Counter-migrating waves propagate along edge of the spot and precursors to fruiting bodies form at perimeter. The inset is a higher magnification showing dark bands correspond to ridges of cells (b and c). Results from numerical simulations of travelling waves shown in a contour plot (b) and perspective plot (c). Waves tilt due to a density gradient of cells from the top region (representing the interior of the disk) to the bottom (periphery of disk). Right moving waves tilt to the right while left moving waves tilt to the left. The intersection of the two counter propagating waves move in a downwards direction, towards the periphery. Darker regions correspond to higher cell density.

The computational model was able to reproduce the spatial patterns from movies made of experiments and show how contact-mediated signalling without any diffusible morphogens could produce these patterns. The work also suggested several testable hypotheses such as the existence of a refractory period, nonlinear density dependence, and wave tilt being determined by the transverse density gradient.

Ripples may be induced by lateral signaling that induces reversals

Subsequent models for rippling used agent-based modelling approaches and further refined the intercellular interactions that mediate wave emergence (Sliusarenko *et al.*, 2006; Börner *et al.*, 2002; Holmes *et al.*, 2009; Zhang *et al.*, 2012). The latest model by Zhang *et al.* (2012), inspired by experimental observations of Mauriello *et al.* (2009), is a discrete, stochastic model. This model showed that three features are needed to generate rippling behaviour: (i) side-to-side signalling between two counter-moving cells that causes one of the cells to reverse, (ii) a minimal time period after each reversal during which cells cannot reverse again, and (iii) physical interactions that cause the cells to locally align. To explain observations suggesting that rippling appears as a consequence of the presence of prey, the model postulated that prey-associated macromolecules induce ripples indirectly by stimulating side-to-side signalling. In addition to the simulations, *M. xanthus* predatory rippling behaviour was experimentally observed with time-lapse microscopy and quantified. An equation expressing the wavelength in terms of reversal time, cell velocity and width of the wave crest has been predicted by the simulations and confirmed by the experimental data. Furthermore, the results suggest that the physiological role of rippling behaviour during *M. xanthus* predation is to increase the rate of spreading over prey cells and to allow predatory cells to remain on the prey for a longer period of time.

Fruiting body development

Development involves the coordinated movement of tens of thousands of cells over several days. While often represented as a single-stage process defined by an endpoint (i.e. a field of fruiting bodies containing spores), recent results using time-lapse microcinematography suggest extraordinary complexity in the choreography of cell movement. It is now clear that development must be represented as a multi-stage process, and separate hypotheses must be formulated and tested for each stage. The stage of development that has garnered the most attention recently has been aggregation, when the cells cluster at centres that will eventually become fruiting bodies (each fruiting body contains approximately 10^5 cells) (Kuner and Kaiser, 1982; Curtis *et al.*, 2007). The mechanism of aggregation is unknown, but first physical principles suggest that cells might synchronize their movement by sensing and responding to biochemical and physical cues (Konovalova *et al.*, 2010). Therefore, a full grasp of development requires tracking cell movement throughout aggregation. In the following section, statistical image analysis is used to explore the underlying mechanisms of fruiting body formation. This is followed by a discussion on computational modelling approaches that study the fruiting body formation.

Image analysis used to quantify fruiting body development

Time-lapse microcinematography is used to observe cell movement and corresponding cell-density patterns during aggregation (Shi *et al.*, 1996; Jelsbak and Sogaard-Andersen, 2003; Kaiser and Welch, 2004; Curtis *et al.*, 2007; Sliusarenko *et al.*, 2007; Taylor and Welch, 2010). Beginning with an approximately uniform cell density on non-nutritive agar, sequences of images describing changes of cell distribution resulting from movements of individual cells are recorded and analysed (Curtis *et al.*, 2007; Taylor and Welch, 2010). The movement of cells within the swarm can also be observed by mixing a relatively small (0.5–2%) number of fluorescent cells into the population and then tracking their movements using fluorescence microscopy while observing cell density patterns using differential interference contrast (DIC), phase contrast or bright field microscopy (Igoshin *et al.*, 2001; Welch and Kaiser, 2001; Sliusarenko *et al.*, 2007; Berleman *et al.*, 2008). These methods are sufficient as long as the cells move within a uniformly dense swarm in two dimensions. However, once cells begin to ascend into nascent aggregates, confocal microscopy must be used to track them in three dimensions (Lux *et al.* 2004; Hu *et al.*, 2012). In addition, once cells begin to form aggregates, they also begin to produce large quantities of extracellular matrix (ECM), which limits the ability to track individual cells or even visualize cell layers. Overcoming this limitation is possible by using *M. xanthus* strains that produce much less ECM but still develop, thus permitting the visualization of cell movement in the fruiting body throughout much of aggregation (Curtis *et al.*, 2007).

Aggregation can be divided into a series of stages (Jelsbak and Sogaard-Andersen, 2003; Kaiser and Welch, 2004; Curtis *et al.*, 2007). The first visible signs of aggregation begin 6–8 h after amino acid limitation with a dramatic increase in cell movement. Small towers of cells several layers thick and about 25 μm in diameter emerge from the more uniform surface of the biofilm and move in seemingly random ways, sometimes sinking back into the biofilm, and sometimes joining with or splitting off from other towers. This first stage of aggregation is characterized by extraordinary horizontal movement of these small towers across the surface of the biofilm. With multiple fusion events, the towers form broader, more stable structures that develop into aggregates 2–4 times broader than the initial small towers (Fig. 8.6).

During the second stage of aggregation the horizontal movement of aggregates diminishes and is replaced by vertical movement of cells within the aggregate. Close examination of the ECM defective strain DK10410 (Δ*pilA*) shows that the maturing aggregate gains height through the addition of successive layers to the top. Each tier begins when cells rise from one or two points in the lower tier and begin to spread evenly across the aggregate surface. The tier continues to expand

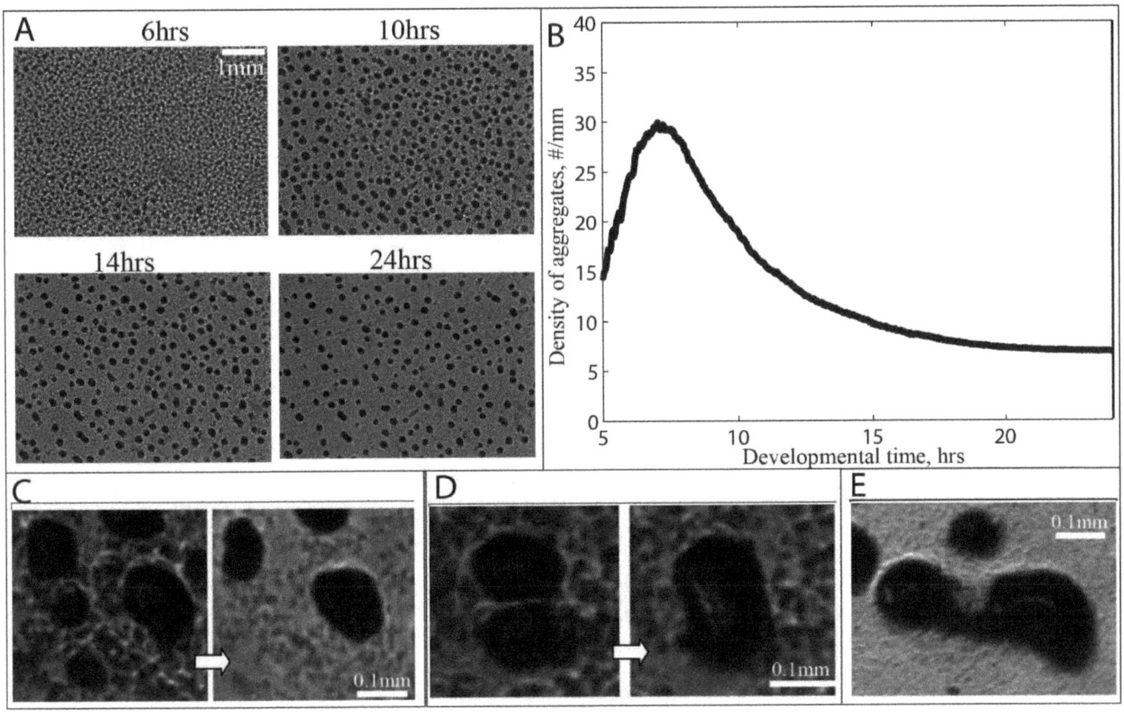

Figure 8.6 Progression of aggregation. (A) The number and distribution of aggregates changes with time during development as shown by select frames of a time lapse movie at 6, 10, 14 and 24h. (B) Change in aggregate number over 1mm2 area during development for a representative experiment. (A) and (B) are adapted from (Zhang *et al.*, 2011). (C–E) Various fates of transient aggregates during development adapted from (Xie *et al.*, 2011). (C) Aggregates persist or disperse. (D) Two aggregates merge to form one fruiting body. (E) One aggregate is beginning to split into two smaller ones.

to the boundary of the fruiting body even as the next layer begins (Curtis *et al.*, 2007).

As with the small towers that began aggregation, many of these aggregates are not stable. During the third stage of aggregation, which partially overlaps with the second, the field of aggregates is culled and tailored to produce fruiting bodies of similar size with a regular spatial distribution. Some aggregates grow in size while others diminish. Those that disappear do so by reversing the aggregation process and disappearing one layer at a time beginning with the uppermost layer. Roughly half of the aggregates disappear, resulting in a more regular distribution of the remaining structures (Fig. 8.6A, B and C). Two other fates await nascent aggregates, though these are much less frequent than aggregate disappearance. Occasionally an aggregate moves across the surface to fuse with another aggregate (Fig. 8.6D). Sometimes an aggregate splits into two smaller aggregates (Fig. 8.6E). Eventually each stable aggregate adopts a regular dome shape then differentiate myxospores to become mature fruiting bodies (Curtis *et al.*, 2007). The dynamic changes observed by time-lapse microcinematography await a fuller explanation as to mechanism and purpose (Xie *et al.*, 2011; Zhang *et al.*, 2011).

A quantitative metric of developmental patterns in *M. xanthus* is necessary, not only to compare experimentally observed aggregation patterns to those produced by models, but to quantitatively compare various mutant phenotypes (Sawai *et al.*, 2007; Xie *et al.*, 2011; Zhang *et al.*, 2011). To this end, digital image processing and statistical analysis of information in images were developed and adapted to study *M. xanthus* development. Methods were developed to filter and segment time-lapse images, to separate aggregates from the surrounding swarm of cells, to track these aggregates in space and time, and to automatically extract the properties related to the individual aggregates and determine their global distribution (Xie *et al.*, 2011; Zhang *et al.*, 2011). Analyses of this type also may demonstrate which properties are most informative when trying to determine aggregate behaviour and fate. To this end, methods of pattern recognition and spatial statistics can be used to extract information from microcinematography data to test various hypotheses. Such techniques are widely used in other areas of biology such as high-content imaging (Murphy, 2010) in cell biology (Ronneberger *et al.*, 2008; Nurgaliev *et al.*, 2010) and medical imaging (Kassner and Thornhill, 2010; Noble, 2010; Kriegeskorte, 2011; Petitjean and Dacher, 2011), but these have not been widely used to study microbial pattern formation.

The first step in image analysis of development involves separating each time-lapse image into individual aggregates and surrounding spaces using a combination of standard segmentation algorithms. Segmentation is possible because aggregates are darker than the surrounding field. Pixels within aggregates are labelled one while the remaining pixels are labelled zero. Assuming that labels vary gradually everywhere except at the aggregate boundaries, segmentation proceeds by optimizing energy function. Discontinuity is penalized and correlations between greyscale and binary intensity are rewarded. The segmented images are thereafter converted to binary images with black aggregates surrounded by white pixels (Xie *et al.*, 2011). Segmentation is applied to time frame images by detecting overlapping aggregates. This approach permits unambiguous identification of dispersing, merging, or splitting aggregates. Methods of the statistical analyses of the images are used to decrypt properties of individual aggregates that influence their fate. Thirty-three parameters were devised to characterize each aggregate according to its geometry (size and shape), proximity to other aggregates, and size of the nearest neighbours. All 33 features were automatically identified for each aggregate.

Only about half of the aggregates mature into fruiting bodies. Most aggregates recede into the biofilm. Presumably the cells reemerge from the biofilm in successful aggregates, though their travels have never been documented. What determines which aggregates survive and which disappear? Can we come to an understanding of the mechanism determining aggregate fate? To determine which features of aggregates (size, proximity to other aggregates, size of nearest neighbours, and shape) correlate with fate, mutual information was calculated for all 33 features. The results indicated that the primary factor correlated with aggregate survival is large size. Those aggregates exceeding 6,000 μm^2 at 13.5 h were most likely to mature (Fig. 8.7). Neither proximity with neighbours nor neighbour size were adequate predictors of survival frequency (Fig. 8.7). The results suggest that the aggregate monitors its size or growth rate to regulate maturation.

Aggregate fusion and aggregate splitting are rare events under the conditions of this assay. Close proximity seems necessary but not sufficient for aggregate merging. Aggregates failed to merge over distances in excess of 150 µm. However, roughly 20% of those aggregates within that distance did not merge (Xie *et al.*, 2011). Both aggregate perimeter and aggregate shape are correlated with splitting. In general, those aggregates that are large and asymmetric tended to split (Xie *et al.*, 2011).

Computational models used to study fruiting body development

Mathematical and computational modelling has been used to test various hypotheses about how fruiting bodies form (Othmer and Stevens, 1997; Igoshin *et al.*, 2001, 2004c; Sozinova *et al.*, 2005; Holmes *et al.*, 2010; Harvey *et al.*, 2012). Thus far, such models have uncovered sets of intercellular interactions that lead to steady-state patterns similar to those observed experimentally. Models of differing complexity and with different formalisms have been used to describe aggregation, but most of these share a common underlying mechanism; cells tend to slow down or stop when entering regions of high cell density (Igoshin *et al.*, 2004a,c; Sozinova *et al.*, 2005; Sliusarenko *et al.*, 2007). This leads to a positive feedback loop as more cells jam into the region, further increasing cell density. Importantly, mathematical simulation of a 'traffic jam' model was able to form aggregates without any diffusible morphogens or chemoattractants. More recent models have also emphasized the importance of polysaccharide (slime) production, and the formation of and following of

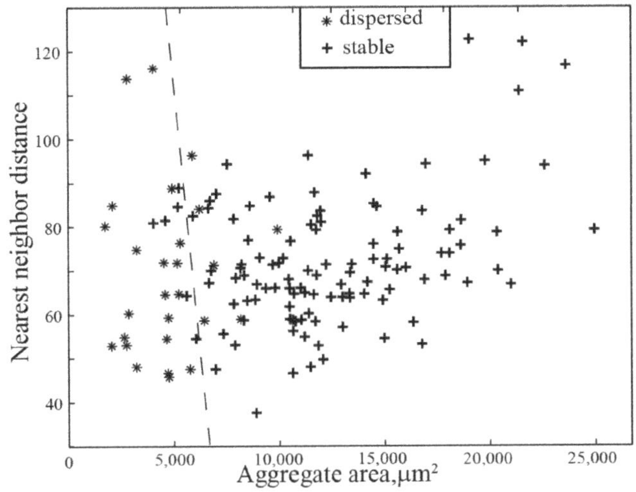

Figure 8.7 Scatter plot of aggregate area vs. nearest-neighbour distance for stable and dispersing aggregates. The results indicate a strong correlation of dispersion with small aggregate size but no correlation between aggregate dispersion and its proximity to the nearest neighbour. The dashed line shows a support-vector machine model separating stable and dispersed aggregates based on a training data set corresponding to 50% of the data points. The line nearly parallel to the y-axis further indicates that distance to a neighbouring aggregate does not significantly affect aggregate stability. (Adapted from Xie *et al.*, 2011. Proc. Natl. Acad. Sci. U.S.A. *108*, 5915–5920.)

'slime trails' during aggregation (Hendrata *et al.*, 2011). Notably, almost all modelling research has focused on the formation of initial large aggregates rather than dynamic processes observed during the earlier stages of organization. Lacking quantitative parameters that describe various stages of aggregation, the comparison between microcinematography and in silico aggregation simulations was mostly qualitative. Two examples of computational modelling used to study fruiting body development are described below.

Lattice gas cellular automata model of aggregation

This model examines how cell movement patterns influence aggregation. A three-dimensional (3D), stochastic, lattice-gas cellular automata (LGCA) model was used by Sozinova *et al.*, (2005) to study two proposed stages of aggregation: (1) a field of motile cells producing traffic jams that act as aggregation centres, and (2) streams of cells expand the aggregation centres into hemispherical mounds. In this discrete model, each cell is a collection of pixels on the underlying 3D hexagonal lattice. The ellipsoidal shape for each cell does not bend, and contains polar regions that are designated as the head and tail. Cell movement is determined by local conditions where the directional movement of a cell aligns with other cells. Cells move along the lattice and accumulate C-signal when the head of a cell overlaps with the tail of another cell.

The simulations captured several features seen in movies of cell aggregation. Traffic jams occurred in particular regions and created high-density zones where cells move rotationally around the site as a 'motile skirt'. Between the traffic jams, streams of cells formed where cells follow one another. This study demonstrated that cell polarity is important because spherical cells did not aggregate like rod-shaped cells.

This computational model was used to study spore formation within fruiting bodies in Sozinova *et al.*, (2006). There, the C-signal level in a cell was used to determine when a cell differentiated into a spherical spore. The authors found that streams could deliver sufficient cells to aggregate into spherical mounds. Furthermore, citing confocal studies showing two density domains in fruiting bodies (Sager and Kaiser, 1993), the authors suggested that the distribution of spores within the mound agreed with experimental findings.

Off-lattice Monte Carlo model of aggregation

A different computational approach was used to study how the spatial organization of cells in a fruiting body is affected by the physical properties of cells. Holmes *et al.* (2010) used off-lattice Monte Carlo simulations to study the spatial organization of cells during fruiting body development. These simulations used a Cellular Potts Model (CPM) representation for simulated bacterial cells. In this model, cells are represented as eight connected segments that are not confined to a lattice. Each segment is a collection of 27 segment nodes arranged in a $(3 \times 3 \times 3)$ cube. The authors point out that lattice models restrict the degrees of freedom in cell shape and flexibility. In this off-lattice model, bending and stretching energies are used to control the shapes of cells and a propulsive force is applied so that cells move along their long axis. Additionally, energies that account for gravity, collisions, and adhesion are included in the model in order to ensure that movement in 3D is realistic. Adhesion between cells is modelled as an attractive force between neighbouring cells which models the effect of the extracellular polysaccharide (EPS) surrounding cells. Cells accumulate C-signal during collisions and differentiate into non-motile spores at a C-signal threshold. The large number of cells needed to model fruiting body aggregation can be computationally prohibitive. Rather than model every cell over a large area, a smaller region filled with cells was used and new cells were added to the domain. As streams of cells moved inward from the perimeter of the simulation domain, a cell influx parameter was defined which determined the rate of new cells created at the perimeter of the simulation domain.

The authors point out that previous aggregation models (Sozinova *et al.* 2005, 2006; Kaiser and Welch, 2004) utilized artificially induced centres where cells aggregate. The simulations using this model were able to study how cells form aggregation centres spontaneously and which physical properties contribute to the emerging fruiting body. A climbing feature was introduced that created an upwards force when cells experience a high number of head on collisions, which allowed second layers to form from a dense monolayer of cells. Without a high level of cell influx, simulations showed that mounds would form transiently but not become stable spherical fruiting bodies. With a high level of cell influx, the density of cells at the aggregation centre was sufficient to create a stable, spherical fruiting body. Simulations showed that stable 3D spherical aggregates could be formed when the high cell density led to cells climbing into higher layers and the adhesive force due to the EPS held cells within the mound. The high cell density in the aggregate led to more collisions and higher signalling, which created non-motile spores. In the simulations, the spores contributed to the stability of the aggregate by limiting the movement of the motile cells thus expanding the size of the traffic jam. An example of motile cells forming a spherical aggregate based on this model can be seen in Fig. 8.8, where each cell is visualized as a string of eight spheres representing the segments. The large hemispherical mound, the interaction between motile and non-motile cells, and the highly aligned streams of cells were some of the experimentally observed phenomena that the authors identified as being reproduced in the simulations. This model demonstrates how computational simulations can study the impact of the physical properties of cells on the spatial organization of cells.

Conclusions and prospects

In this chapter, we demonstrated the path from observation to modelling and predictive simulations for three processes: swarming, travelling waves and fruiting body development. In addition, we showed how statistical image analysis can tease out relationships that are not otherwise obvious. For example, a quantitative metric of features characterizing each aggregate suggests that aggregate fate is determined by behaviour of individual bacteria rather than long-range chemical gradients or biased cell exchange. This approach has directed our attention to internal sensing mechanisms, but has stopped short of defining specific genetic or biochemical pathways that might be involved.

The chapter also provided examples of continuous and discrete models as well as deterministic and stochastic models used to study myxobacteria. Different cell-based approaches are particularly

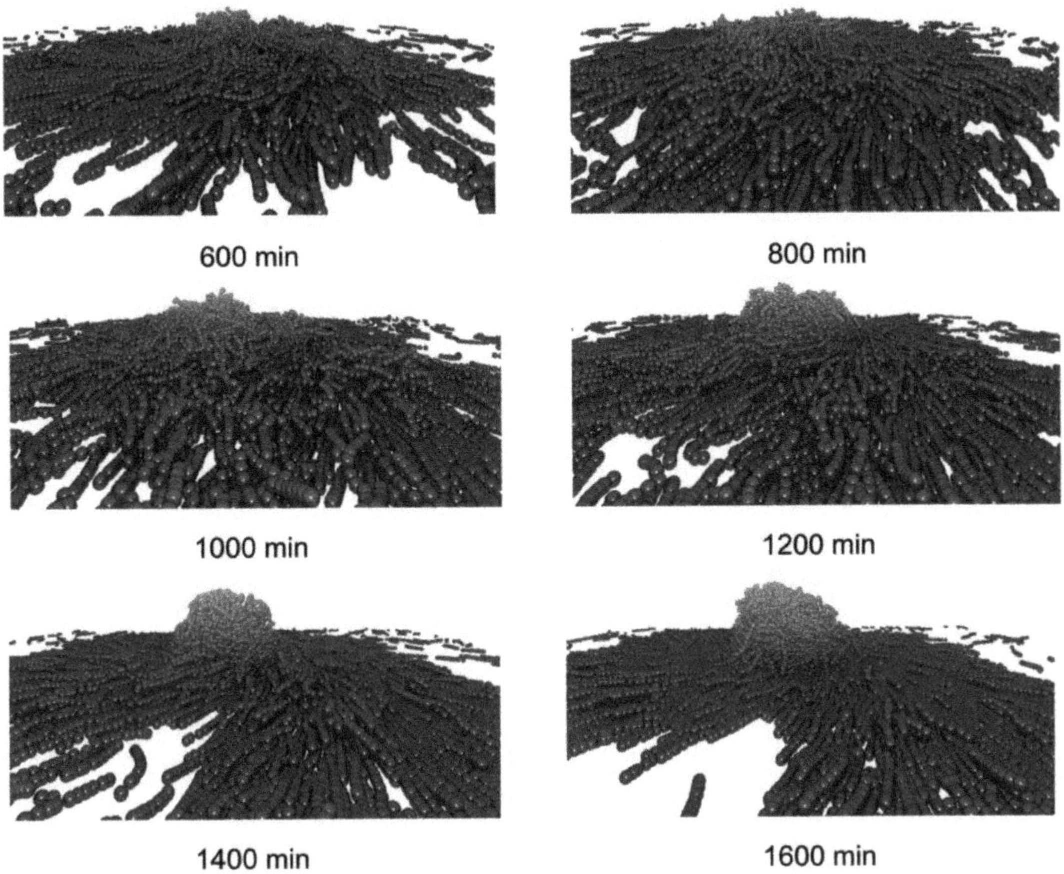

Figure 8.8 (Fig. 6 from Holmes *et al.*, 2010.) Fruiting body aggregation simulation. This sequence of images is taken from the off-lattice Monte Carlo simulation where cells form a 3D spherical aggregate. Streams of cells collide at an aggregation site to form a stable spherical mound of motile cells. Individual cells are represented by strings of eight spheres.

well suited for studying individual cell behaviour and physical properties during cell–cell interactions. Such approaches usually require a trade-off in the level of biological detail incorporated into the model and the number of cells that a simulation can contain. Detailed representation of cells often focuses on populations with hundreds to a few thousands of cells. For patterns that emerge in large populations of cells (tens of thousands to millions), continuum models describing cell density instead of individual cells provide a good option for the modeler. Experimentalists must make these kinds of tradeoffs as well when, for example, they choose to image cell movement through a low magnification lens to see a large area rather than a high magnification lens to observe more detailed behaviour of individual cells.

Continuum models describe the movement of bacteria in a more general way to capture group behaviour. Discrete approaches can incorporate more details about motility mechanisms and interactions between cells and the environment. Ultimately, either approach can be used to compare specific models of cell movement with experimental data. The chapter also describes how pathways involving protein–protein interactions can be studied with a deterministic, continuous model that makes predictions about the roles of proteins in the Frz pathway.

While the ideal modelling process would lead to new experiments, it is not always possible to test a model prediction. Sometimes, simulations are used when running experiments is difficult or expensive. Modelling efforts often rely on outcomes of known experiments in order to 'calibrate' the model (not unlike testing meteorological models or financial market models with historical data). Ideally, detailed and biologically relevant simulations should be run in parallel with specifically designed and novel experiments. More and more, researchers are beginning to utilize an interdisciplinary approach by forming collaborations between modelers and biologists, and training students at this exciting interface.

Many predictions from simulations are specific to patterns of movement that depend on cell properties that can be modified through genetic mutation. Some such predictions require improved imaging and tracking techniques in order to get statistical information about individual cells and groups of cells for comparison. Other predictions about swarm expansion rate or cell density dependence on nutrient concentrations could be tested in the lab with currently available methods. As more and more biologists begin to engage with modelers and use newly developed tools of computational modelling, new and exciting questions can be explored in the lab. Essential to the modelling process is the input of biologists who can inform and critique the biological assumptions that are built into a model and help interpret any biological significance in the observations that emerge from simulations.

A number of important questions regarding emergent behaviour can benefit from modelling including a better understanding of the mechanism of A-motility, signalling during development and biochemical pathways driving direction reversals, among others. While all computational models necessarily omit certain biological details known about myxobacteria, and in this sense do not completely simulate reality, they provide critical insight and predictive power that is becoming an important complementary research tool. As researchers continue to improve computational methods and modelling techniques implemented on high performance computer clusters, collaborations between experimentalists and computational modelers will lead to novel insights and discoveries.

Acknowledgements

This work was supported by the National Institute of General Medical Science of the National Institutes of Health under award number R01GM095826 to LJS, by the National Science Foundation under award number MCB0742976 to LJS, and NSF CAREER Grant MCB-0845919 to O.A.I. C.H. and M.A. were partially supported by the National Institute of General Medical Science of the National Institutes of Health under award number R01GM100470.

Glossary of terms

Agent-based model (ABM) *see* Cell-based model.

Cell-based model (CBM) Also called an agent-based model or individual-based model. Models that use autonomous cells (agents/individuals) and a set of rules to determine how the cells are updated each step of a simulation. Rules can vary from simple to highly complex where they may be influenced by the presence of neighbouring cells (agents/individuals). A cell can be represented as either a point or as having finite size with a defined area or volume.

Cellular Potts model (CPM) A lattice-based, stochastic model where cells are represented by a collection of sites in 2- or 3-dimensional lattices. A lattice site can be assigned a value corresponding to one of multiple types of cells or the medium containing the cells. Update to a value at the randomly chosen lattice site is based on the values of the neighbouring sites and is accepted with probability depending on the change of energy of the system.

Continuum model Model that describes dynamics of cell and chemical densities in a homogenous or heterogeneous environment by using ordinary or partial differential equations. The variables in continuum models may assume any value over some continuous range, e.g. a fractional concentration of chemicals, the density of cells in a swarm, the size of a tumour in a continuous cancer model, or the fraction of the population infected in epidemiology models.

Deterministic model A mathematical model whose outcome does not depend on any random variation. The initial values of the variables for the model always yield the exact same results (values) in the end of a simulation based on the equations defining the process being modelled. Varying the input of a deterministic model and parameter values allows the modeler to see how sensitive the outcome is to the input.

Discrete model A model that allows variables in the system, spatial positions or time to have non-continuous values, e.g. a particle moving on a 2D grid has a position (x,y) where x and y can only be integer numbers. A population of cells can be discretely modelled by having each cell tracked in space instead of the cell density.

Lattice (or grid) The subdivision of the spatial domain of the model (in 1D, 2D, or 3D space) into units (triangles, squares, cubes, etc.). The nodes that make up a lattice (grid) are referred to as the lattice sites (grid points). For example, a 2D square domain with each side having length L divided into an $n \times n$ grid, where each lattice unit is a square of size $L/n \times L/n$ would be called a rectangular lattice (or rectangular grid).

Mean field model A continuous model involving many interacting particles. A mean field approach uses averages over all particles to create continuous fields that allow each particle to interact with the field, dramatically reducing the computation to a one-body problem from the many-body problem. Since the field is the result of average values, it is referred to as a mean field. This approach can be used to model a variety of systems from magnetic materials to networks to swarming of bacteria or insects. The field may describe the mean value of all magnetic spins in the system, the population density, the mean velocity field of self-propelled objects or the mean concentration of molecules on a cell.

Monte Carlo A stochastic method that relies on randomly sampling from a probability distribution in order to determine an outcome of an event during a simulation.

Phase Velocity For a periodic variable (a variable whose value repeats after a given period of time, an example is the sine function), the phase refers to the current value of the variable, and the phase velocity refers to the rate at which the value changes. In myxobacteria models that use periodic variables to model the Frz pathway in order to determine directional reversals, the phase velocity is how quickly a cell's internal 'clock' is running which will determine who quickly it reverses direction.

State The state of a model refers to the current values of all variables in the system (e.g. the positions of all cells, the concentration levels of proteins, the phase of cycles, etc. at the fixed moment in time). Simulations evolve a system from one state to the next in either a deterministic or stochastic fashion. May also refer to a condition of a particular lattice site in a lattice model. For example, a particular lattice site can have the current state of being a 'cell' or a 'substrate' depending on the current state of the simulation.

Stochastic model Model that uses probabilistic rules in order to evolve the system from its current state to a new state. (It is also called stochastic process.) This is the counter to the deterministic model. Simulations beginning in the same state can evolve differently. Consider the example of a population of cells that are dividing. To determine how many cells divide, a simulation may iterate over all cells in the simulation and let them divide according to some probability. If the number of cells at a given time step is N, then the next step can have N' equal to some number between N (if no cells divided) and 2N (if all cells divided).

Subcellular element mode A type of discrete model that divides cells into several sub-regions' (subcellular elements (SCEs)) that interact via intracellular interactions. The intracellular interactions maybe modelled as springs between SCEs that fix the geometry of the SCEs or a more simple volume excluding interaction that prevent SCEs from overlapping (attractive interaction between SCEs within a cell are used to maintain the cell shape if spring-like interactions are not used). Depending on the nature of the intracellular interaction, cells can take on simple spherical geometry, or more complex shapes like rods, discs, etc. These interactions determine the shape of a cell or the functions of specific regions represented by the SCEs represent. Interactions between cells are determined by intercellular interactions with neighbouring cells by mechanical or chemical means. This representation can add biological details to a cell-based model.

References

Austin, M. (2007). Species distribution models and ecological theory: A critical assessment and some possible new approaches. Ecol. Model. *200*, 1–19.

Berleman, J.E., Chumley, T., Cheung, P., and Kirby, J.R. (2006). Rippling is a predatory behavior in *Myxococcus xanthus*. J. Bacteriol. *188*, 5888–5895.

Berleman, J.E., Scott, J., Chumley, T., and Kirby, J.R. (2008). Predataxis behavior in *Myxococcus xanthus*. Proc. Natl. Acad. Sci. U.S.A. *105*, 17127–17132.

Blackhart, B.D., and Zusman, D.R. (1985). 'Frizzy' genes of *Myxococcus xanthus* are involved in control of frequency of reversal of gliding motility. Proc. Natl. Acad. Sci. U.S.A. *82*, 8767–8770.

Börner, U., Deutsch, A., Reichenbach, H., and Bar, M. (2002). Rippling patterns in aggregates of myxobacteria arise from cell–cell collisions. Physical Review Letters *89*, 1–4.

Burchard, R. (1974). Growth of surface colonies of the gliding bacterium *Myxococcus xanthus*. Arch. Microbiol. *96*, 247–254.

Chandler, D. (1987). Introduction to Modern Statistical Mechanics. (New York, NY: Oxford University Press.)

Curtis, P.D., Taylor, R.G., Welch, R.D., and Shimkets, L.J. (2007). Spatial organization of *Myxococcus xanthus* during fruiting body formation. J. Bacteriol. *189*, 9126–9130.

Du, H., Xu, Z., Shrout, J.D., and Alber, M. (2011). Multiscale modeling of *Pseudomonas aeruginosa* swarming. Math.l Models Methods Appl. Sci. *21 (Suppl)*, 939–954.

Gallegos, A., Mazzag, B., and Mogilner, A. (2006). Two continuum models for the spreading of myxobacteria swarms. Bull. Math. Biol. *68*, 837–861.

Gardner, Martin (1970). Mathematical Games – The fantastic combinations of John Conway's new solitaire game 'life'. Sci. Am. *223*, 120–123.

Graner, F., and Glazier, J.A. (1992). Simulation of biological bell sorting using a two-dimensional extended Potts Model. Phys. Rev. Lett. *69*, 2013–2016.

Harvey, C.W., Morcos, F., Sweet, C.R., Kaiser, D., Chatterjee, S., Liu, X., Chen, D.X., and Alber, M. (2011). Study of elastic collisions of *Myxococcus xanthus* in swarms. Phys. Biol. *8*, 026016.

Harvey, C.W., Du, H., Xu, Z., Kaiser, D., Aranson, I., and Alber, M. (2012). Interconnected cavernous structure of bacterial fruiting bodies. PLoS Computational Biology *8*, e1002850.

Hendrata, M., Yang, Z., Lux, R., and Shi, W. (2011). Experimentally guided computational model discovers important elements for social behavior in myxobacteria. PLoS One *6*, e22169.

Holmes, A.B., Kalvala, S., and Whitworth, D.E. (2009). Myxobacteria motility: a novel 3D model of rippling behaviour in *Myxococcus xanthus*. Communications of the Systematics and Informatics World Network. *6*, 65–70.

Holmes, A.B., Kalvala, S., and Whitworth, D.E. (2010). Spatial simulations of myxobacterial development. PLoS Comput. Biol. *6*, e1000686.

Holmes, E.E., Lewis, M.A., Banks, J.E., and Veit, R.R. (1994). Partial Differential Equations in Ecology: Spatial Interactions and Population Dynamics. Ecol. *75*, 17–29.

Hu, W., Yang, Z., Lux, R., Zhao, M., Wang, J., He, X., and Shi, W. (2012). Direct visualization of the interaction between pilin and exopolysaccharides of *Myxococcus xanthus* with eGFP-fused PilA protein. FEMS Microbiol. Lett. *326*, 23–30.

Igoshin, O.A., Mogilner, A., Welch, R.D., Kasier, D., and Oster, G. (2001). Pattern formation and traveling waves in myxobacteria: theory and modeling. Proc. Natl. Acad. Sci. U.S.A. *98*, 14913–14918.

Igoshin, O.A., Neu, J., and Oster, G. (2004a). Developmental waves in myxobacteria: a distinctive pattern formation mechanism. Phys. Rev. E *70*, 1–11.

Igoshin, O.A., Goldbeter, A., Kaiser, D., and Oster, G. (2004b). A biochemical oscillator explains several aspects of *Myxococcus xanthus* behavior during development. Proc. Natl. Acad. Sci. U.S.A. *101*, 15760–15765.

Igoshin, O.A., Welch, R., Kasier, D., and Oster, G. (2004c). Waves and aggregation patterns in myxobacteria. Proc. Natl. Acad. Sci. U.S.A. *101*, 4256–4261.

Janulevicius, A., van Loosdrecht, M., Simone, A., and Picioreanu, C. (2010). Cell flexibility affects the alignment of model myxobacteria. Biophys. J. *99*, 3129–3138.

Jelsbak, L., and Sogaard-Andersen, L. (1999). The cell surface-associated intercellular C-signal induces behavioral changes in individual *Myxococcus xanthus* cells during fruiting body morphogenesis. Proc. Natl. Acad. Sci. U.S.A. *96*, 5031–5036.

Jelsbak, L., and Sogaard-Andersen, L. (2003). Cell behavior and cell–cell communication during fruiting body morphogenesis in *Myxococcus xanthus*. J. Microbiol. Methods. *55*, 829–839.

Kaiser, D. (2008). *Myxococcus*-from single-cell polarity to complex multicellular patterns. Annu. Rev. Genet. *42*, 109–130.

Kaiser, D., and Warrick, H. (2011). *Myxococcus xanthus* swarms are driven by growth and regulated by a pacemaker. J. Bacteriol. *193*, 5898–5904.

Kaiser, D., and Welch, R. (2004). Dynamics of fruiting body morphogenesis. J. Bacteriol. *186*, 919–927.

Kaiser, D., and Yu, R. (2005). Reversing cell polarity: evidence and hypothesis. Curr. Opin. Microbiol. *8*, 216–221.

Kassner, A., and Thornhill, R.E. (2010). Texture analysis: a review of neurologic MR imaging applications. AJNR Am. J. Neuroradiol. *31*, 809–816.

Konovalova, A., Petters, T., and Sogaard-Andersen, L. (2010). Extracellular biology of *Myxococcus xanthus*. FEMS Microbiol. Rev. *34*, 89–106.

Kriegeskorte, N. (2011). Pattern-information analysis: from stimulus decoding to computational-model testing. Neuroimage. *56*, 411–421.

Kuner, J.M., and Kaiser, D. (1982). Fruiting body morphogenesis in submerged cultures of *Myxococcus xanthus*. J. Bacteriol. *151*, 458–461.

Lux, R., Li, Y., Lu, A., and Shi, W. (2004). Detailed three-dimensional analysis of structural features of *Myxococcus xanthus* fruiting bodies using confocal laser scanning microscopy. Biofilms *1*, 293–303.

Leonardy, S., Freymark, G., Hebener, S., Ellehauge, E., and Sogarrd-Andersen, L. (2007). Coupling of protein localization and cell movements by a dynamically localized response regulator in *Myxococcus xanthus*. EMBO J. *26*, 4433–4444.

Leonardy, S., Miertzschke, M., Bulyha, I., Sperling, E., Wiitinghofer, A., and Sogaard-Andersen, L. (2010). Regulation of dynamic polarity switching in bacteria by a Ras-like G-protein and its cognate GAP. EMBO J. *29*, 2276–2289.

Macal, C.M., and North, M.J. (2010). Tutorial on agent-based modelling and simulation. J. Simulat. *4*, 151–162.

Mauriello, E.M., Astling, D.P., Silusarenko, O., and Zusman, D.R. (2009). Localization of a bacterial cytoplasmic receptor is dynamic and changes with cell–cell contacts. Proc. Natl. Acad. Sci. U.S.A. *106*, 4852–4857.

Murphy, R.F. (2010). Communicating subcellular distributions. Cytometry A. *77*, 686–692.

Neumann, J.V. (1966). Theory of self-reproducing automata. (A. W. Burks, Ed.) Mathematics of Computation. University of Illinois Press.

Newman, S., Christley, S., Glimm, T., Hentschel, H.G.E., Kazmierczak, B., Zhang,Y.T., Zhu. J., and Alber, M. (2008). Multiscale models for vertebrate limb development. Curr. Top. Dev. Biol. *81*, 311–340.

Newman, T.J. (2005). Modeling multi-cellular systems using sub-cellular elements. Math. Biosci. Eng. *2*, 613–624.

Noble, J.A. (2010). Ultrasound image segmentation and tissue characterization. Proc. Inst. Mech. Eng. H. *224*, 307–316.

Nurgaliev, D., Gatanov, T., and Needleman, D.J. (2010). Automated identification of microtubules in cellular electron tomography. Methods Cell Biol. *97*, 475–495.

Othmer, H.G., and Stevens, A. (1997). Aggregation, blowup, and collapse: The ABC's of taxis in reinforced random walks. Siam Journal on Applied Mathematics *57*, 1044–1081.

Parker, S. (2003). McGraw-Hill Dictionary of Scientific and Technical Terms. 6th edn. (McGraw-Hill).

Pekalski, A. (2004). A short guide to predator-prey lattice models. Comp. Sci. Eng. *6*, 62- 66.

Pelling, A.E., Li, Y., Shi, W., and Gimzewski, J.K. (2005). Nanoscale visualization and characterization of *Myxococcus xanthus* cells with atomic force microscopy. Proc. Nat. Acad. Sci. U.S.A. *102*, 6484–6489.

Pelling, A.E., Li, Y., Cross, S.E., Castaneda, S., Shi, W., and Gimzewski, J.K. (2006). Self-organized and highly ordered domain structures within swarms of *Myxococcus xanthus*. Cell Motil. Cytoskeleton. *63*, 141–148.

Petitjean, C., and Dacher, J.N. (2011). A review of segmentation methods in short axis cardiac MR images. Med. Image. Anal. *15*, 169–184.

Reynolds, C.W. (1987). Flocks, herds, and schools: a distributed behavioral model. Comput. Graphics *21*, 25–34.

Riddel, J.P., Aouizerat, B.E., Miaskowski, C., and Lillicrap, D.P. (2007). Theories of blood coagulation. J. Pediatr. Oncol. Nurs. *24*, 123–131.

Ronneberger, O., Baddeley, D., Scheipi, F., Vereer, P.J., Burkhardt, H., Cremer, C., Fahrmeir, L., Cremer, T., and Joffe, B. (2008). Spatial quantitative analysis of fluorescently labeled nuclear structures: problems, methods, pitfalls. Chromosome Res. *16*, 523–562.

Sager, B., and Kaiser, D. (1993). Two cell-density domains within the *Myxococcus xanthus* fruiting body. Proc. Natl. Acad. Sci. U.S.A. *90*, 3690–3694.

Shi, W., Ngok, F.K., and Zusman, D.R. (1996). Cell density regulates cellular reversal frequency in *Myxococcus xanthus*. Proc. Natl. Acad. Sci. U.S.A 93, 4142–4146.

Sliusarenko, O., Neu, J., Zusman, D.R., and Oster, G. (2006). Accordion waves in *Myxococcus xanthus*. Proc. Natl. Acad. Sci. U.S.A. *103*, 1534–1539.

Sliusarenko, O., Zusman, D.R., and Oster, G. (2007). Aggregation during fruiting body formation in *Myxococcus xanthus* is driven by reducing cell movement. J. Bacteriol. *189*, 611–619.

Sozinova, O., Jiang, Y., Kaiser, D., and Alber, M. (2005). A three-dimensional model of myxobacterial aggregation by contact–mediated interactions. Proc. Natl. Acad. Sci. U.S.A. *102*, 11308–11312.

Sozinova O., Jiang, Y., Kaiser, D., and Alber, M. (2006). A three-dimensional model of myxobacterial fruiting-body formation. Proc. Natl. Acad. Sci. U.S.A. *103*, 17255–17259.

Spormann, A.M. (1999). Gliding motility in bacteria: insights from studies of *Myxococcus xanthus*. Microbiol. Mol. Biol. Rev. *63*, 621–641.

Spormann, A.M., and Kaiser, A.D. (1995). Gliding movements in *Myxococcus xanthus*. J. Bacteriol. *177*, 5846–5852.

Starruß, J., Bley, Th., Sogaard-Andersen, L., and Deutsch, A. (2007). A new mechanism for collective migration in *Myxococcus xanthus*. J. Stat. Phys. *128*, 269–286.

Taylor, R.G., and Welch, R.D. (2010). Recording multicellular behavior in *Myxococcus xanthus* biofilms using time-lapse microcinematography. J. Vis. Exp. *42*, pii: 2038.

Thakar, J., Poss, M., Albert, R., Long, G.H., and Zhang, R. (2010). Dynamic models of immune responses: what is the ideal level of detail? Theor. Biol. Med. Model. *7*, 35–42.

Ward, M.J., and Zusman, D.R. (1997). Motility in *Myxococcus xanthus* and its role in developmental aggregation. Curr. Opin. Microbiol. *2*, 624–629.

Welch, R.D., and Kaiser, D. (2001). Cell behavior in traveling wave patterns of myxobacteria. Proc. Natl. Acad. Sci. U.S.A. *98*, 14907–14912.

Wu, Y., Jiang, Y., Kaiser, D., and Alber, M. (2007). Social interactions in myxobacterial swarming. PLoS Comput. Biol. 3, e253.

Wu, Y., Kaiser, A.D., Jiang, Y., and Alber, M. (2009). Periodic reversal of direction allows myxobacteria to swarm. Proc. Natl. Acad. Sci. U.S.A. *106*, 1222–1227.

Wu, Y., Jiang, Y., Kaiser, A.D., and Alber, M. (2011). Self-organization in bacterial swarming: lessons from myxobacteria. Phys. Biol. *8*, 055003.

Xie, C., Zhang, H., Shimkets, L.J., and Igoshin, O.A. (2011). Statistical image analysis reveals features affecting fates of *Myxococcus xanthus* developmental aggregates. Proc. Natl. Acad. Sci. U.S.A. *108*, 5915–5920.

Xu, Z., Kim, O., Kamocka, M., Rosen, E., and Alber, M. (2012a). Models of Thrombogenesis, Wiley Interdiscip Rev Syst Biol Med. 4(3):237–46.

Xu, Z., Christley, S., Lioi, J., Harvey, C., Sun, W., Rosen, E., and Alber, M. (2012b). Multiscale modeling of fibrin accumulation on thrombus surface and platelet dynamics. Methods Cell Biol. *110*, 367–388.

Zhang, H., Angus, S., Tran, M., Xie, C., Igoshin, O.A., and Welch, R.D. (2011). Quantifying aggregation dynamics during *Myxococcus xanthus* development. J. Bacteriol. *193*, 5164–5170.

Zhang, H., Vaksman, Z., Litwin, D.B., Shi, P., Kaplan, H.B., and Igoshin, O.A. (2012). The mechanistic basis of *Myxococcus xanthus* rippling behavior and its physiological role during predation. PLoS Compt. Biol. *8*, e1002715.

Zhang, Y., Guzzo, M., Ducret, A., Yue-Zhong, L., and Mignot, T. (2012). A dynamic response regulator protein modulates G-protein-dependent polarity in the bacterium *Myxococcus xanthus*. PLoS Genet, *8*, e1002872.

The Mechanism of A-Motility

Jennifer Luciano, Beiyan Nan, David R. Zusman and Tâm Mignot

9

Abstract

Myxococcus xanthus is able to move over solid surfaces as large coordinated groups (social (S) motility) or as single isolated cells (adventurous (A) motility). While S-motility was shown to rely on polar Type-IV pili in the early 2000s, the mechanism of A-motility has long remained mysterious because it occurs in absence of detectable extracellular organelles. In the past, many models have been proposed to explain A-motility. In recent years, with the advent of high-resolution fluorescence microscopy and phylogenomics, it has been possible to identify the A-motility machinery and propose a motility mechanism. In this chapter, we discuss the state-of-the-art in A-motility research and future directions to decipher this long mystery of modern bacteriology.

Myxococcus uses two distinct motility systems: A- and S-motility

In 1979, Hodgkin and Kaiser showed that *M. xanthus* uses two genetically separable motility systems (Hodgkin and Kaiser, 1979a). In this seminal study, a large number of mutants were isolated, indicating that motility is controlled by two distinct sets of genes. One set of genes controlled social (S) motility, the movement of cells in groups, whereas the other set of genes was required for adventurous (A) motility, the movement of individual cells. In their experiments, cells only became completely non-motile when both motility systems were inactivated (Fig. 9.1). However, one locus *mglA*, inactivated both motility systems when mutated. Subsequent molecular studies revealed that A- and S-motility result from two molecular engines. S-motility is a form of twitching motility and results from the cooperation of Type IV pili (T4P) and surface exopolysaccarides (EPS) (Li *et al.*, 2003; Wu and Kaiser, 1995). S-motility is discussed in details in Chapter 10. The mechanism of A-motility remained unknown for a long time, mostly because it functions in the absence of any visible extracellular organelles (Burchard, 1981). In this chapter, we present recent findings that greatly increase our understanding of A-motility.

Characterizing the motility machinery

Searching for A-motility genes

Initial attempts to understand the mechanisms of A- and S-motility involved genetic screens to identify motility mutants. While this approach was successful at showing that S-motility is powered by T4P, it was more difficult to understand the A-motility machinery using this approach (Hodgkin and Kaiser, 1979b; Youderian *et al.*, 2003; Yu and Kaiser, 2007). The early work by Hodgkin and Kaiser identified a collection of A^-S^+ mutants that were non-motile as isolated cells but still able to move by S-motility in groups (Hodgkin and Kaiser, 1977, 1979b). Importantly, these mutants could be classified in two subclasses: *agl* and *cgl*. Intriguingly, *cgl* (contact-transient gliding) mutant were A^-, but motility could be restored transiently when they were mixed with wild type cells or other A^- mutant cells from a different complementation group. Subsequent

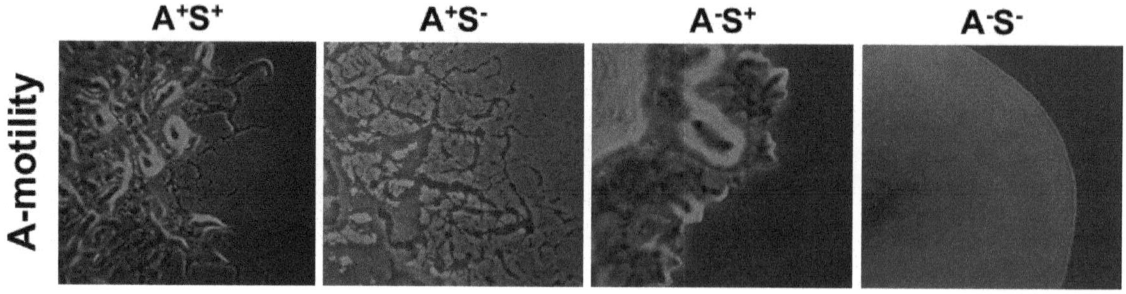

Figure 9.1 *Myxococcus xanthus* utilizes two independent motility systems. Motility phenotype of strains carrying mutations in either motility systems. Wild type *M. xanthus* strain DZ2 with both A- and S- motilities (A^+S^+), cells lacking S-motility (A^+S^-), cells lacking A-motility (A^-S^+) and cells lacking both motility systems (A^-S^-) are shown. A-motility assays were performed on hard (1.5%) agar and single cells swarming out from the edge of colonies were monitored as an indicator of A-motility after 48h at 32°C. See Figure 10.1 in Chapter 10 for S-motility assays on soft agar. A^+S^+ wild type cells show normal motility on both agar concentrations, A^+S^- cells move primarily as individuals, A^-S^+ cells move only as groups and A^-S^- cells are non-motile.

characterization of *cgl* mutations revealed that they caused defects in outer membrane lipoproteins and that rescue results from extracellular complementation involving protein transfer between cells (Hodgkin and Kaiser, 1977; Pathak and Wall, 2012; Pathak *et al.*, 2012) (see also Chapter 5). In contrast, *agl* (adventurous gliding) mutants could not be rescued in cell mixing experiments. In 2003, Youderian *et al.* (Youderian *et al.*, 2003) used the transposable element *mariner* to isolate additional A-motility mutants. This screen identified approximately 40 genes required for gliding motility, providing an important advance in the field. Unfortunately, many of these genes encode polypeptides with homology to proteins with unknown function or involved in general cellular processes, rendering further interpretations of A-motility mechanisms difficult.

Identification of the core A-motility machinery encoding genes

In an effort to identify the motility machinery, the 51 putative gliding motility genes of *M. xanthus* identified by genetic screens were re-annotated and their taxonomic distribution, genomic localization, and phylogeny examined (Luciano *et al.*, 2011). Based on this analysis, fifteen genes were identified to code for the core A-motility machinery based on their presence in other gamma- and delta-proteobacterial genomes (Fig. 9.2A). Remarkably, while these fourteen genes localize in three main gene clusters scattered in the *Myxococcus* genome, seven of them, *gltD, E, G, C* and *aglR, Q, S*, formed a single genetic cluster in some delta- and in many gamma-proteobacteria. The conservation of these genes across diverse phyla suggests that it encodes a common structure or function. The phylogenetic relationships between the genes suggested that motility emerged by functional specialization of this core system and through the acquisition of seven additional genes *gltF, H, I, J, K, B* and *A*) (Fig. 9.2B).

The motility machinery forms a complex in the cell envelope

Three lines of evidence below suggest that *agl* and *glt* genes encode the components of a motility machinery:

1. It has been shown (Luciano *et al.*, 2011; Nan *et al.*, 2010; Sun *et al.*, 2011) that individual deletion of the *agl* and *glt* genes rendered the cells completely deficient in A-motility, whereas S-motility was not affected (A^-S^+).
2. Phylogenomics indicate that *glt* and *nfs* genes were derived by a recent duplication event (Luciano *et al.*, 2011), raising the possible existence of an intriguing link between motility and sporulation. Nevertheless, since the *nfs* genes encode a cell envelope complex required for assembly of the spore coat (Müller *et al.*, 2012), this close relationship suggests that the *glt* genes also encode a cell envelope complex.

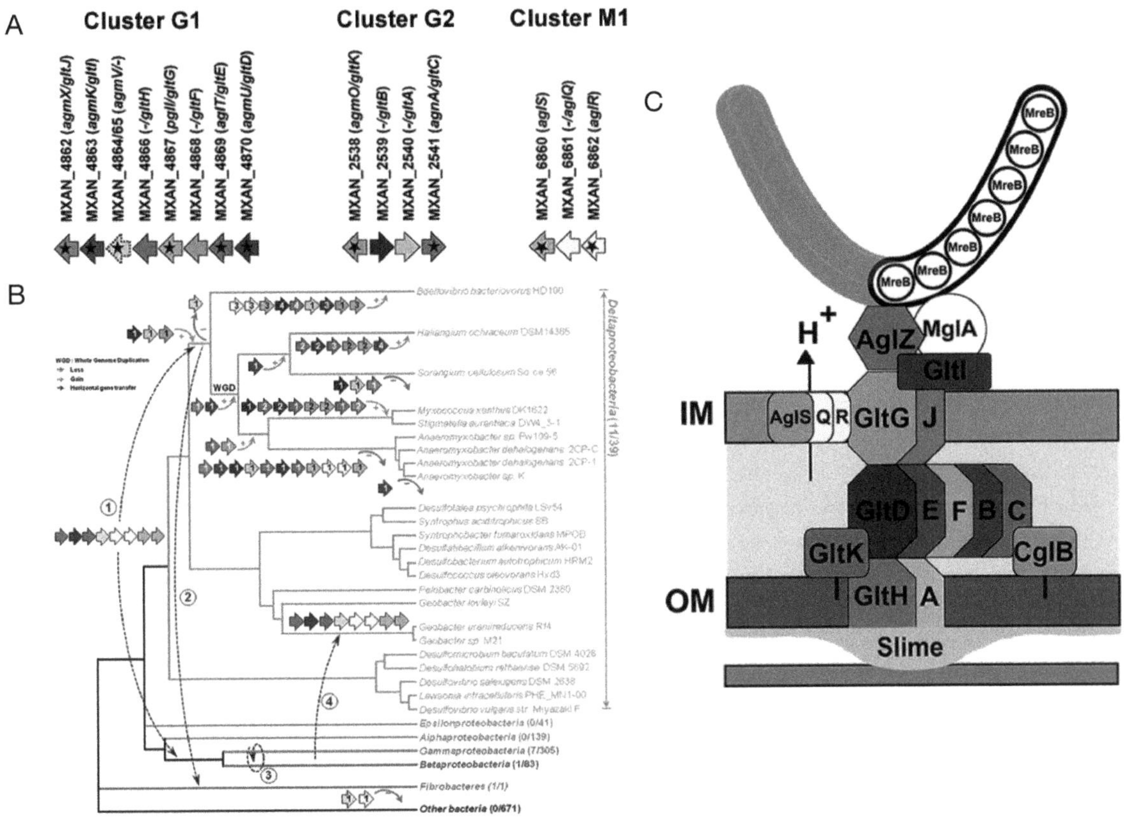

Figure 9.2 A-motility machinery is a PMF-energized envelope-spanning protein complex. (A) Genetic organization of the 15 genes composing the G1, G2 and M1 clusters encoding the core components of the A-motility machinery in *Myxococcus xanthus.* Stars indicate the genes that were hit by transposons in the Youderian *et al.* (2003) and Yu and Kaiser (2007) transposon screens. (B) Evolutionary scenario describing the emergence and evolution of the A-motility machinery in *M. xanthus*. Each gene colour is based on the colour code of the Figure 9.2A. Green and red arrows respectively indicate gene acquisition and gene loss. The number of gene copies that were acquired or lost is indicated within arrows. The purple dotted arrows represent horizontal gene transfer events of one or several components. WGD marks the putative whole genome duplication event that occurred in the ancestor of Myxococcales. (C) Architecture of the *Myxococcus* A-motility machinery. The proposed diagram is compiled from bioinformatic predictions, mutagenesis, pull down and localization studies. Note that direct physical interactions have only been shown between MglA and AglZ, AglR and GltG. The peptidoglycan layer is not shown because its layer with machinery components is not presently clear. Outer membrane proteins belonging to the system are shown integral or exposed in the periplasm based on sequence prediction, including a predicted beta-barrel fold for GltH and lipoprotein anchors for CglB and GltK. MreB may constitute the helical track but it cannot be excluded that the Glt proteins themselves form the track. It is unknown which surface protein(s) may contact the slime. A colour version of this figure is located in the plate section at the back of the book.

3 Combined work by Nan *et al.* (2010), Luciano *et al.* (2011) and Sun *et al.* (2011) showed that *aglR, aglQ, aglS, gltD, E, F, G* and *H* encode proteins of the cell envelope. Pull-down experiments identified a motility complex formed by the GltD, GltE, GltI and GltJ proteins (Nan *et al.*, 2010). AglR was shown to interact with GltG by two-hybrid experiments.

While it seems clear that *agl-glt* genes encode the structural motility machinery, many additional proteins may also be involved. For example, in their pull down experiments Nan *et al.* (2010) found also CglB, AglW and AglZ. CglB is an outer membrane lipoprotein and may constitute a surface-exposed component of the motility complex (Nudleman *et al.*, 2005). AglW is a TolB-like protein and is encoded in an operon encoding a complete Tol-Pal system. In Gram- negative bacteria, the Tol-Pal system is ubiquitous and composed of five core components, TolQ, TolR, TolA, TolB and Pal that span the entire cell envelope to control membrane integrity, cell division and macromolecule transport across the cell envelope (Cascales *et al.*, 2007; Lloubes *et al.*, 2001). The connection between AglW (TolB) and the motility machinery remains unclear but it is possible that AglW co-purified with the motility proteins due to its function in cell envelope integrity.

The AglZ protein was originally identified by the Hartzell laboratory and is essential for A-motility (Yang *et al.*, 2004). However, work by Mauriello *et al.* (2009) has recently shown that AglZ defines a distinct class of A-motility proteins with a regulatory role rather than a structural function. This conclusion came from the finding that the motility defect of the *aglZ* mutant is suppressed when *frzCD*, encoding the receptor protein of the Frz signal transduction system, is disrupted (Mauriello *et al.*, 2009; see also Chapter 11). The function of AglZ is thus clearly distinct from Glt proteins because *frz* mutations did not restore the motility of any of the *glt* deletions (Luciano *et al.*, 2011).

The A-motility machinery may be linked to the MreB cytoskeleton

AglZ and the small G-protein MglA may bridge the motility machinery to the MreB cytoskeleton in the cytoplasm. In bacteria, MreB is an actin homologue that forms polymers on the inner side of the plasma membrane to facilitate the peptidoglycan synthesis by guiding the cell wall synthetic machinery (Domínguez-Escobar *et al.*, 2011). Other functions suggested for MreB include chromosome segregation and the control of cell polarity (Carballido-Lopez, 2006). Mauriello *et al.* (2010) investigated a potential function of *M. xanthus* MreB in A-motility. Immunofluorescence microscopy showed that MreB in *M. xanthus* forms apparent helical structures similar to those observed in other bacteria (Mauriello *et al.*, 2010). The direct involvement of MreB in A-motility was demonstrated using A22, an MreB-specific perturbing agent. It was observed that this agent caused a rapid arrest of gliding in moving cells and disrupted the assembly of the motility complex as observed with an AglZ-YFP fusion (see below, Mauriello *et al.*, 2010). Finally, cross-linking experiments showed that AglZ interacts directly with MreB (Mauriello *et al.*, 2010). Importantly, the Mauriello *et al.* (2010) study also showed that the MglA protein binds to the motility complex in an MreB-dependent way. MglA is a bacterial Ras-like small GTPase that has long been known to be important for A-motility (Hartzell and Kaiser, 1991). Together with AglZ, MglA could bridge the motility complex to the MreB cytoskeleton.

A-motility is propelled by flagellar stator (MotAB) homologues

What proteins energize motility? The *aglRQS* genes encode proteins with homologies to MotA and MotB as well as TolQR/ExbBD. These proteins represent a class of motor proteins that use the proton motive force (PMF) directly to energize various processes in bacteria: flagella rotation, macromolecular transport across the cell envelope, and cell division (Gerding *et al.*, 2007; Minamino *et al.*, 2008; Postle and Larsen, 2007). More specifically, in the case of MotAB or the TolQR system, MotA homologues associate with MotB homologues to form a proton channel in the inner membrane; in both cases conformational changes induced by proton flow generate a power stroke that causes the flagellar rotor to rotate or the TolA protein to contact the outer membrane (Cascales *et al.*, 2000; Wadhams and Armitage, 2004). Based on sequence analysis and the conservation of critical residues, AglR, the MotA homologue, is expected to associate with AglQ and AglS, two MotB homologues, to form active H^+-gated channels in the plasma membrane (Sun *et al.*, 2011). Experiments using PMF-uncoupling drugs (CCCP, Nigericin) demonstrated that A-motility is driven directly by the proton gradient (Nan *et al.*, 2011; Sun *et al.*, 2011). Moreover, a point mutation in a conserved residue of AglQ (D28N), which is predicted to abolish H^+ binding within the lumen of the proton channel, blocked A-motility and paralysed the complex (Sun *et al.*, 2011). Finally, AglR interacts directly with GltG suggesting that the activity of the Agl channel is transduced by a direct interaction with the Glt proteins (Luciano *et al.*, 2011; Sun *et al.*, 2011). How mechanical work from the AglRQS motor is transduced to cell surface to produce movement will be discussed in the next section.

Based on sequence analysis and published experiments, a low-resolution representation of the motility machinery may be presented: The AglRQS motor and GltG may constitute an inner membrane platform linked to the MreB cytoskeleton via proteins such as AglZ and MglA on the cytosolic side. The connection with the substrate would involve a complex of GltA-K spanning the cell envelope (Fig. 9.2C). At the molecular level, the exact topology of the complex remains to de determined. Characterizing the structure and the interactions between individual components of the machinery should tremendously increase our understanding of the motility mechanism.

The motility mechanism

Location of the A-motility engines

Early models of A-motility suggested that it was powered by the directed secretion of slime, an uncharacterized polysaccharide that forms visible trails in the wake of motile cells (Wolgemuth *et al.*, 2002). It was first proposed that slime secretion could directly propel gliding motility if secretory organelles were located at the lagging pole of the cells. According to this model, propulsion would occur when carbohydrate chains hydrate as they are transported through the outer membrane (Wolgemuth *et al.*, 2002). This model stimulated several experiments that eventually showed that this hypothesis was unlikely: First, studies of cephalexin treated, artificially elongated cells suggested that motility was relatively insensitive to increased cell length, which would be expected if the A-motility motors were located at the back of the cells (Sun *et al.*, 1999). Second, in some elongated cells, movement was observed in the anterior portion of the cell body, while the posterior portion remained stationary (Sliusarenko *et al.*, 2007). Third, slime secretion was observed directly in live motile cells and shown to be secreted all along the cell body ((Ducret *et al.*, 2012), see below). These experimental observations are inconsistent with a model of polar engines at the lagging cell pole for A-motility.

An alternative motility model, which is now gaining wide acceptance, proposed that A-motility is powered by motor units along the cell body. This idea was initially suggested by cytological work showing that in motile cells, AglZ localized to clusters distributed along the length of the cell (Mignot *et al.*, 2007). Time-lapse video microcopy revealed that these clusters were initially assembled at the leading pole and appeared to retain a fixed position relative to the substratum as the cell moved forward; the clusters were dispersed when they reached the lagging cell pole (Fig. 9.3A). Later experiments showed similar localization patterns for additional proteins of the motility machinery, such as AglQ, GltD and GltF (Nan *et al.*, 2010, 2011; Luciano *et al.*, 2011; Sun *et al.*, 2011), supporting the model of distributed A-motility motors.

A-motility is propelled by helical rotary assemblies

More recent experiments on the localization of AglZ-YFP directly showed that traction forces are exerted at sites where motility components accumulate (Sun *et al.*, 2011). In these experiments, polystyrene beads were applied to the surface of cells with an optical tweezer and then observed to be transported forward when they collided with an AglZ cluster. As expected, transport was dependent on the activity of the AglRQS complex and was abolished by PMF uncouplers, showing that the Agl–Glt complex drives bead transport (Sun *et al.*, 2011, see below).

How does the A-motility machinery promote cell movement? Dynamic studies of a functional GltD-mCherry fusion showed that this protein co-localizes with AglZ in clusters that contact the surface (Luciano *et al.*, 2011; Nan *et al.*, 2010). However, deconvolution microscopy further revealed that GltD-mCherry also forms a closed helical loop that spans the length of the cell (Fig. 9.3B; Nan *et al.*, 2011). Time-lapse experiments showed that the helix rotates during movement and that rotation depends both on the PMF and the MreB cytoskeleton.

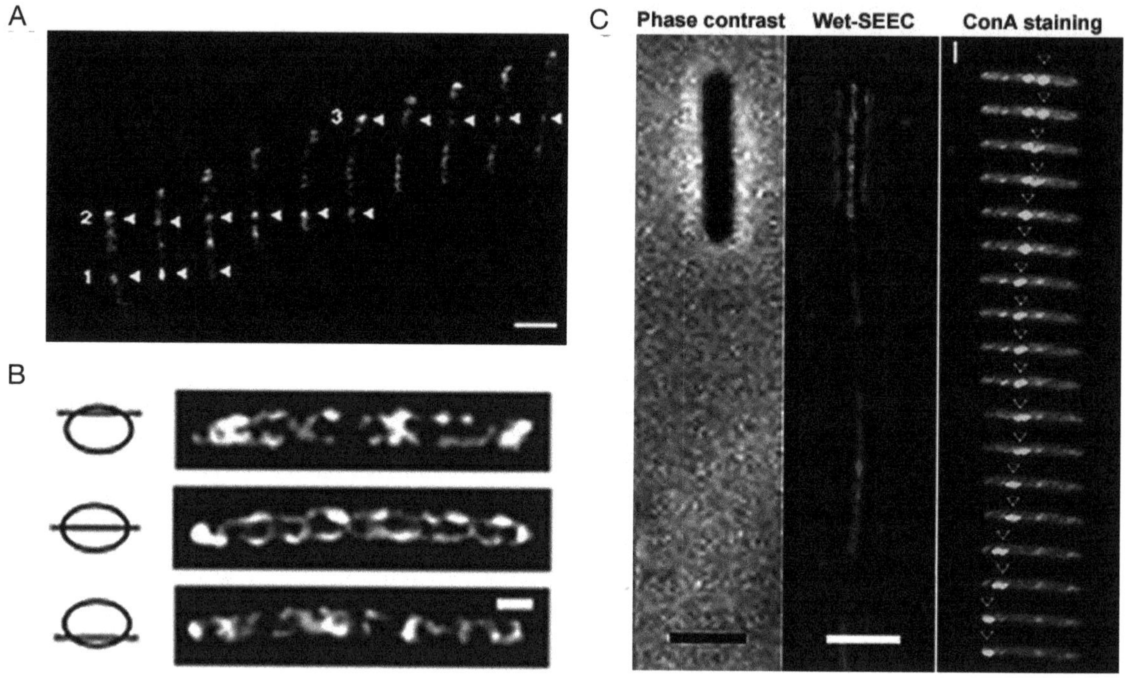

Figure 9.3 A-motility is powered by distributed motors. (A) Time-lapse fluorescence microscopy showing fixed AglZ-YFP clusters in a moving *M. xanthus* cell. Reprinted with permission from Mignot *et al.* (2007). (B) Deconvolution reveals that GltD-mCherry forms a closed loop helical structure in the cell. Reprinted with permission form Nan *et al.* (2011). Scale bar, 1 μm. (C) Slime deposition by moving cells as observed by Wet-SEEC and concanavalin-A staining. Phase contrast and Wet-SEEC images of slime deposition in the wake of a motile cell (left and middle panel). Extended slime segments separated by patchy deposits are only visible by Wet-SEEC. Right: moving Con-A stained clusters at the cell surface. Reprinted with permission from Ducret *et al.* (2012). Scale bars, 1 μm.

Three additional lines of evidence argue that the rotation of GltD is linked to the motility mechanism:

1 Fluorescence Recovery After Photobleaching (FRAP) experiments showed that the GltD-mCherry signal recovered within seconds in both directions suggesting that motility complex proteins move in both directions, consistent with them travelling along a looped helical track, (Nan *et al.*, 2011). Importantly, this result also suggests that GltD itself binds to but does not form the closed loop.
2 When cells were placed on soft agar (or water) where they are non-motile, GltD did not form clusters but the GltD helices kept rotating (Nan *et al.*, 2011). In contrast, cells on hard surfaces showed almost stationary GltD clusters (hard surfaces cause motor proteins to slow down and stall; Nan *et al.*, 2010). This suggests that trafficking GltD-mCherry forms clusters because they stall when they encounter the underlying substrate.
3 TIRF microscopy and AFM probing of the surface indicated cell surface undulations, suggesting the existence of helical depressions (the surface is uneven) (Nan *et al.*, 2011; Pelling *et al.*, 2005).

Together, these findings provide convincing evidence that A-motility is powered by distributed protein complexes that accumulate and provide force at the cell body–substrate interface. According to this model, A-motility is propelled by a PMF-driven helical rotor that exerts force on the substrate as it moves on a cytoskeletal filament.

The current A-motility model

The independent identification of the motility genes by several groups, the deduced phylogenetic history of these genes and interaction data strongly suggest that the A-motility machinery has been identified, at last (Luciano *et al.*, 2011; Nan *et al.*, 2010; Youderian *et al.*, 2003). Furthermore, the dynamics studies discussed above may be compiled into a coherent mechanochemical model to explain A-motility.

The core of the model proposes that motility is propelled by a rotary helical structure that spans the entire cell length. While proteins (MreB, Glt proteins or other proteins) that form the loop remain be identified, AglRQS-motor units may associate with the loop and use PMF as an energy source to track along this structure. In this system, motility would arise from pole-specific regulations of the motility complex. The motility machinery may be activated at the leading cell pole, for example by loading the AglRQS motor with Glt proteins. This way, the Agl motor would transport the complexes along the loop and propel motility when the complexes aggregate in contact with the substratum and adhere to it or create a drag force (Fig. 9.4; see below). At the lagging cell pole, the interaction with the substratum must be disengaged for motility to proceed. Unloading of the Agl motor spatially, for example by removing its interaction with the Glt proteins could alleviate connection to the substratum and allow movement forward. Because the motors would still traffic, their constant movement along the loop would direct them back to the leading cell pole where they would be available for a new cycle (Fig. 9.4). The localization of critical A-motility regulators, MglA and AglZ at the leading cell pole (Mauriello *et al.*, 2010) is consistent with assembly of the motility machinery at the leading cell pole.

Interaction between the motility machinery and the substratum

While the existence of the helical rotor is strongly suggested by the dynamics of the internal parts of the machinery, how the machinery connects with the surface remains unclear. The initial hypothesis by Mignot *et al.* (2007) suggested that A-motility is powered by complexes that span the entire

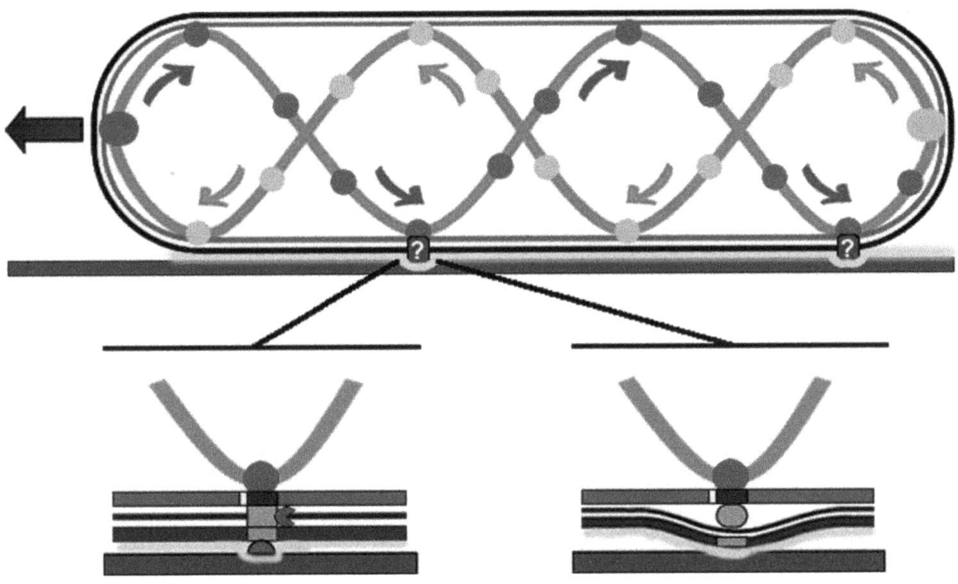

Figure 9.4 A-motility model. The purple arrow represents the direction of movement. Motility motors carrying high-drag (red circles) or low-drag (yellow circles) cargoes translocate along an endless closed loop. Only the high-drag cargoes, presumably active Glt units, are proficient for movement. The machinery could afford the rigid PG by two different ways: The motility complex may span the entire cell envelope and a PG-hydrolase may facilitate insertion of the complex through the PG (left). Alternatively, the motility complex could deform the PG, creating surface depression and drag. Outer membrane proteins may in this system reinforce local contacts at the depressions. A colour version of this figure is located in the plate section at the back of the book.

cell envelope. According to this mechanism, motility would arise as the outer parts of the motility machinery contact the substratum and the rigid helical structure gets threaded through AglRQS motor units. While this is supported by the presence of motility proteins in all envelope layers (with a large number of outer membrane proteins), a difficulty with this hypothesis is the necessity for the complex to span the rigid peptidoglycan at all times. This problem could be solved by motility complex-associated peptidoglycan hydrolases, digesting the cell wall locally and thus allowing penetration of the machinery. However, such hydrolases have yet to be found and no protein of the Glt complex has a predicted hydrolase function.

There is an alternative possibility: Nan *et al.* (2011) suggested that the motility complex does not penetrate the cell wall but rather distorts it and pushes against the outer membrane to create local drag forces against the substratum. In this scenario, the motility proteins appear fixed when observed by fluorescence microscopy because they aggregate at the cell body–substratum interface due to friction forces. TIRF microscopy revealed contact zones between the cell body and the substratum, which could be consistent with the existence of motility machinery-driven cell envelope deformations (Nan *et al.*, 2011).

Whatever the exact mechanism by which the motility machinery the contacts the substratum, it seems to involve slime, a self deposited polysaccharide. Recently, the use of a novel contrast-enhancing microscopy technique (Wet-SEEC: surface-enhanced ellipsometric contrast microscopy in wet condition) showed for the first time that slime is deposited underneath the cell body at approximately constant rates and that the deposition patterns are dependent on motility (Ducret *et al.*, 2012). ConA (concanavalin A)-staining of slime on the surface of moving cells revealed that slime patches are associated with the A-motility machinery (as observed using AglZ-YFP and AglQ-mCherry) and transported along the cell body (Fig. 9.3C). Based on these results, Ducret *et al.* (2012) proposed that the machinery binds slime as a self-generated substratum. In this process, a specific element of the machinery or travelling deformations may enrich slime locally to facilitate energy transduction to the underlying substratum. In the future, it will be crucial to determine the exact composition of slime to elucidate its exact contribution to the motility process.

Conclusion

Significant progress has been made over the last five years, including the first characterization of the motility machinery, its observation by fluorescence microscopy in live cells, and the first experiment-driven motility model. However, many questions still need to be answered. How is the mechanical force generated by motors transduced by the motility apparatus? What is the exact role of the MreB cytoskeleton? How is directionality ensured? What is the composition of slime and how is it recognized by the gliding machinery? How does the machinery contact the cell wall and what is the role of the outer membrane proteins? In the future, high-resolution fluorescence microscopy and biophysical approaches will probably be critical to solving these problems.

References

Burchard, R.P. (1981). Gliding motility of prokaryotes: ultrastructure, physiology, and genetics. Annu. Rev. Microbiol. *35*, 497–529.

Carballido-Lopez, R. (2006). The bacterial actin-like cytoskeleton. Microbiol. Mol. Biol. Rev. *70*, 888–909.

Cascales, E., Gavioli, M., Sturgis, J.N., and Lloubès, R. (2000). Proton motive force drives the interaction of the inner membrane TolA and outer membrane pal proteins in *Escherichia coli*. Mol. Microbiol. *38*, 904–915.

Cascales, E., Buchanan, S.K., Duche, D., Kleanthous, C., Lloubes, R., Postle, K., Riley, M., Slatin, S., and Cavard, D. (2007). Colicin biology. Microbiol. Mol. Biol. Rev. *71*, 158–229.

Domínguez-Escobar, J., Chastanet, A., Crevenna, A.H., Fromion, V., Wedlich-Söldner, R., and Carballido-López, R. (2011). Processive Movement of MreB-Associated Cell Wall Biosynthetic Complexes in Bacteria. Science. 333, 225–228.

Ducret, A., Valignat, M.-P., Mouhamar, F., Mignot, T., and Theodoly, O. (2012). Wet-surface-enhanced ellipsometric contrast microscopy identifies slime as a major adhesion factor during bacterial surface motility. Proc. Natl. Acad. Sci. U.S.A. *109*, 10036–10041.

Gerding, M.A., Ogata, Y., Pecora, N.D., Niki, H., and De Boer, P.A.J. (2007). The trans-envelope Tol-Pal complex is part of the cell division machinery and required for proper outer-membrane invagination during cell constriction in *E. coli*. Mol. Microbiol. *63*, 1008–1025.

Hartzell, P., and Kaiser, D. (1991). Function of MglA, a 22-kilodalton protein essential for gliding in *Myxococcus xanthus*. J. Bacteriol. *173*, 7615–7624.

Hodgkin, J., and Kaiser, D. (1977). Cell-to-cell stimulation of movement in nonmotile mutants of *Myxococcus*. Proc. Natl. Acad. Sci. U.S.A *74*, 2938–2942.

Hodgkin, J., and Kaiser, D. (1979a). Genetics of Gliding Motility in *Myxococcus xanthus* (Myxobacterales) – 2 Gene Systems Control Movement. Molecular & General Genetics *171*, 177–191.

Hodgkin, J., and Kaiser, D. (1979b). Genetics of Gliding Motility in *Myxococcus xanthus* (Myxobacterales) – Genes-Controlling Movement of Single Cells. Molecular & General Genetics *171*, 167–176.

Li, Y., Sun, H., Ma, X., Lu, A., Lux, R., Zusman, D., and Shi, W. (2003). Extracellular polysaccharides mediate pilus retraction during social motility of *Myxococcus xanthus*. Proc. Natl. Acad. Sci. U.S.A. *100*, 5443–5448.

Lloubes, R., Cascales, E., Walburger, A., Bouveret, E., Lazdunski, C., Bernadac, A., and Journet, L. (2001). The Tol-Pal proteins of the *Escherichia coli* cell envelope: an energized system required for outer membrane integrity? Res. Microbiol. *152*, 523–529.

Luciano, J., Agrebi, R., Le Gall, A.V., Wartel, M., Fiegna, F., Ducret, A., Brochier-Armanet, C., and Mignot, T. (2011). Emergence and modular evolution of a novel motility machinery in bacteria. PLoS Genet. *7*, e1002268.

Mauriello, E.M., Nan, B., and Zusman, D.R. (2009). AglZ regulates adventurous (A-) motility in *Myxococcus xanthus* through its interaction with the cytoplasmic receptor, FrzCD. Mol. Microbiol. *72*, 964–977.

Mauriello, E.M., Mouhamar, F., Nan, B., Ducret, A., Dai, D., Zusman, D.R., and Mignot, T. (2010). Bacterial motility complexes require the actin-like protein, MreB and the Ras homologue, MglA. Embo J. *29*, 315–326.

Mignot, T., Shaevitz, J.W., Hartzell, P.L., and Zusman, D.R. (2007). Evidence that focal adhesion complexes power bacterial gliding motility. Science *315*, 853–856.

Minamino, T., Imada, K., and Namba, K. (2008). Molecular motors of the bacterial flagella. Curr. Opin. Struct. Biol. *18*, 693–701.

Müller, F.D., Schink, C.W., Hoiczyk, E., Cserti, E., and Higgs, P.I. (2012). Spore formation in *Myxococcus xanthus* is tied to cytoskeleton functions and polysaccharide spore coat deposition. Mol. Microbiol. *83*, 486–505.

Nan, B., Mauriello, E.M., Sun, I.H., Wong, A., and Zusman, D.R. (2010). A multi-protein complex from *Myxococcus xanthus* required for bacterial gliding motility. Mol. Microbiol. *76*, 1539–1554.

Nan, B., Chen, J., Neu, J.C., Berry, R.M., Oster, G., and Zusman, D.R. (2011). Myxobacteria gliding motility requires cytoskeleton rotation powered by proton motive force. Proc. Natl. Acad. Sci. U.S.A *108*, 2498–2503.

Nudleman, E., Wall, D., and Kaiser, D. (2005). Cell-to-cell transfer of bacterial outer membrane lipoproteins. Science *309*, 125–127.

Pathak, D.T., and Wall, D. (2012). Identification of the cglC, cglD, cglE, and cglF genes and their role in cell contact-dependent gliding motility in *Myxococcus xanthus*. J. Bacteriol. *194*, 1940–1949.

Pathak, D.T., Wei, X., Bucuvalas, A., Haft, D.H., Gerloff, D.L., and Wall, D. (2012). Cell contact-dependent outer membrane exchange in myxobacteria: genetic determinants and mechanism. PLoS Genet. *8*, e1002626.

Pelling, A.E., Li, Y., Shi, W., and Gimzewski, J.K. (2005). Nanoscale visualization and characterization of *Myxococcus xanthus* cells with atomic force microscopy. Proc. Natl. Acad. Sci. U.S.A. *102*, 6484–6489.

Postle, K., and Larsen, R.A. (2007). TonB-dependent energy transduction between outer and cytoplasmic membranes. Biometals *20*, 453–465.

Sliusarenko, O., Zusman, D.R., and Oster, G. (2007). The motors powering A-motility in *Myxococcus xanthus* are distributed along the cell body. J. Bacteriol. *189*, 7920–7921.

Sun, H., Yang, Z., and Shi, W. (1999). Effect of cellular filamentation on adventurous and social gliding motility of *Myxococcus xanthus*. Proc. Natl. Acad. Sci. U.S.A. *96*, 15178–15183.

Sun, M., Wartel, M., Cascales, E., Shaevitz, J.W., and Mignot, T. (2011). Motor-driven intracellular transport powers bacterial gliding motility. Proc. Natl. Acad. Sci. U.S.A. *108*, 7559–7564.

Wadhams, G.H., and Armitage, J.P. (2004). Making sense of it all: bacterial chemotaxis. Nat. Rev. Mol. Cell Biol. *5*, 1024–1037.

Wolgemuth, C., Hoiczyk, E., Kaiser, D., and Oster, G. (2002). How myxobacteria glide. Curr. Biol. *12*, 369–377.

Wu, S.S., and Kaiser, D. (1995). Genetic and functional evidence that Type IV pili are required for social gliding motility in *Myxococcus xanthus*. Mol. Microbiol. *18*, 547–558.

Yang, R.F., Bartle, S., Otto, R., Stassinopoulos, A., Rogers, M., Plamann, L., and Hartzell, P. (2004). AglZ is a filament-forming coiled-coil protein required for adventurous gliding motility of *Myxococcus xanthus*. J. Bacteriol. *186*, 6168–6178.

Youderian, P., Burke, N., White, D.J., and Hartzell, P.L. (2003). Identification of genes required for adventurous gliding motility in *Myxococcus xanthus* with the transposable element mariner. Mol. Microbiol. *49*, 555–570.

Yu, R., and Kaiser, D. (2007). Gliding motility and polarized slime secretion. Mol. Microbiol. *63*, 454–467.

Type IV Pili and Exopolysaccharide-dependent Motility in *Myxococcus xanthus*

10

Zhaomin Yang, Chengyun Li, Carmen Friedrich and Lotte Søgaard-Andersen

Abstract

Social (S) motility of *Myxococcus xanthus* was discovered in the late 1970s. In the last few decades, especially since the mid-1990s, great strides have been made in our understanding of the molecular mechanisms underlying this form of motility. It is now well established that *M. xanthus* S-motility as well as twitching motility in other bacteria rely on the retraction of the type IV pilus (T4P) to bring about cell translocation. In fact, T4P retraction generates the strongest force ever measured for a biological motor. In this chapter, we review the landmark studies leading to the conclusion that T4P retraction powers these forms of bacterial surface locomotion. The structure of the T4P system in comparison with the type II secretion system is discussed as they are evolutionally related. We additionally describe the requirement of *M. xanthus* S-motility for exopolysaccharides. We will conclude by highlighting some of the unanswered questions, with special attention to *M. xanthus* T4P-dependent motility.

Introduction

The locomotion of myxobacteria on solid surfaces was noted by Thaxter in 1892 when he first described his discovery of these bacteria (Thaxter, 1892). Their two motility systems, known as social (S) and adventurous (A), were revealed many decades later when sophisticated genetics was applied to the model species *Myxococcus xanthus* (Hodgkin and Kaiser, 1977). With ingenuity and some serendipity, the Kaiser group obtained an *M. xanthus* mutant partially deficient in S-motility (Hodgkin and Kaiser, 1977; Wall *et al.*, 1999). Using this strain as the parent, motility mutants with mutations in A-motility genes were isolated (Hodgkin and Kaiser, 1979a). Generalized transduction of these A^- mutants using a wild-type (WT) as the donor allowed the construction of fully S-motile or A^-S^+ strains (Hodgkin and Kaiser, 1979a). These A^-S^+ strains in turn enabled the isolation of non-motile mutants with mutations in S-motility genes (Hodgkin and Kaiser, 1979b). Morphologically (Fig. 9.1 and Fig. 10.1), the advancing colony edge of the WT displays both isolated cells and large cell groups with the latter often referred to as rafts. In contrast, the colony edge of A^-S^+ strains shows only large rafts of cells whereas that of the A^+S^- strains displays predominantly single cells and small cell groups. In the early 1990s (Shi and Zusman, 1993), it was revealed that A-motility, while effective in translocation on hard agar surfaces, is largely non-functional on soft agar surfaces. In contrast, S-motility is particularly effective on a soft surface (Fig. 10.1), and this provides a convenient functional assay for the study of the S-motility system. The two motility systems may allow *M. xanthus* cells to move on very different surfaces in their natural environment.

This chapter will focus on S-motility, which is now well established to depend on the retraction of type IV pili (T4P) and the presence of extracellular polysaccharides or exopolysaccharides (EPS). We will first provide a historical perspective of early advances, followed by a discussion of our current understanding and unresolved questions for future inquiry. Please consult Chapter 9 for a discussion on A-motility. Refer to Chapter 5 for information on the involvement of cell-to-cell

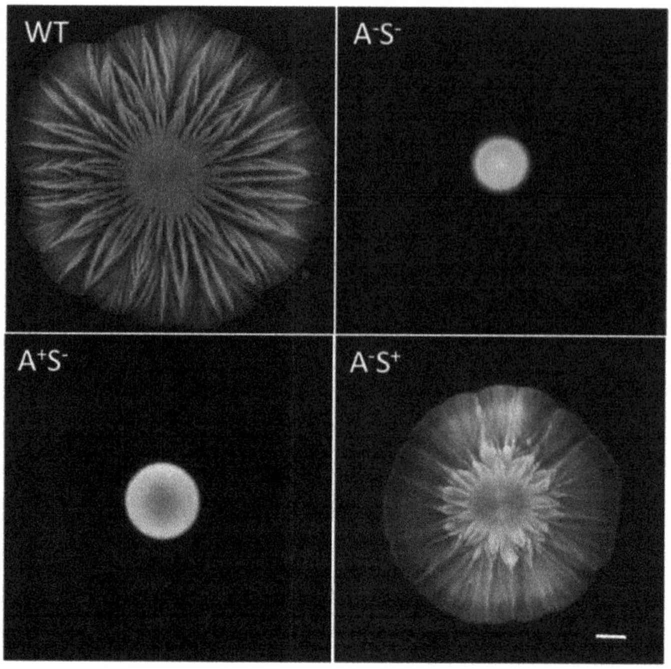

Figure 10.1 Swarming or spreading of *M. xanthus* on soft agar surfaces. Cell suspensions of strains with indicated motility phenotypes were placed in the centre of a nutrient plate with 0.4% agar. Photographs were taken after 5 days of incubation at 32°C. Scale bar, 5 mm.

protein transfer in T4P-dependent motility and Chapter 11 for the sensory regulation of *M. xanthus* motility systems.

Early advances: pili and exopolysaccharides in *Myxococcus xanthus* S-motility

Investigations prior to the mid-1990s uncovered critical pieces of the S-motility puzzle, one of which was the correlation of S-motility defects with the absence of pili. Once S-motility was discovered genetically (Hodgkin and Kaiser, 1979b), this was quickly followed by the observation that many S-motility mutants lack a thin pilus at their cell poles (Kaiser, 1979). Rosenbluh and Eisenbach (Rosenbluh and Eisenbach, 1992) extended this association by mechanically depiliating cells. They observed that depiliated cells of genetically S^+ strains showed motility behaviours similar to those of S^- mutants. These early results, while circumstantial, strongly argued for the direct involvement of polar pili in *M. xanthus* S-motility.

The second important piece to the puzzle was the requirement of S-motility for a cell surface material initially termed extracellular matrix (ECM) fibrils or the matrix material (Arnold and Shimkets, 1988; Shimkets, 1986). It is now known that the functionally relevant constituent of the ECM in S-motility is EPS. Major contributions came from the studies of an S^- mutant known as *dsp* (for *dis*persed growth) by Shimkets (Shimkets, 1986). This *dsp* mutant, besides the lack of S-motility, was found to be severely defective in cellular cohesion by agglutination assays (Shimkets, 1986). Studies by scanning electron microscopy (EM) indicated that WT *M. xanthus* cells existed in clusters in an ECM in which cells are linked by a network of fibrils of variable dimensions (Arnold and Shimkets, 1988; Behmlander and Dworkin, 1991). In contrast, *dsp* mutant cells rarely formed clusters and they lacked this matrix or network of fibrils. Because *dsp* mutant cells were piliated at least to the same extent as the WT (Arnold and Shimkets, 1988), this matrix material was the second extracellular component implicated in S-motility. It should be noted that the fibrils were likely an artefact of sample preparations. More recent studies indicate that live *M. xanthus* cells are covered by a coat of ECM (Merroun *et al.*, 2003; Palsdottir *et al.*, 2009; Pelling *et al.*, 2005). The ECM consists of EPS and proteins in a roughly 1:1 ratio (Behmlander and Dworkin, 1994). The proteins in the ECM have diverse functions (for a review see Konovalova *et al.*, 2010). The focus here will be on EPS as it pertains to S-motility.

Because the formation and maintenance of cell groups was considered a prerequisite for the manifestation of S-motility, the roles of both the polar pili and the 'ECM fibrils' in cellular cohesion dominated the thinking of how S-motility might function in the early days. Based on the observation that piliated *M. xanthus* cells tend to clump in liquid culture, it was suggested that pili might bind cells together or sense the presence of another cell to enable S-motility (Kaiser, 1979). Along the same line, it was suggested that pili could provide means of cooperativity or coordination among cells in S-motility (Rosenbluh and Eisenbach, 1992). Since the presence of EPS correlated with cellular cohesion (Arnold and Shimkets, 1988; Behmlander and Dworkin, 1991; Dana and Shimkets, 1993), which correlated with S-motility (Shimkets, 1986), it was assumed that EPS could provide the major cohesive force required for S-motility (Arnold and Shimkets, 1988). These ideas would explain the mechanisms of cell clustering or group formation, but the mechanisms of S-motility or force generation remained the key issue yet to be resolved at that time.

Current model for S-motility

It is now well established that the type IV pilus (T4P) in *M. xanthus* and other bacteria can function as a motor by pulling cells forward once a T4P tethers or attaches to a biotic or abiotic anchor through its distal end (Mattick, 2002). Under laboratory conditions, the attachment site for *M. xanthus* T4P may be the EPS on an adjacent cell or a suitable surface. In the current model for S-motility (Fig. 10.2), a cell first extends its polar T4P. The extended T4P may then adhere to the EPS on an adjacent cell or other appropriate surfaces in the environment nearby. Due to this adherence, retraction resulting from the disassembly of the T4P at the proximal end pulls a cell forward. The continued cycles of extension and retraction explain how T4P allows a cell to move and why T4P-dependent motility only functions when cells are placed on a surface. *M. xanthus* cells can reverse their direction of movement periodically. Reversals involve the disassembly of T4P at one pole and their reassembly at the opposite pole and their subsequent retraction. The following few paragraphs will highlight some of the landmark studies that laid the foundation for realizing that T4P powers S-motility and EPS functions as a trigger for T4P retraction in *M. xanthus*.

T4P is a retractile motor

An important turning point in S-motility studies was the discovery that the polar pili required for *M. xanthus* S-motility belong to the family of T4P (Wu and Kaiser, 1995). It was well known by then that many Gram-negative bacteria including *Pseudomonas aeruginosa* depended on their T4P to perform a form of surface translocation called twitching motility (Henrichsen, 1975a; Mattick *et al.*, 1996). The revelation that S-motility had a similar requirement for T4P was critical because it merged the studies of *M. xanthus* S-motility with twitching and made the extensive literature on

Figure 10.2 Current model for S-motility. The distal end of an extended T4P from one cell attaches to the EPS on a neighbouring cell or the agar surface with or without EPS deposited. This attachment triggers the retraction of T4P at its proximal end to result in cell movement. The direction of the arrow below or above the T4P indicates T4P retraction (pointing left) or extension (pointing right).

bacterial twitching applicable to *M. xanthus* (Wall and Kaiser, 1999). The key, as stated by Wu and Kaiser (Wu and Kaiser, 1995), is that this discovery brought to the forefront the very possibility that the T4P system could provide the engine that powers *M. xanthus* S-motility.

Here, we briefly review some of the early literature on twitching motility and its association with the T4P. In a 1972 review (Henrichsen, 1972), twitching was defined as a flagellum-independent form of surface translocation resulting in intermittent and jerky movement. The link between pili, or fimbriae, with bacterial twitching was first observed in *Moraxella* species (Henrichsen *et al.*, 1972) and subsequently extended to other twitching bacteria (Mattick *et al.*, 1996). While the reasons for this requirement could only be speculated in the motility field early on, a separate line of research on phages by Bradley (1972) proved instrumental. He provided the earliest evidence for pilus retraction as he observed the shortening of *P. aeruginosa* polar pili when phage particles attached to these pili as their host receptors. Most of his phage resistant mutants were unpiliated as expected. However, a fortuitous resistant isolate was found to be piliated (Bradley, 1974). In comparison with the WT, this phage-resistant mutant possessed more pili, but they did not shorten upon phage infection. Bradley concluded that while the structure of these pili was not altered, they were paralysed or non-retractile (NR). Later the connection between pilus retraction and motility was established when it was observed that the *P. aeruginosa* NR mutants failed to twitch just as non-piliated strains (Bradley, 1980). He concluded that pilus retraction, and not its mere physical presence or construction (Henrichsen, 1975b), was required for *P. aeruginosa* to twitch (Bradley, 1980). Moreover, it was proposed that twitching motility was the results of pilus extension and retraction and that cell movement was produced by pilus adhesion followed by retraction. By the early 1990s, it was known that Bradley's NR mutants harboured mutations in *pilT*, which encodes a homologue of ATPases required for protein export (Whitchurch *et al.*, 1991). This discovery spurred the idea that PilT might hydrolyse ATP to power T4P retraction and twitching. The hypothesis of T4P retraction as the driving force for twitching, while yet to be proven by the 1990s, was well supported by circumstantial evidence and widely accepted as one of the most viable among different models (Wall and Kaiser, 1999).

The association of T4P with bacterial surface translocation received support from additional studies in *M. xanthus* (Rodriguez-Soto and Kaiser, 1997; Wu and Kaiser, 1996, 1997; Wu *et al.*, 1997, 1998) and *Neisseria gonorrhoeae*, another important bacterium that moves by twitching motility (Drake and Koomey, 1995; Wolfgang *et al.*, 1998a,b). Of particular significance was the discovery that *pilT* mutants of *M. xanthus* and pathogenic *Neisseria* were piliated but defective in surface translocation (Carbonnelle *et al.*, 2006; Wolfgang *et al.*, 1998a; Wu *et al.*, 1997). *M. xanthus pilT* mutants were reported to be hyperpiliated (Li *et al.*, 2003; Wu *et al.*, 1997) as expected if PilT functions in T4P retraction. Because T4P are peritrichously arranged in *Neisseria*, it was unclear if *pilT* mutant cells of these diplococci were more piliated than WT (Biswas *et al.*, 1989; Wolfgang *et al.*, 1998a), but *pilT* mutations restored piliation to certain T4P$^-$ *pil* mutants in *N. gonorrhoeae* and *N. meningitidis* (Carbonnelle *et al.*, 2006; Wolfgang *et al.*, 1998b). These observations, which all supported the hypothesis that PilT functions as an ATPase to stimulate T4P retraction, brought the T4P motility field near the cusp of a breakthrough.

That breakthrough came in early 2000 by the application of advanced biophysics tools and microscopy. Using laser tweezers, it was demonstrated that *N. gonorrhoeae* T4P could indeed tether and retract to move *N. gonorrhoeae* cells with pulling forces exceeding 80 pN (Merz *et al.*, 2000). In fact, T4P retraction generates the strongest force measured so far for any known biological motors (Clausen *et al.*, 2009; Holz *et al.*, 2010; Maier *et al.*, 2002). It was also determined that the speed of T4P retraction correlated with that of cell translocation by twitching motility. While the piliated *pilT* mutant cells tethered to biotic or abiotic surfaces by its pilus, they displayed no or little detectable retraction. More traditional video microscopy provided indirect evidence for the involvement of T4P retraction in *M. xanthus* S-motility (Sun *et al.*, 2000). Finally, in 2001, the remarkable direct visualization of T4P retraction was achieved in *P. aeruginosa* (Skerker and Berg, 2001). Using total internal reflection fluorescence microscopy, Skerker and Berg observed T4P extension and retraction as well as retraction-mediated cell movement. In total, these results proved unequivocally that T4P retraction, which is PilT dependent, can generate enough force to move cells over a solid surface in T4P-dependent bacterial motility, be it twitching or *M. xanthus* S-motility (Merz *et al.*, 2000; Skerker and Berg, 2001; Sun *et al.*, 2000).

EPS as a receptor and trigger for T4P retraction

The early discovery that S-motility requires EPS (Arnold and Shimkets, 1988; Behmlander and Dworkin, 1991; Shimkets, 1986) was further solidified by later investigations of additional S^- mutants. Of particular interest were the discoveries following the identification of the Dif chemosensory system and its role in EPS regulation (Yang *et al.*, 1998; Yang *et al.*, 2000b) (see below). With the knowledge that T4P retractions were possibly responsible for S-motility (Wu and Kaiser, 1995; Wu *et al.*, 1997), it was speculated that *M. xanthus* EPS could function as a signal to activate the S-motility specific motor in a contact-dependent manner (Yang *et al.*, 2000b). More importantly, it was shown that purified *M. xanthus* EPS interact with T4P and this interaction may trigger T4P retraction in a PilT-dependent manner (Li *et al.*, 2003). Interestingly, *M. xanthus* EPS^- cells displayed EPS-independent but T4P-dependent movement at the single cell level in a viscous environment over certain solid surfaces (Hu *et al.*, 2011), indicating a certain degree of receptor promiscuity in T4P tethering. However, this EPS-independent movement under a viscous layer of liquid may be artificial physiologically; a critical hallmark of *M. xanthus* S-motility is the macroscopic movement over an air–solid interface by this strictly aerobic microbe (Fig. 10.1). Regardless, there is no doubt that a T4P can attach to a relatively stationary and suitable receptor and the ensuing retraction will move cells over a solid surface in either twitching or *M. xanthus* S-motility. In the latter case, T4P may preferentially attach to EPS on a neighbouring cell to allow concerted movement of cells in large groups or rafts. There have been conflicting reports for the requirement of lipopolysaccharide (LPS) O-antigen by *M. xanthus* S-motility (Bowden and Kaplan, 1998; Fink and Zissler, 1989; Yang *et al.*, 2000a). Because some of the O-antigen mutants are clearly proficient in S-motility (Fink and Zissler, 1989; Yang *et al.*, 2000a), O-antigen per se is therefore not critical for S-motility. Rather, some of the O-antigen biosynthetic genes may be required for EPS production, which explains the S-motility defects of selected O-antigen mutants.

Structure of T4P in comparison with T2SS

It has long been recognized that T4P forms a supramolecular structure evolutionally related to that of the type II secretion system (T2SS) (Burrows, 2012; Korotkov *et al.*, 2012; Peabody *et al.*, 2003; Planet *et al.*, 2001). T2SS is well known for its role in protein transport across the outer membrane in Gram-negative bacteria (Korotkov *et al.*, 2012). Consequently, T2SS has been the subject of extensive scientific inquiry, which in turn has contributed to the understanding of T4P. In recent years, high resolution structures of T4P proteins along with their T2SS counterparts have become available (Table 10.1). These structural studies together with other approaches have provided insights into the organization of the supramolecular structure that enables T4P extension and retractions. This section will focus on the structural organization of the proteins required for T4P assembly and function with references to the homologous T2SS when appropriate (Fig. 10.3A).

Based on the conservation of proteins, there are two T4P subtypes, **a** and **b** or T4Pa and T4Pb (Burrows, 2012; Craig *et al.*, 2003; Pelicic, 2008; Shaw and Taylor, 1990). Generally, the **a** subtypes are dynamic with cycles of T4P extension and retraction capable of powering bacterial surface motility; T4Pb only extends and does not retract (Burrows, 2012; Pelicic, 2008). The T4P in *M. xanthus* as well as in *P. aeruginosa* and *Neisseria* species is of T4Pa and is the focus of the discussion here. All T4Pa systems share a core set of 10 structural proteins, most of which have

Table 10.1 Conserved T4P proteins and their T2SS counterparts

T4P proteins[a]					T2SS proteins[b]	
M&P	Ngo	Tt	PDB ID [c]	Function/localization [d]	Name	PDB ID[c]
PilA	PilE	PilA4	1OQW (*Pa*)	Pilus subunit/IM and pilus	GspG	3G20 (*Ec*)
PilB	PilF	PilF		Assembly ATPase/cytoplasmic	GspE	1P9R (*Vc*)
PilC	PilG	PilG	2WHN (*Tt*)	Unknown/IM; multi-TM	GspF	2VMA (*Vc*)
PilD	PilD	PilD		Prepilin peptidase/IM; multi-TM	GspO	
Tgl[a]	PilW	?	2VQ2 (*Ngo*)	Pilotin/OM; lipoprotein	GspS	3SOL (*Ec*)
PilM	PilM	PilM	2YCH (*Tt*)	Unknown/cytoplasmic	$GspL_{cyto}$	2BH1 (*Vc*)
PilN	PilN	PilN		Unknown/IM; single TM	$GspL_{peri}$	2W7V (*Vp*)
PilO	PilO	PilO	2RJZ (*Pa*)	Unknown/IM; single TM	GspM	1UV7 (*Vc*)
PilP	PilP	PilW	2LC4 (*Pa*)	Secretin associated/IM; lipoprotein	GspC	3OSS (*Ec*)
PilQ	PilQ	PilQ		Secretin/OM	GspD	3EZJ (*Ec*)
PilT	PilT	PilT	3JVV (*Pa*)	Retraction ATPase/cytoplasmic	N/A	N/A

[a]T4P proteins are listed for *M. xanthus* (*Mx*) and *P. aeruginosa (Pa)* in the first column (M&P). *M. xanthus* Tgl and *P. aeruginosa* PilF are homologues. The second and third columns list the corresponding proteins in *N. gonorrhoeae (Ngo)* and *Thermus thermophilus (Tt)*. Please note the differences in nomenclature. N/A: not applicable.
[b]Counterparts of T4P proteins in T2SS is indicated by its General Secretion Pathway (Gsp) nomenclature. The subscripts 'cyto' and 'peri' indicates the cytoplasmic and the periplasmic part of GspL, respectively.
[c]The Protein Data Bank (PDB) ID for the high-resolution structure of a protein is provided. If multiple structures are available for the same protein, only one is provided with preference given to one with a higher resolution or of a more complete protein. Please see a structure at http://www.pdb.org and the reference therein. The source organism for each protein is provided in parenthesis following each ID: *Escherichia coli* (*Ec*), *Vibrio cholerae* (*Vc*), and *Vibrio parahaemolyticus* (*Vp*).
[d]IM, inner membrane; OM, outer membrane; TM, transmembrane helix/helices.

counterparts in T2SS (Table 10.1 and Fig. 10.3A) (Burrows, 2012; Pelicic, 2008). In *M. xanthus*, genes for 9 of these 10 proteins are found in the *pil* locus with *tgl* located elsewhere (Wall and Kaiser, 1999). Various studies and comparisons with T2SS (Burrows, 2012; Korotkov *et al.*, 2012; Mattick, 2002) suggest that these 10 proteins form an envelope-spanning protein complex that traverses the cytoplasm, inner membrane (IM), periplasm, outer membrane (OM) and beyond (Fig. 10.3A). Nine of these proteins are required for T4P assembly with PilT required only for retraction.

The T4P filament, which is largely composed of a major pilin subunit, is a thin and somewhat flexible structure about 5–8 nanometres in diameter and several micrometres in length (Craig and Li, 2008; Pelling *et al.*, 2005; Skerker and Berg, 2001). Mature pilins, which are processed from the prepilin PilA by the transmembrane (TM) prepilin peptidase PilD (Nunn and Lory, 1991; Wu *et al.*, 1998), are distributed in the IM possibly as monomers when not assembled into the pilus (Fig. 10.3A) (Morand *et al.*, 2004; Skerker and Berg, 2001; Yang *et al.*, 2010). Assembly or extension of the T4P filament requires the polymerization of these IM-embedded pilin molecules whereas retraction results from their depolymerziation and redistribution into the IM (Morand *et al.*, 2004; Skerker and Berg, 2001). As such, both the polymerization (extension) and the depolymerization (retraction) occur at the base of the pilus. In T2SS, a pilus-like structure is also assembled from mature pseudopilins processed from pre-pseudopilins by a PilD-like peptidase, GspO (Fig. 10.3A) (Ayers *et al.*, 2010; Korotkov *et al.*, 2011). This so-called pseudopilus is thought to serve as a piston-like structure to push secreted proteins through the OM secretion channel (Korotkov *et al.*, 2012). While the pseudopilus does not extend visibly beyond the cell envelope in a WT, overexpression of the major pseudopilin did lead to the formation of a T4P-like surface filament (Sauvonnet *et al.*, 2000), further validating

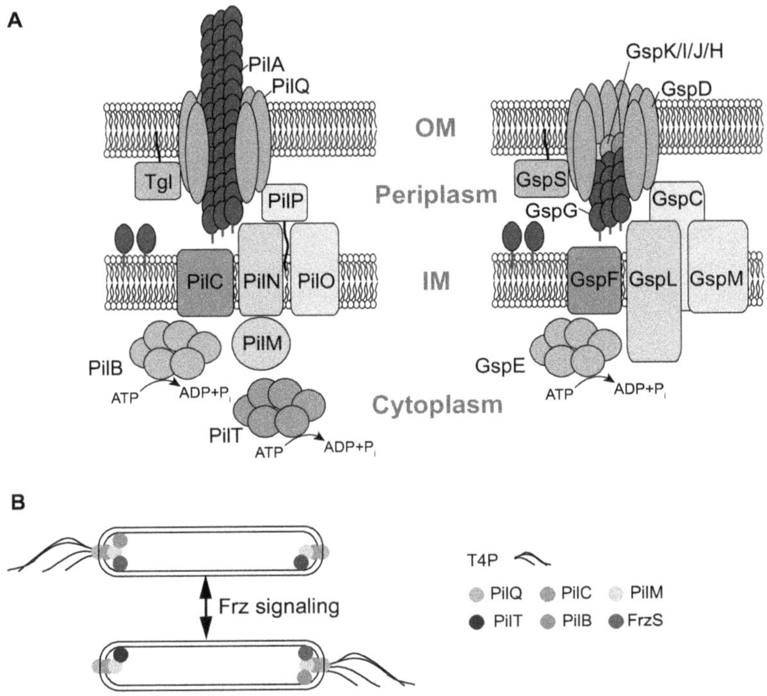

Figure 10.3 Organization of T4P and T2SS envelope structures. (A) Models for T4P (left) and T2SS (right) are shown. Homologous proteins and/or structural analogues are indicated by the same colours in the two models. Tgl and PilP are lipoproteins with attached black lines indicating their lipid moiety. For convenience, PilA here represents mature pilin instead of prepilin. See text for details. (B) Regulation of T4P localization. The two cells demonstrate the localization of T4P proteins before and after a Frz-induced reversal. The leading cell pole is the one with T4P. Shown are the cellular locations of six proteins as colour coded in the legend on the right. A colour version of this figure is located in the plate section at the back of the book.

the close evolutionary relationships of T2SS and T4P.

The pilus filament traverses the OM through a channel formed by the multimerization of the secretin PilQ (Burkhardt *et al.*, 2011; Collins *et al.*, 2001; Helm *et al.*, 2007) (Fig. 10.3A). The OM lipoprotein Tgl [PilF in *P. aeruginosa* (Koo *et al.*, 2008)] facilitates secretin multimerization (Koo *et al.*, 2008; Nudleman *et al.*, 2006). The secretin multimer is proposed to form the OM conduit or channel for T4P (Fig. 10.3A), which may be connected to the IM by the IM lipoprotein PilP, (Drake *et al.*, 1997). It has been suggested additionally that PilP may control the gating of the PilQ channel (Balasingham *et al.*, 2007). The structural homologues of PilQ, Tgl and PilP in the T2SS are GspD, GspS and GspC (Table 10.1 and Fig. 10.3A), respectively. It is noted that GspC is a TM protein; this is in contrast to PilP which is a lipoprotein anchored to the IM by a lipid moiety. The overall interactions and domain structures between these proteins in T4P and T2SSS are conserved and they are proposed to interact in a similar fashion in both systems (Korotkov *et al.*, 2012) (Fig. 10.3A).

PilC, PilN and PilO are the three integral IM T4P proteins, (Fig. 10.3A). PilC is the only T4P protein proposed to traverse the IM multiple times. While its exact membrane topology and interacting partners remain to be clarified (Arts *et al.*, 2007; Blank and Donnenberg, 2001; Georgiadou *et al.*, 2012; Karuppiah *et al.*, 2010; Py *et al.*, 2001), its null mutations results in lack of T4P and motility defects in *M. xanthus*, *P. aeruginosa* and *Neisseria* (Nunn *et al.*, 1990; Tonjum *et al.*, 1995; Wu *et al.*, 1997). This underscores the vital role of PilC in T4P assembly and/or in the modulation of T4P dynamics. The other two, PilN and PilO, have a single TM helix each and they interact with each other through their periplasmic domains (Sampaleanu *et al.*, 2009). This PilN-PilO IM complex may additionally interact with the IM lipoprotein PilP, which is also suggested to interact with the OM secretin channel PilQ as discussed above (Georgiadou *et al.*, 2012; Tammam *et al.*, 2011). These observations would suggest a complex consisting of PilN, PilO and PilP in the IM and PilQ in the OM (Fig. 10.3A). It is possible that this multicomponent complex is linked to PilC via PilO (Georgiadou *et al.*, 2012). The periplasmic portion of the assembled pilus may also mediate additional interactions between the T4P components in the IM and OM (Jain *et al.*, 2011). The structural homologue of PilC in the T2SS is GspF, which likely has three TM helices (Korotkov *et al.*, 2012) (Fig. 10.3A). GspM and the periplasmic C-terminal domain of GspL ($GspL_{peri}$) in T2SS are structural analogues of PilO and PilN, respectively (Ayers *et al.*, 2010; Korotkov *et al.*, 2011) (Table 10.1 and Fig. 10.3A). The interactions between GspM and $GspL_{peri}$ have been demonstrated (Johnson *et al.*, 2007; Sandkvist *et al.*, 1999), supporting the interaction of PilN-PilO in T4P as shown in Fig. 10.3A.

Three cytoplasmic T4P proteins, PilM, PilB and PilT, are central to T4P function and dynamics (Fig. 10.3A). PilM is an actin- and FtsA-like protein homologous to the cytoplasmic N-terminal domain of GspL ($GspL_{cyto}$). Recently, PilM was crystalized in a complex with a short cytoplasmic N-terminus of PilN, indicating an intimate interaction between PilM and PilN (Georgiadou *et al.*, 2012; Karuppiah and Derrick, 2011). This interaction is consistent with the observation that their counterparts in T2SS are fused into a single protein, GspL (Fig. 10.3A) (Ayers *et al.*, 2010; Korotkov *et al.*, 2011). PilB and PilT are both ATPases of the secretion ATPase superfamily (Peabody *et al.*, 2003; Planet *et al.*, 2001). Structural and biochemical studies indicate that PilT forms a dynamic hexamer with ATPase activity (Jakovljevic *et al.*, 2008; Satyshur *et al.*, 2007). For all intents and purposes, PilT is the ATPase that powers T4P retraction whereas PilB, whose structure has yet to be determined, likely provides energy for T4P assembly (Ayers *et al.*, 2010; Jakovljevic *et al.*, 2008; Korotkov *et al.*, 2012; Misic *et al.*, 2010). In comparison, T2SS only contains a single ATPase, GspE, which is homologous to PilB (Korotkov *et al.*, 2012) (Fig. 10.3A). Here, the cytoplasmic ATPase GspE interacts with and is recruited to the IM by the biotopic membrane protein GspL (Abendroth *et al.*, 2005; Sandkvist *et al.*, 1995; Shiue *et al.*, 2006). GspE ATPase activity is stimulated by $GspL_{cyto}$ and acidic phospholipids *in vitro* (Camberg *et al.*, 2007). Once the GspE–GspL complex has formed, GspF, the PilC homologue, is thought to interact with this complex to facilitate the assembly of the T2SS apparatus (Py *et al.*, 2001). Furthermore, because GspG (the major pseudopilin) and GspL can be cross-linked *in vivo* (Gray *et al.*, 2011), it has been proposed that GspL is a coupling protein between the GspE ATPase and the pseudopilin. In this capacity, GspL may serve to transmit conformational changes in GspE induced by ATP hydrolysis across the IM to promote pseudopilus assembly and protein secretion (Gray *et al.*, 2011; McLaughlin *et al.*, 2012). Although interaction between either PilB or PilT with PilC and PilM has yet to be demonstrated in T4P, PilB and PilT are presumed to interact with PilC and the PilM–PilN complex to assembly the T4P multicomponent structure and/or to extend and retract T4P in bacterial motility (Table 10.1 and Fig.10.3A).

In summary, various studies discussed above led to the model in Fig. 10.3A. PilM, PilN, PilO, PilP and PilQ are proposed to assemble into a complex that spans from the cytoplasm, through the IM and periplasm, to the OM. PilM, PilN and PilO form part of an IM complex that may connect to the PilC assembly platform through PilO. In analogy to GspL in T2SS, the PilM and PilN binary complex may act to couple the energy from ATP hydrolysis by PilB and PilT to T4P assembly and disassembly. PilP is proposed to bridge the IM complex to the OM PilQ multimer and may control the closing and opening of this PilQ channel. Overall, this integrated T4P structure makes possible the energy transformation and transduction necessary for T4P to function as the strongest biological motor known to date.

Dynamics of the T4P motor

As has been discussed, T4P-powered surface motility depends on the repetitive cycles of T4P extension and retraction to translocate bacterial cells. A major question in the T4P field is how the switch between T4P extension and retraction is brought about. In the rod-shaped *M. xanthus* cells, T4P is at the leading cell pole in a moving cell (Mignot *et al.*, 2005; Sun *et al.*, 2000). Interestingly, PilB and PilT mostly localize to opposite poles in unexpected patterns (Bulyha *et al.*, 2009). Using immunofluorescence microscopy, it was observed that PilB was unipolar in 40% of the cells and bipolar in the remaining cells with 35% showing asymmetric and 25% showing symmetric bipolar localization. The unipolar and the larger of the two asymmetric PilB clusters localize to the pole opposite of an A-motility protein that localizes to the lagging pole. This led to the conclusion that PilB predominantly localizes to the leading cell pole. PilT localization was followed by live cell fluorescence microscopy using a YFP-tagged functional PilT (Bulyha *et al.*, 2009). Surprisingly, PilT was found to mostly localize to the lagging pole and with only occasional accumulation at the leading cell pole. Based on these observations, it was proposed that the spatial separation of PilB and PilT would allow the temporal separation of T4P extension and retraction (Bulyha *et al.*, 2009). In this model, T4P only assembles at the leading pole which possesses the predominant or the only PilB cluster. The occasional accumulation of PilT at the leading pole would result in T4P retraction to move cells forward.

M. xanthus cells can reverse their direction of movement periodically with WT cells doing so every 7 to 15 min on average (Blackhart and Zusman, 1985). Reversals are regulated predominantly by the Frz chemosensory system (see details in Chapter 11). Without the Frz system, cells seldom reverse their direction of movement. During a reversal, the previous leading pole becomes the new lagging pole and vice versa. It was observed that T4P localizes to the leading cell pole, and during a reversal, the assembly of T4P must switch poles (Mignot *et al.*, 2005; Sun *et al.*, 2000). The A- and S-motility systems in *M. xanthus* have been shown to function synergistically (Kaiser and Crosby, 1983); this brings up the question of how the two motility systems are coordinated, which is addressed in Chapter 11. Another question is how the assembly of T4P switches from one pole to the other during a reversal, which is discussed below.

Localization studies of *M. xanthus* proteins required for T4P function suggest that a subset of T4P proteins play critical roles in switching the poles of T4P assembly (Fig. 10.3B). FrzS is required for *M. xanthus* S-motility (Ward *et al.*, 2000). Using an active FrzS–GFP fusion, it was shown that FrzS predominantly localizes to the leading pole in moving cells, and during a reversal, it relocates to the new leading cell pole in a Frz-dependent manner (Mignot *et al.*, 2005). More importantly, PilT and PilB, which localize in either a unipolar or an asymmetric bipolar pattern as discussed above (Bulyha *et al.*, 2009), switch poles during a cell reversal in a Frz-dependent manner (Bulyha *et al.*, 2009). In addition, a FLaSH-tagged Tgl protein was shown to localize in a unipolar pattern (Nudleman *et al.*, 2006). However, not all T4P proteins show unipolar or asymmetric biopolar localization. The OM secretin PilQ localizes to both cell poles with similar sized clusters (Bulyha *et al.*, 2009; Nudleman *et al.*, 2006). Moreover, the IM protein PilC and the cytoplasmic protein PilM, the latter followed using an active PilM–YFP fusion, were also found to localize in a bipolar, symmetric pattern (Bulyha *et al.*, 2009). These observations indicate that the localization of T4P proteins falls into two categories in relationship to cellular reversal: some are more stationary such as PilC, PilM and PilQ which localize to both poles symmetrically; others are more dynamic such as PilT and PilB which are either unipolar or asymmetrically bipolar. Based on these observations, it has been suggested that *M. xanthus* cells contain at both poles preassembled complexes of T4P proteins that include PilM, PilC, PilQ and possibly others. On the other hand, it is the dynamically localized proteins such as FrzS, PilB and PilT that determine at which pole T4P assembles (Bulyha *et al.*, 2009) (Fig. 10.3B). In other words, during a reversal, and in a Frz-dependent manner, a previously functional T4P assembly complex at the previous leading cell pole is disabled and a new one is enabled to function at the new leading pole thanks to those dynamically localized proteins.

Regulation of EPS accumulation in *Myxococcus xanthus*

The accumulation of EPS in *M. xanthus* is regulated by an extensive network of signal transduction pathways (Fig. 10.4). While the only known EPS biosynthesis genes are mapped to the single *eps* locus (Lu *et al.*, 2005), many loci house genes that regulate EPS accumulation. Here, we will discuss in more detail the Dif chemosensory system and the SgmT-DigR two-component system, which are the best studied *M. xanthus* EPS regulators.

Five genes at the *dif* locus encode proteins homologous to the core components of a chemosensory system (Fig. 10.4) (Black and Yang, 2004; Yang *et al.*, 1998). DifA, DifC and DifE, which are the respective orthologues of MCP (methyl-accepting chemoreceptor proteins), CheW and CheA, are required for *M. xanthus* EPS accumulation (Bellenger *et al.*, 2002; Yang *et al.*, 2000b). DifD and DifG, which are the respective orthologues of the CheY response regulator and the CheC phosphatase, negatively regulate EPS accumulation (Black and Yang, 2004). Based on results from genetic, biochemical and molecular biology studies, a model has been proposed to explain how the Dif proteins interact to regulate EPS in *M. xanthus* (Fig. 10.4). In this model, DifA, DifC and DifE form a membrane-anchored signalling complex that stimulates EPS accumulation (Black *et al.*, 2010; Yang and Li, 2005). The histidine protein kinase DifE is the most downstream of the three and it is proposed to stimulate EPS by the phosphorylation of an unknown regulator, DifX (Black *et al.*, 2006, 2010; Black and Yang, 2004). DifD, a substrate of the DifE kinase, is thought to function as a phosphate sink for the phosphoryl group on DifE to inhibit EPS accumulation. DifG, which dephosphorylates DifD~P, may augment the function of DifD as a phosphate-sink to inhibit EPS accumulation. In addition, it was observed that T4P$^-$ mutants accumulate reduced amounts of EPS and that a *pilT* mutant accumulates increased amounts of EPS. Partially because *dif* mutations are epistatic to *pil* mutations in EPS accumulation, T4P are proposed to serve as a sensor for the Dif system (Black *et al.*, 2006).

sgmT and *digR*, two orphan genes at separate chromosomal loci, negatively regulate EPS

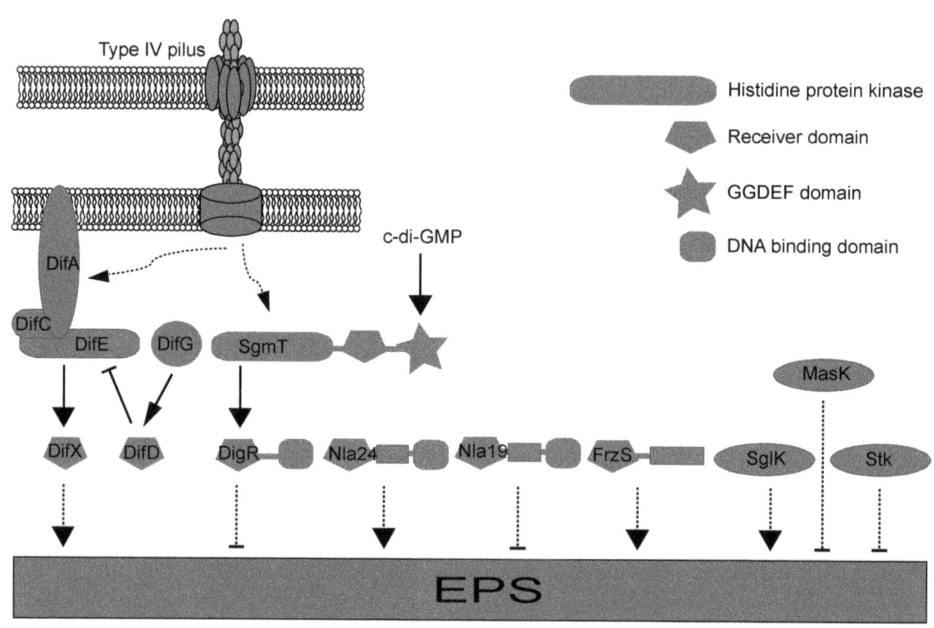

Figure 10.4 Regulation of EPS accumulation in *M. xanthus*. A T4P is shown spanning the cell envelope with both the Dif pathway and the SgmT-DigR system downstream in EPS regulation. Shapes in the upper right are the legend for protein domains as discussed in the text. Rectangles in the figure refer to domains not discussed. Arrows and bars indicate stimulatory and inhibitory effects, respectively. Solid lines indicate proven physical interactions whereas dashed lines designate indirect or inferred interactions.

accumulation in *M. xanthus* as their mutants have more EPS and are defective in S-motility (Overgaard *et al.*, 2006; Petters *et al.*, 2012) (Fig. 10.4) (see also Chapter 7). SgmT is a histidine protein kinase with a complex architecture (Fig. 10.4). Besides the conserved kinase domains, SgmT contains a GAF domain at its N-terminus as well as a receiver domain followed by a GGDEF domain at the C-terminus (Petters *et al.*, 2012). DigR is a response regulator with a DNA binding domain. Biochemical studies demonstrated that SgmT and DigR form a genuine two-component system as SgmT is able to phosphorylate DigR *in vitro*. In other bacterial species, the second messenger c-di-GMP regulates the switch from a planktonic, motile state to a surface-associated biofilm state by stimulating EPS accumulation and inhibiting motility (Hengge, 2009; Sondermann *et al.*, 2012). c-di-GMP is synthesized by proteins containing a so-called GGDEF domain and degraded by phophodiesterases that either contain an EAL-domain or an HD-GYP domain. Not all GGDEF domains have enzymatic activity and they function instead as c-di-GMP receptors (Hengge, 2009; Jenal and Malone, 2006; Schirmer and Jenal, 2009). The SgmT GGDEF domain likely functions as a receptor as it binds c-di-GMP but is likely inactive enzymatically (Petters *et al.*, 2012). While lack of the GGDEF domain in SgmT does not result in any overt phenotypes under laboratory conditions, it is involved in localizaing SgmT to a cluster between midcell and a cell pole. GAF domains are known to have regulatory functions in many proteins and systems. Since the SgmT GAF domain is required for the function of this protein (Petters *et al.*, 2012), this domain is proposed to sense cytoplasmic signals to influence SgmT kinase activity for the regulation of EPS accumulation. In totality, results from genetics, biochemistry and transcriptional profiling show that SgmT and DigR form a cognate two-component system that negatively regulate EPS accumulation in *M. xanthus* (Petters *et al.*, 2012). In addition, a *digR* mutation restored EPS accumulation to a *pilA* mutant but not to a *difE* mutant (Yang, unpublished results). These findings suggest that SgmT and DigR function downstream of T4P in EPS regulation, possibly in a parallel pathway with the Dif systems (Fig.10.4).

Unresolved questions and future perspectives

Much progress has been made on the identification and studies of T4P proteins and their genes. There has been progress on deciphering the pairwise interactions among T4P proteins as well. Likely, the interactions between T4P proteins are highly dynamic in order to give rise to the repeated cycles of extension and retraction. It is largely unknown how the T4P proteins and their interactions change dynamically during extensions and retractions. While it is generally accepted that PilB and PilT are the T4P extension and retraction motor ATPases, respectively, the mechanisms of energy transmission from the cytoplasm across the cell envelope are largely unknown. Regarding the dynamics of T4P, it remains an open question whether PilC and/or the PilM-PilN complex participate in both T4P assembly and disassembly (Carbonnelle *et al.*, 2006; Georgiadou *et al.*, 2012; Nunn *et al.*, 1990; Tonjum *et al.*, 1995; Wu *et al.*, 1997). Another major question in the T4P field relates to how the switch between T4P extension and retraction is regulated. It has been suggested that the localization of PilB and PilT is important for this regulation in *M. xanthus* (Bulyha *et al.*, 2009). As discussed in Chapter 11, the polarity of T4P is regulated by the Frz chemosensory system. The exact mechanisms by which the Frz system and the downstream GTPase MglA regulate the cellular polarity of T4P and the A-motility motors will remain areas of intense research.

Similarly, many open questions remain on how EPS is produced and its accumulation regulated. Even for the two best studied pathways, the SgmT-DigR two-component and the Dif signalling systems, it is not understood how they regulate EPS accumulation. It appears that neither pathway affect EPS by modulating the transcription of known *eps* genes (Lu *et al.*, 2005; Petters *et al.*, 2012). It therefore remains to be determined which step(s) of EPS accumulation is affected by these two pathways. Moreover, there are other known EPS regulators and signal transduction pathways not discussed in any details in this chapter (Fig. 10.4). MasK is a putative tyrosine protein kinase with a critical role in EPS regulation (Thomasson *et al.*, 2002). Two NtrC-like transcriptional activators, Nla19 and Nla24, affect EPS accumulation antagonistically: an *nla19* mutant shows increased EPS accumulation (Lancero *et al.*, 2005), while an *nla24* (also referred to as *epsI*; Lu *et al.*, 2005) mutant accumulates reduced levels (Lancero *et al.*, 2004) (Fig. 10.4). SglK and Stk, two DnaK-like proteins, also affect EPS in opposite directions: *sglK* mutants lack detectable EPS (Weimer *et al.*, 1998), whereas *stk* mutants accumulate increased amounts (Dana and Shimkets, 1993) (Fig. 10.4). It is interesting to note that *sglK* and *dif* genes map to the same region as certain *dsp* mutations (Lancero *et al.*, 2002). More recently, FrzS was shown to positively affect EPS accumulation (Berleman *et al.*, 2011) (Fig. 10.4). Last but not least, there are numerous other S-motility genes (Youderian and Hartzell, 2006, 2007) whose function we know little about. Much work remains to elucidate how the different EPS regulators and pathways interact with one another to coordinately affect EPS accumulation in *M. xanthus*. It will also be a daunting task to understand how ~30 additional S-motility genes (Youderian and Hartzell, 2006, 2007) contribute to *M. xanthus* T4P-mediated surface motility.

References

Abendroth, J., Murphy, P., Sandkvist, M., Bagdasarian, M., and Hol, W.G. (2005). The X-ray structure of the type II secretion system complex formed by the N-terminal domain of EpsE and the cytoplasmic domain of EpsL of *Vibrio cholerae*. J. Mol. Biol. *348*, 845–855.

Arnold, J.W., and Shimkets, L.J. (1988). Cell surface properties correlated with cohesion in *Myxococcus xanthus*. J. Bacteriol. *170*, 5771–5777.

Arts, J., de Groot, A., Ball, G., Durand, E., El Khattabi, M., Filloux, A., Tommassen, J., and Koster, M. (2007). Interaction domains in the *Pseudomonas aeruginosa* type II secretory apparatus component XcpS (GspF). Microbiology *153*, 1582–1592.

Ayers, M., Howell, P.L., and Burrows, L.L. (2010). Architecture of the type II secretion and type IV pilus machineries. Future Microbiol. *5*, 1203–1218.

Balasingham, S.V., Collins, R.F., Assalkhou, R., Homberset, H., Frye, S.A., Derrick, J.P., and Tonjum, T. (2007). Interactions between the lipoprotein PilP and the secretin PilQ in *Neisseria meningitidis*. J. Bacteriol. *189*, 5716–5727.

Behmlander, R.M., and Dworkin, M. (1991). Extracellular fibrils and contact-mediated cell interactions in *Myxococcus xanthus*. J. Bacteriol. *173*, 7810–7821.

Behmlander, R.M., and Dworkin, M. (1994). Biochemical and structural analyses of the extracellular matrix fibrils of *Myxococcus xanthus*. J. Bacteriol. *176*, 6295–6303.

Bellenger, K., Ma, X., Shi, W., and Yang, Z. (2002). A CheW homologue is required for *Myxococcus xanthus* fruiting body development, social gliding motility, and fibril biogenesis. J. Bacteriol. *184*, 5654–5660.

Berleman, J.E., Vicente, J.J., Davis, A.E., Jiang, S.Y., Seo, Y.E., and Zusman, D.R. (2011). FrzS regulates social motility in *Myxococcus xanthus* by controlling exopolysaccharide production. PLoS ONE *6*, e23920.

Biswas, G.D., Lacks, S.A., and Sparling, P.F. (1989). Transformation-deficient mutants of piliated *Neisseria gonorrhoeae*. J. Bacteriol. *171*, 657–664.

Black, W.P., and Yang, Z. (2004). *Myxococcus xanthus* chemotaxis homologs DifD and DifG negatively regulate fibril polysaccharide production. J. Bacteriol. *186*, 1001–1008.

Black, W.P., Xu, Q., and Yang, Z. (2006). Type IV pili function upstream of the Dif chemotaxis pathway in *Myxococcus xanthus* EPS regulation. Mol. Microbiol. *61*, 447–456.

Black, W.P., Schubot, F.D., Li, Z., and Yang, Z. (2010). Phosphorylation and dephosphorylation among Dif chemosensory proteins essential for exopolysaccharide regulation in *Myxococcus xanthus*. J. Bacteriol. *192*, 4267–4274.

Blackhart, B.D., and Zusman, D.R. (1985). 'Frizzy' genes of *Myxococcus xanthus* are involved in control of frequency of reversal of gliding motility. Proc. Natl. Acad. Sci. U.S.A. *82*, 8771–8774.

Blank, T.E., and Donnenberg, M.S. (2001). Novel topology of BfpE, a cytoplasmic membrane protein required for type IV fimbrial biogenesis in enteropathogenic *Escherichia coli*. J. Bacteriol. *183*, 4435–4450.

Bowden, M.G., and Kaplan, H.B. (1998). The *Myxococcus xanthus* lipopolysaccharide O-antigen is required for social motility and multicellular development. Mol. Microbiol. *30*, 275–284.

Bradley, D.E. (1972). Shortening of *Pseudomonas aeruginosa* pili after RNA-phage adsorption. J. Gen. Microbiol. *72*, 303–319.

Bradley, D.E. (1974). The adsorption of *Pseudomonas aeruginosa* pilus-dependent bacteriophages to a host mutant with nonretractile pili. Virology *58*, 149–163.

Bradley, D.E. (1980). A function of *Pseudomonas aeruginosa* PAO polar pili: twitching motility. Can. J. Microbiol. *26*, 146–154.

Bulyha, I., Schmidt, C., Lenz, P., Jakovljevic, V., Hone, A., Maier, B., Hoppert, M., and Søgaard-Andersen, L. (2009). Regulation of the type IV pili molecular machine by dynamic localization of two motor proteins. Mol. Microbiol. *74*, 691–706.

Burkhardt, J., Vonck, J., and Averhoff, B. (2011). Structure and function of PilQ, a secretin of the DNA transporter from the thermophilic bacterium *Thermus thermophilus* HB27. J. Biol. Chem. *286*, 9977–9984.

Burrows, L.L. (2012). *Pseudomonas aeruginosa* twitching motility: Type IV pili in action. Annu. Rev. Microbiol. *66*, 493–520.

Camberg, J.L., Johnson, T.L., Patrick, M., Abendroth, J., Hol, W.G., and Sandkvist, M. (2007). Synergistic stimulation of EpsE ATP hydrolysis by EpsL and acidic phospholipids. Embo J. *26*, 19–27.

Carbonnelle, E., Helaine, S., Nassif, X., and Pelicic, V. (2006). A systematic genetic analysis in *Neisseria* meningitidis defines the Pil proteins required for assembly, functionality, stabilization and export of type IV pili. Mol. Microbiol. *61*, 1510–1522.

Clausen, M., Jakovljevic, V., Sogaard-Andersen, L., and Maier, B. (2009). High-force generation is a conserved property of type IV pilus systems. J. Bacteriol. *191*, 4633–4638.

Collins, R.F., Davidsen, L., Derrick, J.P., Ford, R.C., and Tonjum, T. (2001). Analysis of the PilQ secretin from *Neisseria meningitidis* by transmission electron microscopy reveals a dodecameric quaternary structure. J. Bacteriol. *183*, 3825–3832.

Craig, L., and Li, J. (2008). Type IV pili: paradoxes in form and function. Curr. Opin. Struct. Biol. *18*, 267–277.

Craig, L., Taylor, R.K., Pique, M.E., Adair, B.D., Arvai, A.S., Singh, M., Lloyd, S.J., Shin, D.S., Getzoff, E.D., Yeager, M., *et al.* (2003). Type IV pilin structure and assembly: X-ray and EM analyses of *Vibrio cholerae* toxin-coregulated pilus and *Pseudomonas aeruginosa* PAK pilin. Mol. Cell *11*, 1139–1150.

Dana, J.R., and Shimkets, L.J. (1993). Regulation of cohesion-dependent cell interactions in *Myxococcus xanthus*. J. Bacteriol. *175*, 3636–3647.

Drake, S.L., and Koomey, M. (1995). The product of the *pilQ* gene is essential for the biogenesis of type IV pili in *Neisseria gonorrhoeae*. Mol. Microbiol. *18*, 975–986.

Drake, S.L., Sandstedt, S.A., and Koomey, M. (1997). PilP, a pilus biogenesis lipoprotein in *Neisseria gonorrhoeae*, affects expression of PilQ as a high-molecular-mass multimer. Mol. Microbiol. *23*, 657–668.

Fink, J.M., and Zissler, J.F. (1989). Defects in motility and development of *Myxococcus xanthus* lipopolysaccharide mutants. J. Bacteriol. *171*, 2042–2048.

Georgiadou, M., Castagnini, M., Karimova, G., Ladant, D., and Pelicic, V. (2012). Large-scale study of the interactions between proteins involved in type IV pilus biology in *Neisseria meningitidis*: characterization of a subcomplex involved in pilus assembly. Mol. Microbiol. *84*, 857–873.

Gray, M.D., Bagdasarian, M., Hol, W.G., and Sandkvist, M. (2011). *In vivo* cross-linking of EpsG to EpsL suggests a role for EpsL as an ATPase-pseudopilin coupling protein in the Type II secretion system of *Vibrio cholerae*. Mol. Microbiol. *79*, 786–798.

Helm, R.A., Barnhart, M.M., and Seifert, H.S. (2007). *pilQ* missense mutations have diverse effects on PilQ multimer formation, piliation, and pilus function in *Neisseria gonorrhoeae*. J. Bacteriol. *189*, 3198–3207.

Hengge, R. (2009). Principles of c-di-GMP signalling in bacteria. Nat. Rev. Microbiol. *7*, 263–273.

Henrichsen, J. (1972). Bacterial surface translocation: a survey and a classification. Bacteriol. Rev. *36*, 478–503.

Henrichsen, J. (1975a). The occurrence of twitching motility among Gram-negative bacteria. Acta Pathol. Microbiol. Scand. [B] *83*, 171–178.

Henrichsen, J. (1975b). On twitching motility and its mechanism. Acta Pathol. Microbiol. Scand. [B] *83*, 187–190.

Henrichsen, J., Froholm, L.O., and Bovre, K. (1972). Studies on bacterial surface translocation. 2. Correlation of twitching motility and fimbriation in colony variants of *Moraxella nonliquefaciens*, *M. bovis*, and *M. kingii*. Acta Pathol. Microbiol. Scand. [B] Microbiol. Immuno.l *80*, 445–452.

Hodgkin, J., and Kaiser, D. (1977). Cell-to-cell stimulation of movement in nonmotile mutants of *Myxococcus*. Proc. Natl. Acad. Sci. U.S.A. *74*, 2938–2942.

Hodgkin, J., and Kaiser, D. (1979a). Genetics of gliding motility in *Myxococcus xanthus* (Myxobacterales): genes controlling movement of single cells. Mol. Gen. Genet. *171*, 167–176.

Hodgkin, J., and Kaiser, D. (1979b). Genetics of gliding motility in *Myxococcus xanthus*: two gene systems control movement. Mol. Gen. Genet. *171*, 177–191.

Holz, C., Opitz, D., Greune, L., Kurre, R., Koomey, M., Schmidt, M.A., and Maier, B. (2010). Multiple pilus motors cooperate for persistent bacterial movement in two dimensions. Phys. Rev. Lett. *104*, 178104.

Hu, W., Hossain, M., Lux, R., Wang, J., Yang, Z., Li, Y., and Shi, W. (2011). Exopolysaccharide-independent social motility of *Myxococcus xanthus*. PLoS ONE *6*, e16102.

Jain, S., Moscicka, K.B., Bos, M.P., Pachulec, E., Stuart, M.C., Keegstra, W., Boekema, E.J., and van der Does, C. (2011). Structural Characterization of Outer Membrane Components of the Type IV Pili System in Pathogenic *Neisseria*. PLoS ONE *6*, e16624.

Jakovljevic, V., Leonardy, S., Hoppert, M., and Søgaard-Andersen, L. (2008). PilB and PilT are ATPases acting antagonistically in type IV pilus function in *Myxococcus xanthus*. J. Bacteriol. *190*, 2411–2421.

Jenal, U., and Malone, J. (2006). Mechanisms of cyclic-di-GMP signaling in bacteria. Annu. Rev. Genet. *40*, 385–407.

Johnson, T.L., Scott, M.E., and Sandkvist, M. (2007). Mapping critical interactive sites within the periplasmic domain of the *Vibrio cholerae* type II secretion protein EpsM. J. Bacteriol. *189*, 9082–9089.

Kaiser, D. (1979). Social gliding is correlated with the presence of pili in *Myxococcus xanthus*. Proc. Natl. Acad. Sci. U.S.A. *76*, 5952–5956.

Kaiser, D., and Crosby, C. (1983). Cell movement and its coordination in swarms of *Myxococcus xanthus*. Cell Motil. 3, 227–245.

Karuppiah, V., and Derrick, J.P. (2011). Structure of the PilM-PilN inner membrane type IV pilus biogenesis complex from *Thermus thermophilus*. J. Biol. Chem. *286*, 24434–24442.

Karuppiah, V., Hassan, D., Saleem, M., and Derrick, J.P. (2010). Structure and oligomerization of the PilC type IV pilus biogenesis protein from *Thermus thermophilus*. Proteins *78*, 2049–2057.

Konovalova, A., Petters, T., and Søgaard-Andersen, L. (2010). Extracellular biology of *Myxococcus xanthus*. FEMS Microbiol. Rev. *34*, 89–106.

Koo, J., Tammam, S., Ku, S.Y., Sampaleanu, L.M., Burrows, L.L., and Howell, P.L. (2008). PilF is an outer membrane lipoprotein required for multimerization and localization of the *Pseudomonas aeruginosa* type IV pilus secretin. J. Bacteriol. *190*, 6961–6969.

Korotkov, K.V., Johnson, T.L., Jobling, M.G., Pruneda, J., Pardon, E., Heroux, A., Turley, S., Steyaert, J., Holmes, R.K., Sandkvist, M., *et al.* (2011). Structural and functional studies on the interaction of GspC and GspD in the type II secretion system. PLoS Pathog. *7*, e1002228.

Korotkov, K.V., Sandkvist, M., and Hol, W.G. (2012). The type II secretion system: biogenesis, molecular architecture and mechanism. Nat. Rev. Microbiol. *10*, 336–351.

Lancero, H., Brofft, J.E., Downard, J., Birren, B.W., Nusbaum, C., Naylor, J., Shi, W., and Shimkets, L.J. (2002). Mapping of *Myxococcus xanthus* social motility *dsp* mutations to the *dif* genes. J. Bacteriol. *184*, 1462–1465.

Lancero, H., Caberoy, N.B., Castaneda, S., Li, Y., Lu, A., Dutton, D., Duan, X.Y., Kaplan, H.B., Shi, W., and Garza, A.G. (2004). Characterization of a *Myxococcus xanthus* mutant that is defective for adventurous motility and social motility. Microbiology *150*, 4085–4093.

Lancero, H.L., Castaneda, S., Caberoy, N.B., Ma, X., Garza, A.G., and Shi, W. (2005). Analysing protein–protein interactions of the *Myxococcus xanthus* Dif signalling pathway using the yeast two-hybrid system. Microbiology *151*, 1535–1541.

Li, Y., Sun, H., Ma, X., Lu, A., Lux, R., Zusman, D., and Shi, W. (2003). Extracellular polysaccharides mediate pilus retraction during social motility of *Myxococcus xanthus*. Proc. Natl. Acad. Sci. U.S.A. *100*, 5443–5448.

Lu, A., Cho, K., Black, W.P., Duan, X.Y., Lux, R., Yang, Z., Kaplan, H.B., Zusman, D.R., and Shi, W. (2005). Exopolysaccharide biosynthesis genes required for social motility in *Myxococcus xanthus*. Mol. Microbiol. *55*, 206–220.

McLaughlin, L.S., Haft, R.J., and Forest, K.T. (2012). Structural insights into the Type II secretion nanomachine. Curr. Opin. Struct. Biol. *22*, 208–216.

Maier, B., Potter, L., So, M., Long, C.D., Seifert, H.S., and Sheetz, M.P. (2002). Single pilus motor forces exceed 100 pN. Proc. Natl. Acad. Sci. U.S.A. *99*, 16012–16017.

Mattick, J.S. (2002). Type IV pili and twitching motility. Annu. Rev. Microbiol. *56*, 289–314.

Mattick, J.S., Whitchurch, C.B., and Alm, R.A. (1996). The molecular genetics of type-4 fimbriae in *Pseudomonas aeruginosa* – a review. Gene *179*, 147–155.

Merroun, M.L., Ben Chekroun, K., Arias, J.M., and Gonzalez-Munoz, M.T. (2003). Lanthanum fixation by *Myxococcus xanthus*: cellular location and extracellular polysaccharide observation. Chemosphere *52*, 113–120.

Merz, A.J., So, M., and Sheetz, M.P. (2000). Pilus retraction powers bacterial twitching motility. Nature *407*, 98–102.

Mignot, T., Merlie, J.P., Jr., and Zusman, D.R. (2005). Regulated pole-to-pole oscillations of a bacterial gliding motility protein. Science *310*, 855–857.

Misic, A.M., Satyshur, K.A., and Forest, K.T. (2010). *P. aeruginosa* PilT structures with and without nucleotide reveal a dynamic type IV pilus retraction motor. J. Mol. Biol. *400*, 1011–1021.

Morand, P.C., Bille, E., Morelle, S., Eugene, E., Beretti, J.L., Wolfgang, M., Meyer, T.F., Koomey, M., and Nassif, X. (2004). Type IV pilus retraction in pathogenic *Neisseria* is regulated by the PilC proteins. Embo J. *23*, 2009–2017.

Nudleman, E., Wall, D., and Kaiser, D. (2006). Polar assembly of the type IV pilus secretin in *Myxococcus xanthus*. Mol. Microbiol. *60*, 16–29.

Nunn, D.N., and Lory, S. (1991). Product of the *Pseudomonas aeruginosa* gene *pilD* is a prepilin leader peptidase. Proc. Natl. Acad. Sci. U.S.A. *88*, 3281–3285.

Nunn, D., Bergman, S., and Lory, S. (1990). Products of three accessory genes, pilB, pilC, and pilD, are required for biogenesis of *Pseudomonas aeruginosa* pili. J. Bacteriol. *172*, 2911–2919.

Overgaard, M., Wegener-Feldbrügge, S., and Søgaard-Andersen, L. (2006). The orphan response regulator DigR is required for synthesis of extracellular matrix fibrils in *Myxococcus xanthus*. J. Bacteriol. *188*, 4384–4394.

Palsdottir, H., Remis, J.P., Schaudinn, C., O'Toole, E., Lux, R., Shi, W., McDonald, K.L., Costerton, J.W., and Auer, M. (2009). Three-dimensional macromolecular organization of cryofixed *Myxococcus xanthus* biofilms as revealed by electron microscopic tomography. J. Bacteriol. *191*, 2077–2082.

Peabody, C.R., Chung, Y.J., Yen, M.R., Vidal-Ingigliardi, D., Pugsley, A.P., and Saier, M.H., Jr. (2003). Type II protein secretion and its relationship to bacterial type IV pili and archaeal flagella. Microbiology *149*, 3051–3072.

Pelicic, V. (2008). Type IV pili: e pluribus unum? Mol. Microbiol. *68*, 827–837.

Pelling, A.E., Li, Y., Shi, W., and Gimzewski, J.K. (2005). Nanoscale visualization and characterization of *Myxococcus xanthus* cells with atomic force microscopy. Proc. Natl. Acad. Sci. U.S.A. *102*, 6484–6489.

Petters, T., Zhang, X., Nesper, J., Treuner-Lange, A., Gomez-Santos, N., Hoppert, M., Jenal, U., and Søgaard-Andersen, L. (2012). The orphan histidine protein kinase SgmT is a c-di-GMP receptor and regulates composition of the extracellular matrix together with the orphan DNA binding response regulator DigR in *Myxococcus xanthus*. Mol. Microbiol. *84*, 147–165.

Planet, P.J., Kachlany, S.C., DeSalle, R., and Figurski, D.H. (2001). Phylogeny of genes for secretion NTPases: identification of the widespread *tadA* subfamily and development of a diagnostic key for gene classification. Proc. Natl. Acad. Sci. U.S.A. *98*, 2503–2508.

Py, B., Loiseau, L., and Barras, F. (2001). An inner membrane platform in the type II secretion machinery of Gram-negative bacteria. EMBO Rep. *2*, 244–248.

Rodriguez-Soto, J.P., and Kaiser, D. (1997). The *tgl* gene: social motility and stimulation in *Myxococcus xanthus*. J. Bacteriol. *179*, 4361–4371.

Rosenbluh, A., and Eisenbach, M. (1992). Effect of mechanical removal of pili on gliding motility of *Myxococcus xanthus*. J. Bacteriol. *174*, 5406–5413.

Sampaleanu, L.M., Bonanno, J.B., Ayers, M., Koo, J., Tammam, S., Burley, S.K., Almo, S.C., Burrows, L.L., and Howell, P.L. (2009). Periplasmic domains of *Pseudomonas aeruginosa* PilN and PilO form a stable heterodimeric complex. J. Mol. Biol. *394*, 143–159.

Sandkvist, M., Bagdasarian, M., Howard, S.P., and DiRita, V.J. (1995). Interaction between the autokinase EpsE and EpsL in the cytoplasmic membrane is required for extracellular secretion in *Vibrio cholerae*. Embo J. *14*, 1664–1673.

Sandkvist, M., Hough, L.P., Bagdasarian, M.M., and Bagdasarian, M. (1999). Direct interaction of the EpsL and EpsM proteins of the general secretion apparatus in *Vibrio cholerae*. J. Bacteriol. *181*, 3129–3135.

Satyshur, K.A., Worzalla, G.A., Meyer, L.S., Heiniger, E.K., Aukema, K.G., Misic, A.M., and Forest, K.T. (2007). Crystal structures of the pilus retraction motor PilT suggest large domain movements and subunit cooperation drive motility. Structure *15*, 363–376.

Sauvonnet, N., Vignon, G., Pugsley, A.P., and Gounon, P. (2000). Pilus formation and protein secretion by the same machinery in *Escherichia coli*. Embo J. *19*, 2221–2228.

Schirmer, T., and Jenal, U. (2009). Structural and mechanistic determinants of c-di-GMP signalling. Nat. Rev. Micro. *7*, 724–735.

Shaw, C.E., and Taylor, R.K. (1990). *Vibrio cholerae O395 tcpA* pilin gene sequence and comparison of predicted protein structural features to those of type 4 pilins. Infect. Immun. *58*, 3042–3049.

Shi, W., and Zusman, D.R. (1993). The two motility systems of *Myxococcus xanthus* show different selective advantages on various surfaces. Proc. Natl. Acad. Sci. U.S.A. *90*, 3378–3382.

Shimkets, L.J. (1986). Correlation of energy-dependent cell cohesion with social motility in *Myxococcus xanthus*. J. Bacteriol. *166*, 837–841.

Shiue, S.J., Kao, K.M., Leu, W.M., Chen, L.Y., Chan, N.L., and Hu, N.T. (2006). XpsE oligomerization triggered by ATP binding, not hydrolysis, leads to its association with XpsL. Embo J. *25*, 1426–1435.

Skerker, J.M., and Berg, H.C. (2001). Direct observation of extension and retraction of type IV pili. Proc. Natl. Acad Sci. U.S.A. *98*, 6901–6904.

Sondermann, H., Shikuma, N.J., and Yildiz, F.H. (2012). You've come a long way: c-di-GMP signaling. Curr. Opin. Microbiol. *15*, 140–146.

Sun, H., Zusman, D.R., and Shi, W. (2000). Type IV pilus of *Myxococcus xanthus* is a motility apparatus controlled by the *frz* chemosensory system. Cur. Biol. *10*, 1143–1146.

Tammam, S., Sampaleanu, L.M., Koo, J., Sundaram, P., Ayers, M., Chong, P.A., Forman-Kay, J.D., Burrows, L.L., and Howell, P.L. (2011). Characterization of the PilN, PilO and PilP type IVa pilus subcomplex. Mol. Microbiol. *82*, 1496–1514.

Thaxter, R. (1892). On the Myxobacteriaceae, a new order of Schizomycetes. Bot. Gaz. *14*, 389–406.

Thomasson, B., Link, J., Stassinopoulos, A.G., Burke, N., Plamann, L., and Hartzell, P.L. (2002). MglA, a small GTPase, interacts with a tyrosine kinase to control type IV pili-mediated motility and development of *Myxococcus xanthus*. Mol. Microbiol. *46*, 1399–1413.

Tonjum, T., Freitag, N.E., Namork, E., and Koomey, M. (1995). Identification and characterization of pilG, a highly conserved pilus-assembly gene in pathogenic *Neisseria*. Mol. Microbiol. *16*, 451–464.

Wall, D., and Kaiser, D. (1999). Type IV pili and cell motility. Mol. Microbiol. *32*, 1–10.

Wall, D., Kolenbrander, P.E., and Kaiser, D. (1999). The *Myxococcus xanthus pilQ (sglA)* gene encodes a secretin homolog required for type IV pilus biogenesis, social motility, and development. J. Bacteriol. *181*, 24–33.

Ward, M.J., Lew, H., and Zusman, D.R. (2000). Social motility in *Myxococcus xanthus* requires FrzS, a protein with an extensive coiled-coil domain. Mol. Microbiol. *37*, 1357–1371.

Weimer, R.M., Creighton, C., Stassinopoulos, A., Youderian, P., and Hartzell, P.L. (1998). A chaperone in the HSP70 family controls production of extracellular fibrils in *Myxococcus xanthus*. J. Bacteriol. *180*, 5357–5368.

Whitchurch, C.B., Hobbs, M., Livingston, S.P., Krishnapillai, V., and Mattick, J.S. (1991). Characterisation of a *Pseudomonas aeruginosa* twitching motility gene and evidence for a specialised protein export system widespread in eubacteria. Gene *101*, 33–44.

Wolfgang, M., Lauer, P., Park, H.S., Brossay, L., Hebert, J., and Koomey, M. (1998a). PilT mutations lead to simultaneous defects in competence for natural transformation and twitching motility in piliated *Neisseria gonorrhoeae*. Mol. Microbiol. *29*, 321–330.

Wolfgang, M., Park, H.S., Hayes, S.F., van Putten, J.P., and Koomey, M. (1998b). Suppression of an absolute defect in type IV pilus biogenesis by loss-of-function mutations in pilT, a twitching motility gene in *Neisseria gonorrhoeae*. Proc. Natl. Acad. Sci. U.S.A. *95*, 14973–14978.

Wu, S.S., and Kaiser, D. (1995). Genetic and functional evidence that Type IV pili are required for social gliding motility in *Myxococcus xanthus*. Mol. Microbiol. *18*, 547–558.

Wu, S.S., and Kaiser, D. (1996). Markerless deletions of *pil* genes in *Myxococcus xanthus* generated by counterselection with the *Bacillus subtilis sacB* gene. J. Bacteriol. *178*, 5817–5821.

Wu, S.S., and Kaiser, D. (1997). Regulation of expression of the *pilA* gene in *Myxococcus xanthus*. J. Bacteriol. *179*, 7748–7758.

Wu, S.S., Wu, J., and Kaiser, D. (1997). The *Myxococcus xanthus pilT* locus is required for social gliding motility although pili are still produced. Mol. Microbiol. *23*, 109–121.

Wu, S.S., Wu, J., Cheng, Y.L., and Kaiser, D. (1998). The *pilH* gene encodes an ABC transporter homologue required for type IV pilus biogenesis and social gliding motility in *Myxococcus xanthus*. Mol. Microbiol. *29*, 1249–1261.

Yang, Z., Geng, Y., Xu, D., Kaplan, H.B., and Shi, W. (1998). A new set of chemotaxis homologues is essential for *Myxococcus xanthus* social motility. Mol. Microbiol. *30*, 1123–1130.

Yang, Z., Guo, D., Bowden, M.G., Sun, H., Tong, L., Li, Z., Brown, A.E., Kaplan, H.K., and Shi, W. (2000a). The *Myxococcus xanthus wbgB* gene encodes a glycosyltransferase homologue required for lipopolysaccharide O-antigen biogenesis. Arch. Microbiol. *174*, 399–405.

Yang, Z., Ma, X., Tong, L., Kaplan, H.B., Shimkets, L.J., and Shi, W. (2000b). *Myxococcus xanthus dif* genes are required for biogenesis of cell surface fibrils essential for social gliding motility. J. Bacteriol. *182*, 5793–5798.

Yang, Z., and Li, Z. (2005). Demonstration of interactions among *Myxococcus xanthus* Dif chemotaxis-like proteins by the yeast two-hybrid system. Arch. Microbiol. *183*, 243–252.

Yang, Z., Lux, R., Hu, W., Hu, C., and Shi, W. (2010). PilA localization affects extracellular polysaccharide production and fruiting body formation in *Myxococcus xanthus*. Mol. Microbiol. *76*, 1500–1513.

Youderian, P., and Hartzell, P.L. (2006). Transposon insertions of magellan-4 that impair social gliding motility in *Myxococcus xanthus*. Genetics *172*, 1397–1410.

Youderian, P., and Hartzell, P.L. (2007). Triple mutants uncover three new genes required for social motility in *Myxococcus xanthus*. Genetics *177*, 557–566.

Sensory Regulation of *Myxococcus xanthus* Motility

11

Emilia M.F. Mauriello, Beiyan Nan and David R. Zusman

Abstract

Myxococcus xanthus cells grow vegetatively on organic nutrients and by killing and digesting prey microorganisms. On hard surfaces, efficient food gathering requires cells to be able to move without flagella. When nutrients become scarce, cells respond by adjusting their movement in order to aggregate into fruiting bodies containing resistant spores. These activities require *M. xanthus* cells to perceive and respond to many signals. To this purpose, *M. xanthus* evolved multiple sensory transduction pathways, including one-component, two-component and complex chemosensory systems. In this chapter, we focus on the *M. xanthus* chemosensory systems and how they control individual cell and group movements, since environmental sensing is essential for survival, permitting bacteria to react to continuously changing external conditions.

Introduction

Myxococcus xanthus cells continuously adjust their behaviours in response to external stimuli. Of particular interest is how *M. xanthus* cells perceive a decrease in nutrients in the growth medium and 'decide' to enter the developmental phase. These decisions involve redirecting cell movements towards aggregation centres (future fruiting bodies) and the differentiation of cells within fruiting bodies into myxospores. In most cases, specialized sensory pathways are engaged to perceive signals and generate specific responses such as the production of needed metabolites, modulation of gene expression and the coordination of cell movements. In this chapter we will specifically focus on how the perception of signals lead to changes in cell movements and how this ultimately affects specific behaviours. We first review how bacterial chemotaxis, the modulation of cell movements in response to chemicals, takes place in the enteric bacterium, *Escherichia coli*, and then describe how *M. xanthus* modulates individual cell movements using similar chemosensory pathways, but highly divergent motility systems. Ultimately, we discuss the impact of the regulated cell movements on *M. xanthus* community behaviours.

The chemotaxis paradigm in *Escherichia coli*

Rotation switches enable a biased random walk

E. coli swims in aqueous environments by rotating 7–14 peritrichous flagella. When the flagella rotate counterclockwise, they form bundles, resulting in forward thrust. In the absence of a chemoattractant gradient, this smooth swimming behaviour is frequently and randomly interrupted by a brief switch in flagellar rotation to the clockwise direction, causing tumbling, which randomizes cellular orientation (Segall *et al.*, 1982). In the presence of a chemical gradient, *E. coli* cells bias the random walk by modulating the frequency of tumbles between runs: when a cell is moving in a favourable direction, it tumbles less; when it moves in a unfavourable direction, it tumbles more frequently, increasing the probability of a changed direction. This biased walk allows *E. coli* cells to move towards a diffusing source of an attractant or away from that of a repellent.

Modified two-component systems regulate the chemotactic response

The chemotaxis pathway of *E. coli* is shown in Fig. 11.1. Membrane chemoreceptor MCPs (methyl-accepting chemotaxis proteins) sense chemicals in the environment, transmit this information to CheA, a cytoplasmic histidine kinase, which autophosphorylates; the phosphoryl group is then transferred to CheY, a single domain response regulator protein (Porter *et al.*, 2011). CheY-P directly binds the flagellar switch triggering clockwise flagella rotation (Sourjik and Berg, 2002a); the phosphate is rapidly removed by the cognate phosphatase, CheZ (Roberts *et al.*, 2010; Yuan *et al.*, 2012). A coupling protein, CheW, is required for the interaction between the MCP and CheA. Attractants inhibit CheA autophosphorylation, whereas repellents stimulate autophosphorylation, increasing tumbling (Porter *et al.*, 2011).

Adaptation in *E. coli* is achieved by modification of the receptors. MCPs are methylated and demethylated on four specific glutamate residues by a methyltransferase (CheR) and methylesterase (CheB), respectively (Fig. 11.1). Feedback occurs as the CheB methylesterase activity increases when phosphorylated by CheA (Porter *et al.*, 2011). MCP methylation and demethylation allow adaptation of the receptor to a persistent stimulus. Ultimately by reducing the CheA activity to pre-stimulus levels, the adaptation mechanisms allow the Che system to respond to changes in the concentration of attractants or repellents instead of the absolute concentration of the stimuli (Roberts *et al.*, 2010).

The *Myxococcus xanthus* chemosensory pathways are structurally similar to that of *E. coli* but the motility systems differ

Myxococcus xanthus motility systems

M. xanthus cells do not produce flagella, in contrast to the enteric paradigm. The cells move on solid surfaces utilizing two genetically distinct motility systems: social (S) motility, powered by Type IV pili (T4P), and Adventurous (A) motility, powered by distributed motor complexes (Li *et al.*, 2003; Luciano *et al.*, 2011; Mauriello *et al.*, 2010a; Sun *et al.*, 2011). T4P are assembled at the leading cell pole (Fig. 11.2). Genetic and behavioural analyses show that, beside T4P (Wall and Kaiser, 1999; Wu and Kaiser, 1995), S-motility also requires extracellular matrix polysaccharide (EPS, also referred to as fibrils). In fact, it has been shown that the binding of *M. xanthus* T4P to the EPS present on the surface of neighbouring cells or in slime trails triggers pilus retraction pulling cells forward (Li *et al.*, 2003). This is also why this type of motility is mostly observed in groups of cells.

Gliding or A-motility is mostly observed in isolated cells and is mediated by membrane spanning multi-protein complexes that are produced at the leading cell pole and redistributed along the cell body by translocation (Fig. 11.2). It is proposed that motility is produced when Agl-Glt motor complexes move on the bacterial cytoskeleton exerting force on the slime covered surface. We invite the reader to refer to Chapter 9 and 10 for detailed descriptions of the A- and S-motility, respectively.

Cells that exhibit only A-motility (A^+S^-) move best on relatively firm, dry surfaces (1.5% agar), but are virtually non-motile on soft wet surfaces (0.3% agar), whereas S motile cells (A^-S^+) move best on relatively soft wet surfaces (0.3% agar) and much more slowly on drier surfaces such as 1.5% agar (Shi and Zusman, 1993). However, on nutrient limiting media, instead of spreading out from the colony centre, cells move inwards to generate aggregating foci that will later mature into fruiting bodies (Fig. 11.3). In the presence of prey cells, such as *E. coli*, *M. xanthus* cells show highly synchronized cell movements, commonly referred to as rippling, generating 'accordion waves' also shown in Fig. 11.3 (Berleman *et al.*, 2008). In the absence of one or both motility systems, aggregation, fruiting body formation, and rippling are defective, indicating that motility is required for these social behaviours. However, as described below, the presence of functional motility apparatuses are not sufficient to generate the

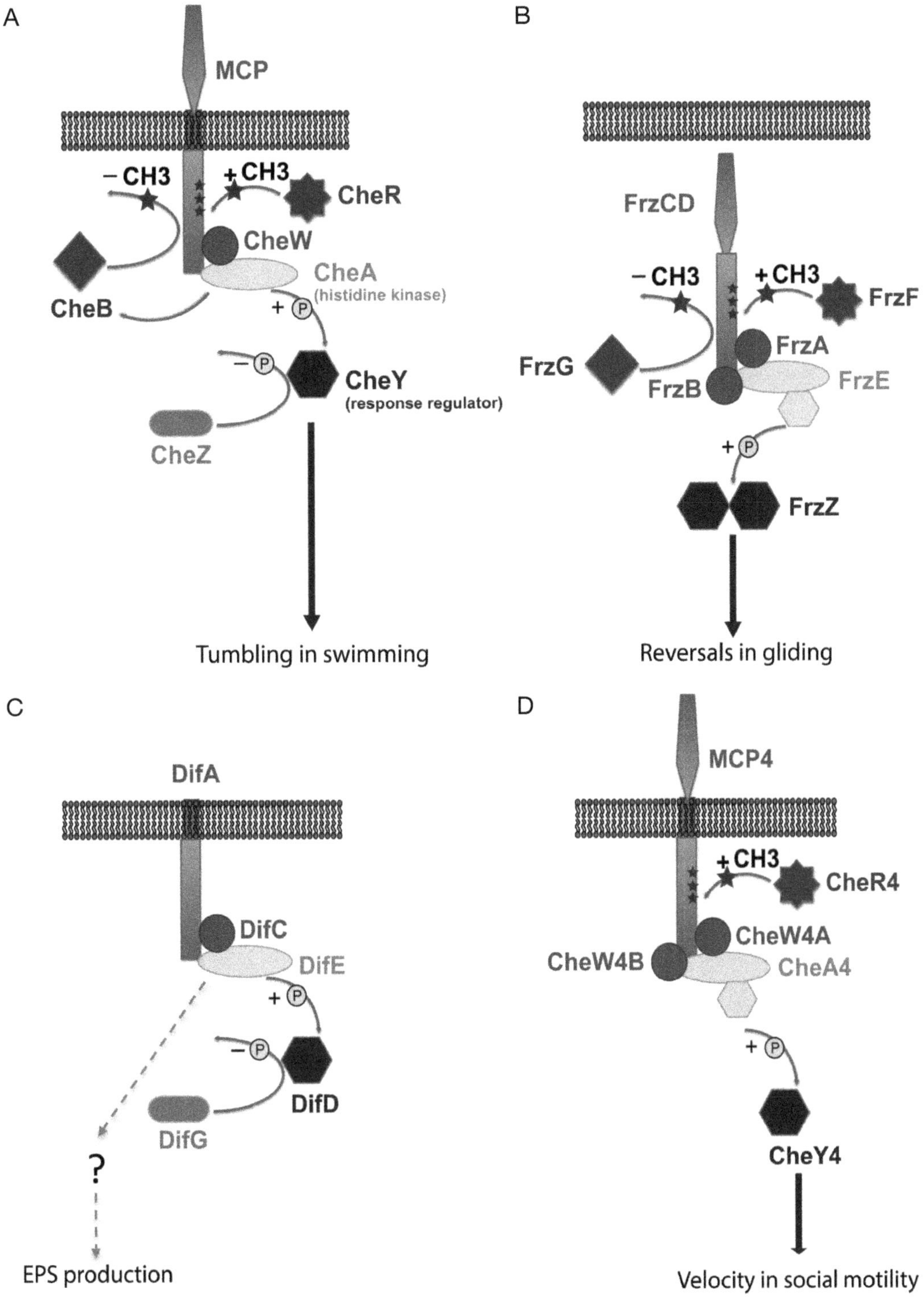

Figure 11.1 Schematic diagrams of the *E. coli* Che and the *M. xanthus* Frz, Dif and Che4 chemotaxis systems. Chemosensory proteins might be arranged similarly to their enteric counterparts. Stars indicate methyl groups.

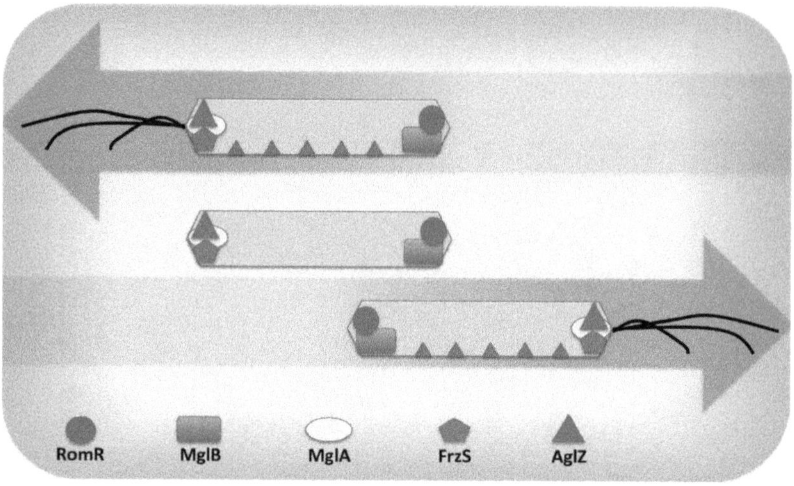

Figure 11.2 Schematic representation of a cell reversal. During a reversal, a *M. xanthus* cell makes a brief pause, A- and S-motility proteins move from one pole to the other and, when polarity and localization are re-established, the cell starts moving in the opposite direction. A colour version of this figure is located in the plate section at the back of the book.

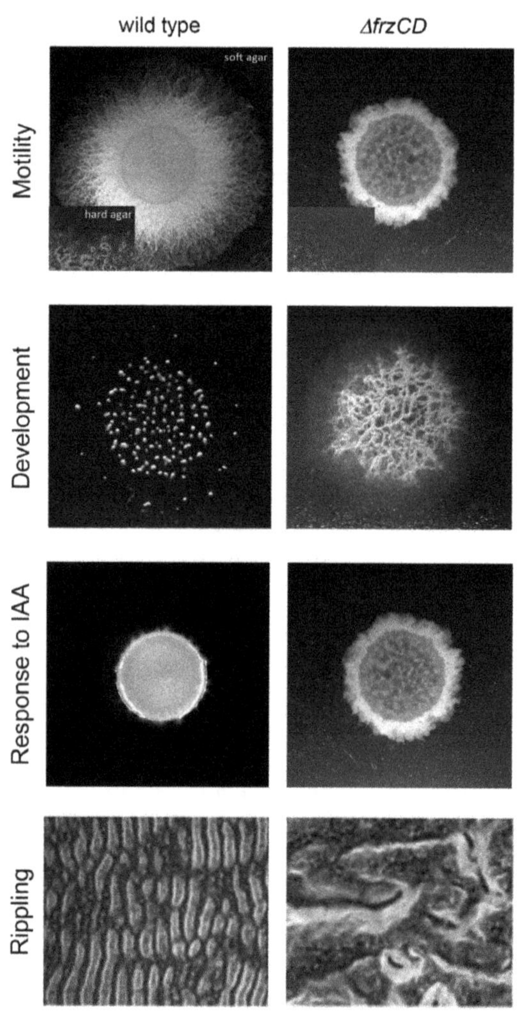

◁ **Figure 11.3** Morphologies of wildtype and *ΔfrzCD* colonies on rich media (Motility, first row); starvation conditions (Development, second row); presence of isoamyl-alchol (Response to IAA, third row); presence of prey (Rippling, fourth row). Motility assays (first row) are performed on plates containing agar concentrations of 0.3% (soft agar, whole colony) or 1.5% agar (hard agar, inset of a colony edge where individual cells can be seen). Response to repellent assays (third row) are performed on 0.3% agar. Fruiting body formation assays (second row) are done on starving media containing 1.5% agar, with incubation for 72h at 32°C. On these media, wildtype cells form fruiting bodies in 48 to 72h. Rippling assays are performed by spotting *M. xanthus* cells next to *E. coli* prey colonies. As *M. xanthus* wildtype cells swarm through and lyse the prey colony, they form the typical wavy pattern shown in the fourth row. *ΔfrzCD* cells are incapable to form fruiting bodies and ripples.

colony patterns described above. Also required is the presence of a chemosensory pathway that modulates directed cell movement (chemotaxis) and coordinates the two motility systems (Figs. 11.1, 11.3 and 11.4; see below).

Directed cell movements in *Myxococcus xanthus* require controlled changes in cell polarity

Individual *M. xanthus* cells move at only a few μm per minute (almost 1000-fold slower than *E. coli*) and periodically reverse the direction of

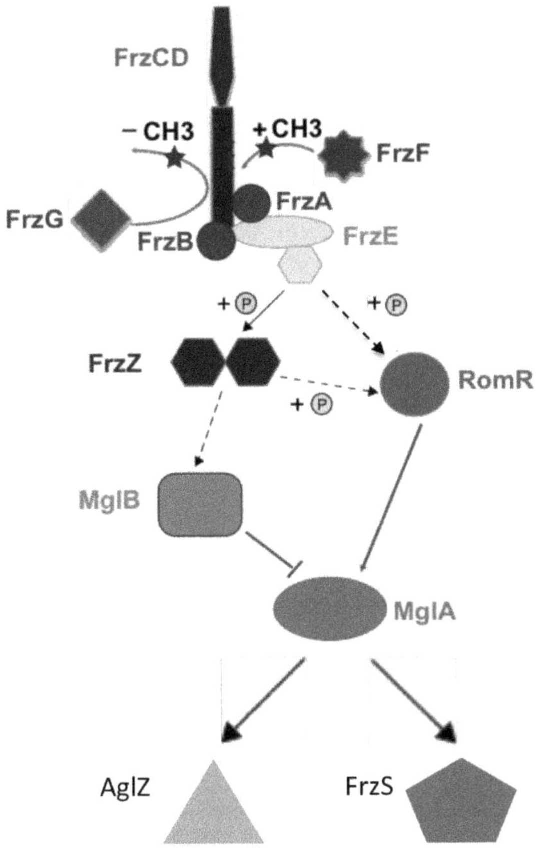

Figure 11.4 Schematic diagram of the Frz pathway and its downstream components RomR, MglB and MglA. Flashes indicate methylation and phosphorylation reactions. Stars indicate methyl groups. Dashed arrows indicates interactions that have not been proven experimentally.

their movement. Wild type cells reverse on the average every 7–14 min, depending on the media and strains. Cell reversals in *M. xanthus* involve a complete inversion of cell polarity: the leading cell poles become the lagging cell poles and the lagging cell poles become the leading cell poles (Keilberg *et al.*, 2012; Leonardy *et al.*, 2010; Mignot *et al.*, 2005; Zhang *et al.*, 2010, 2012) (Fig. 11.2). Thus all of the A- and S-motility proteins and structures that show unipolar localization need to be relocated to the opposite cell pole, in a coordinated manner (see later) (Fig. 11.2). The ability of cells to reverse direction (and polarity) may enable cells to reorient themselves as part of a biased random walk, which would be analogous to controlled tumbles in *E. coli*. Indeed, isolated *M. xanthus* cells respond to attractants or repellents by increasing or decreasing their reversal frequency, which is similar to the responses of *E. coli* that include changes in the switching frequency of flagellar rotation. The regulation of reversal frequency may result in directed cell movement as groups of cells were observed to move towards sources of nutrients (Shi *et al.*, 1993; Taylor and Welch, 2008).

Attractant and repellent responses

Several nutrients including certain lipids, long chain alcohols, casein digests, and yeast extract were shown to increase the methylation of the FrzCD chemoreceptor (see below) and inhibit cellular reversals; these were therefore considered attractants (McBride *et al.*, 1992; Xu *et al.*, 2008). Strong evidence for chemosensory regulation of reversal frequency comes from Kearns and Shimkets showing that individual cells were responsive to certain phosphatidyl-ethanolamine (PE) species (also see below) (Bonner and Shimkets, 2006; Curtis *et al.*, 2006; Kearns and Shimkets, 1998, 2001; Kearns *et al.*, 2001; Xu *et al.*, 2008). They studied the suppression of reversal frequency and showed that it was a function of attractant concentration: reversals were up to four times less frequent in the presence of an attractant (Kearns and Shimkets, 1998). Furthermore, responses to attractants were followed by adaptation. However, *M. xanthus* adaptation times are very long compared to those of *E. coli* (Sourjik and Berg, 2002b), consistent with the much slower movement of *M. xanthus* cells and their response to less diffusible compounds (Reichenbach, 1999).

In the presence of repellents such as short-chain alcohols like isoamyl alchool (IAA) and dimethyl sulphoxide (DMSO), *M. xanthus* cells increase their reversal frequency up to 10-fold, causing cells to reverse in place with little net translocation (Fig. 11.3) (Bustamante *et al.*, 2004). Surprisingly, these responses to repellents are not followed by adaptation (Xu *et al.*, 2007).

Predataxis: the regulation of directed cell movement in the presence of prey

M. xanthus movements in the absence of prey are random with infrequent cellular reversals. However, *M. xanthus* cell groups are highly synchronized in the presence of prey, displaying frequent and regular reversals (rippling) as they move in the direction of the highest concentration of prey (Berleman *et al.*, 2008). This behaviour generates accordion waves that progressively attack and digest the prey (Berleman *et al.*, 2008) (Figs. 11.3 and 11.5A). Reminiscently of adaptation mechanisms, ripple 'wave lengths' increase as cells pass through the prey colony, a behaviour dependent on a functional Frz chemosensory system (see below) (Fig. 11.3). The oriented cell movement, the association with a Che system, and the presence of adaptation has led investigators to refer to this behaviour as *predataxis* (Berleman and Kirby, 2009; Berleman *et al.*, 2008). Rippling also occurs in the presence of dead cells and many macromolecules. Indeed, it is also observed during developmental lysis of some *M. xanthus* strains (Berleman and Kirby, 2009).

Interestingly, *M. xanthus* cells may recognize prey by responding to their membrane PE. (Bonner *et al.*, 2005; Curtis *et al.*, 2006). 18:1 PE is absent in *M. xanthus* membrane but highly represented in the membrane of other bacterial species. Responses to 18:1 PE are growth phase independent and might be involved in the recognition and movement through prey colonies. Conversely, 16:1 PE is a major component of the *M. xanthus* cell membrane (Curtis *et al.*, 2006), while it is rare in other bacterial species (Bonner *et al.*, 2005; Kearns *et al.*, 2001). Responses to 16:1 PE require starvation conditions. It is possible, therefore, that this compound regulates motility during intraspecies behaviours such as fruiting body formation.

Myxococcus xanthus chemosensory systems

The *M. xanthus* genome contains eight operons encoding proteins homologous to *E. coli* Che proteins. Only four of these operons have been characterized (Mauriello *et al.*, 2010a; Zusman *et al.*, 2007). Three of them, Frz, Dif and Che4 are involved in the regulation of cell movement (see below). One system, Che3, is divergent and is implicated in the regulation of gene expression during development (Kirby and Zusman, 2003; Willett and Kirby, 2011). In this section, we will describe the Frz, Dif and Che4 pathway and how they influence the regulation of reversal frequency and motility.

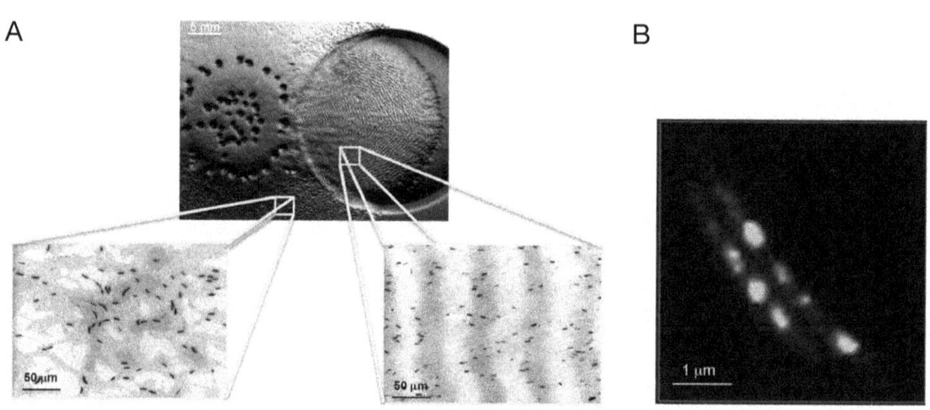

Figure 11.5 Predatory behaviour. (A) When *M. xanthus* cells (left) encounter and then penetrate an *E. coli* colony (right) they align, forming periodic waves (ripples). In contrast, cells that do not encounter *E. coli* cells starve and undergo fruiting body formation (dark structures on the left). Below fluorescence microscopy of wild-type *M. xanthus* cells mixed at a 50:1 ratio with GFP-labelled cells, shows that during rippling behaviour and in the presence of prey (right) individual cells display oriented movement, while in the absence of prey cells do not ripple and move in random directions (left). (B) *M. xanthus* cells expressing FrzCD-GFP. FrzCD forms clusters that align relative to each other when cells make side-to-side contacts.

The Frz pathway, the master regulator of cellular reversal frequency

Reversal frequency is regulated by a set of genes encoding the Frz chemosensory systems (Figs. 11.1 and 11.4) (Bustamante *et al.*, 2004). The *frz* genes were discovered in a screen for mutants defective in aggregation during development because cells failed to form discrete mounds and instead aggregated into 'frizzy' filaments that never matured into fruiting bodies (Fig. 11.3) (Zusman, 1982). The formation of 'frizzy' aggregates was not due to defects in the two motility systems, but rather by the inability of cells to control their reversal frequency during gliding. In fact, individual *frz* null mutant cells reverse very infrequently, about ten times less than wild type; in contrast, constitutively signalling *frzCD*^{c} mutants reverse very frequently, about 5 times more than the parental strain, and show no net swarming (Blackhart and Zusman, 1985; Bustamante *et al.*, 2004).

Sequence analysis of the *frz* genes performed by McBride *et al.* (1989) showed that the *frz* genes encoded proteins that are highly similar to the chemotaxis proteins of enteric bacteria (Fig. 11.1). For example, the *frz* operon encodes FrzCD, a cytoplasmic MCP, and the CheW-like coupling proteins, FrzA and FrzB. Further sequencing showed that the *frz* operon also contains genes encoding FrzE, a CheA (histidine kinase) – CheY (response regulator) fusion protein, FrzF, a CheR (methyltransferase), FrzG, a CheB (methylesterase), and FrzZ, a protein with two CheY domains (Fig. 11.1 and 11.4). All of these genes, when mutated, cause aberrant cellular reversal frequencies, except FrzG whose deletion only causes a mildly increased reversal frequency (see below) (Bustamante *et al.*, 2004). FrzCD, FrzA and FrzE are considered central components of the Frz pathway as (1) they are sufficient for the *in vitro* phosphorylation of FrzZ (Inclán *et al.*, 2007, 2008); (2) only strains lacking one of these three proteins are unresponsive to know repellents (Bustamante *et al.*, 2004).

FrzCD is the MCP receptor associated with the Frz chemosensory system. It is a chemoreceptor lacking the transmembrane and periplasmic domains typical of MCPs (Fig. 11.1) (Bustamante *et al.*, 2004). FrzCD contains two major domains: a unique N-terminal domain that lacks homology to other protein domains, and a C-terminal domain that shares good homology with the methylation and signalling domains of the enteric MCPs (Bustamante *et al.*, 2004). While the unique N-terminal region of FrzCD has been shown to be important for the regulation of A-motility (Mauriello *et al.*, 2009a), its deletion only results in minor defects in S motility and development (Bustamante *et al.*, 2004). However, the methylation of the conserved C-terminal domain of FrzCD has proven to play a central role in regulating the cell reversal frequency, as well as S-motility and development (Astling *et al.*, 2006; McBride and Zusman, 1993; McBride *et al.*, 1992; Scott *et al.*, 2008).

Reversible methylation of MCPs on glutamate residues enables bacteria to sense subtle changes in stimuli and is important for adaptation. FrzCD contains at least seven methylation sites (Astling *et al.*, 2006; Scott *et al.*, 2008). Amino acid substitutions of glutamate to either aspartate or alanine showed that each methylation site plays a unique role in *M. xanthus* single cell behaviours, swarming and fruiting body formation. Moreover, analyses of single, double and triple methylation site mutants revealed that different methylation patterns determine different receptor activities. For example, while certain methylation site mutants cause stimulation of the Frz pathway, leading to hyper-reversing cells, other mutants inhibit Frz signalling and result in cells that rarely reverse (Scott *et al.*, 2008).

Analyses of the localization of the FrzCD receptor showed that the protein is distributed in multiple discrete clusters across the cell (Fig. 11.5B) (Mauriello, 2013; Mauriello *et al.*, 2009b, 2010a). This is in stark contrast to *E. coli*, which has one or two major polar receptor clusters (Maddock and Shapiro, 1993; Sourjik and Berg, 2000). FrzCD localization was also observed to change upon cell–cell contact with other *M. xanthus* cells, events that also induce cell reversals (Mauriello *et al.*, 2009b). It is also possible that FrzCD distributed clusters are used to track contact with prey during predataxis (see above). Movement beyond the prey cell colony, or movement into a crowd of *M. xanthus* cells could both result in a loss of prey contact, triggering cell reversals (Berleman and Kirby, 2009).

FrzF is the enzyme that methylates FrzCD (Shi *et al.*, 1994). It also contains two domains: an N-terminal domain very similar to CheR from *E. coli*, and a C-terminal domain containing three tetratricopeptide repeats (TPRs). These TPR domains are thought to play a role in regulating the methyltransferase activity of FrzF. In fact, *in vitro* assays showed that, while full-length FrzF can only methylate FrzCD at one residue (E182), FrzF lacking the TPR domain could methylate FrzCD at three residues (E168, E175, E182) (Scott *et al.*, 2008). This shows that the TPR domains of FrzF have a regulatory role and modulate methylation at specific sites. How the methylation of FrzCD affects the level of activation of the downstream histidine kinase, FrzE, has yet to be determined. While *frzF* mutants display typical *frizzy* phenotypes in all tested conditions (except the response to repellent), *frzG* mutants only show defects during predation. In predation assays, in fact, *frzG* cells generate the typical wavy pattern (rippling). However, the ripple frequency remains unchanged with respect to changes in prey concentration unlike wildtype cells. These results suggest that adaptation may be more important during predation than during vegetative swarming or fruiting body formation (Berleman *et al.*, 2008; Bustamante *et al.*, 2004).

FrzE is a two-domain protein containing an N-terminal histidine protein kinase similar to CheA and a C-terminal receiver domain (McCleary and Zusman, 1990). As shown in Figs. 11.1 and 11.4, when FrzCD is stimulated (in conjunction with the coupling protein FrzA), it causes FrzE to autophosphorylate at H49 and then transfer phosphoryl groups to FrzZ, a protein consisting of two response regulator domains, at positions D52 and D220 (Inclán *et al.*, 2007, 2008). Interestingly, the receiver domain of FrzE strongly inhibits the autophosphorylation of the CheA domain (Inclán *et al.*, 2008). If FrzZ is the response regulator of the Frz pathway, FrzZ-P should interact with motor components to modulate motility. In *E. coli*, the target of CheY-P, has been shown to be the switch component of the flagellar motor, FliM (Yuan *et al.*, 2012). However, *M. xanthus* does not contain a FliM homologue. Identifying the FrzZ-P interacting partners will supply an important missing component in the signalling pathway.

Cell polarity and reversals

A reversal in *M. xanthus* implies an inversion of cell polarity in which proteins belonging to the A- and S-motility systems translocate from the old leading cell pole to the new leading pole in a synchronous and coordinate manner (Fig. 11.2). Fluorescence microscopy allowed the tracking of several A- and S-motility proteins and studying how they behave before, during and after a reversal. For example, the S-motility protein, FrzS, preferentially accumulates at the leading (piliated) cell pole but it is also present at the lagging cell pole. During movements, the amount of FrzS at the leading pole decreases while FrzS at the lagging pole increases. When cells reverse, the remaining FrzS from the old leading pole rapidly shifts to the new leading pole (Fig.11.2) (Mignot *et al.*, 2005). FrzS has been shown to be important in the regulation of EPS production (Berleman *et al.*, 2011). However, what is the link between its function and subcellular localization remains an open question. Most Agl/Glt A-motility proteins, such as AglZ-YFP, have been shown to localize in a large cluster at the leading pole as well as in smaller clusters distributed in an ordered array spanning the cell length and maintaining fixed positions with respect to the agar surface. Upon cellular reversal, AglZ foci disperse while the polar cluster rapidly relocalize to the new leading pole. When cells begin to move, lateral clusters reappear as well (Fig. 11.2) (Luciano *et al.*, 2011; Mignot *et al.*, 2007; Nan *et al.*, 2010). The frequency of FrzS and AglZ oscillations is under the regulation of the Frz chemosensory system (Mignot *et al.*, 2005, 2007).

Signalling downstream from the Frz pathway

How does Frz contact the A- and S-motility proteins in order to signal reversals? Evidence indicates that this regulation is achieved through the activity of MglA, MglB, and RomR (Fig. 11.2 and 11.4) (Zhang *et al.*, 2010; Zhang *et al.*, 2012; Leonardy *et al.*, 2010; Keilberg *et al.*, 2012). Importantly, MglA, MglB and RomR, which are known to be downstream of the Frz pathway by genetic epistasis, are the only *M. xanthus* proteins known to be essential for both A- and S-motility (Leonardy *et al.*, 2010; Leonardy *et al.*, 2012; Spormann and Kaiser, 1999; Zhang *et al.*, 2010; Zhang *et al.*, 2012). This supports their function of modulating the reversal frequency and motility protein localization in a coordinate manner for the two motility systems. MglB is encoded in an operon with MglA. Genetic analysis showed that MglB acts upstream of MglA and downstream of the Frz pathway in the regulation of the reversal frequency (Fig. 11.4) (Leonardy *et al.*, 2010; Zhang *et al.*, 2010). Moreover, biochemistry showed that while MglA alone binds and hydrolyses GTP at low rates, GTP hydrolysis significantly increases in the presence of MglB, showing that the latter is the cognate GAP (GTPase-activating protein) for MglA (Leonardy *et al.*, 2010; Miertzschke *et al.*, 2011; Zhang *et al.*, 2010). This regulation is analogous to the GTP/GDP cycle catalysed by eukaryotic GTPases, polarity regulators (Charest and Firtel, 2007).

Localization studies showed that MglA forms a cluster at the leading cell pole and that it interacts and co-localize with the A- and S-motility proteins AglZ and FrzS respectively (Fig. 11.2) (Mauriello *et al.*, 2010b). MglB localizes to lagging cell pole and, upon reversals, it relocalizes to the new lagging pole soon after MglA has relocalized to the leading pole. Once the polarity of the two motility motors have been re-established, cells start moving again (Fig. 11.2). It is proposed that MglB prevents MglA accumulation at the lagging cell pole by promoting MglA GTP-to-GDP hydrolysis and keeping GTP-bound MglA at very low concentrations at this site (Zhang *et al.*, 2010). When the slow accumulation of GTP-bound MglA at the lagging pole reaches a certain threshold level, the cell begin a reversal and MglB is pushed to the new lagging pole (Fig. 11.2) (Leonardy *et al.*, 2010; Zhang *et al.*, 2010).

Genetic and cell biology evidence suggests three possible mechanisms by which the Frz pathway might exert its action on MglAB to induce reversals: (i) reversals occur when the MglB inhibitory activity on MglA is released by the Frz system, allowing the MglA relocalization at the new leading cell pole and the switch of the polarity axis; (ii) the Frz system stimulates the accumulation of GTP-bound MglA at the lagging pole; (iii) (i) and (ii) co-occur (Leonardy *et al.*, 2010; Miertzschke *et al.*, 2011; Zhang *et al.*, 2010). Interestingly, *mglA* alleles encoding proteins locked in the GTP-bound form can still reverse in a pendulum-like, unregulated manner. This clearly points to the existence of an additional Frz-modulated regulator of MglA (Leonardy *et al.*, 2010; Miertzschke *et al.*, 2011; Zhang *et al.*, 2010). Indeed, the Frz pathway acts on MglA and MglB also through the RomR protein (Youderian *et al.*, 2003), a response regulator essential for both A- and S-motility (Keilberg *et al.*, 2012; Zhang *et al.*, 2012). Genetic showed that RomR is downstream of Frz and upstream of MglA (Fig. 11.4). It localizes at the lagging cell pole and it drives the positioning of MglA at the leading cell pole by direct interaction (Fig. 11.2) (Leonardy *et al.*, 2007; Keilberg *et al.*, 2012; Zhang *et al.*, 2012). RomR might directly interact with FrzZ, as the two proteins have been recently shown to have a common phylogenetic occurrence (Keilberg *et al.*, 2012).

Genetic and localization studies attempting to show the epistasis between RomR and MglB are controversial rendering such epistasis unclear (Keilberg *et al.*, 2012; Zhang *et al.*, 2012).

The Dif pathway and phosphatidylethanolamine (PE) taxis

The Dif chemosensory system is one of the eight *M. xanthus* chemosensory systems that contains homologues of bacterial chemotaxis proteins (Fig. 11.1) (Yang *et al.*, 2000; Zusman *et al.*, 2007). DifA is similar to bacterial MCP receptors, DifE to CheA kinases, DifD to CheY response regulators, DifC to CheW coupling proteins and DifG to CheC phosphatases. The Dif system does not contain CheR (methyltransferase) or CheB (methylesterase) homologues. Dif regulates the production of exopolysaccharide (EPS) which is essential for the correct functioning of the S-motility machinery (Black and Yang, 2004; Yang *et al.*, 2000).

Besides its role in EPS regulation (see Chapter 10), the Dif system has been shown to modulate the reversal frequency of starved *M. xanthus* cell

(Kearns and Shimkets, 1998). Starved *M. xanthus* cells are also known to respond to certain PE species as chemo-attractants (Bonner and Shimkets, 2006; Kearns and Shimkets, 2001), that is, the reversal frequency of single gliding cells is transiently suppressed by these PE molecules (see above). This transient suppression of reversal, known as the excitation response, requires an intact Dif system. Analysis of *M. xanthus* cell movement in response to PE was performed by tracking isolated cells. Since isolated cells are presumably moving by A-motility, the observed excitation response indicates that the reversal of the A-motility motor is responsive to the Dif chemosensory system.

The response to PE also indicated that the Dif and Frz pathways must interact in order to coordinate their functions in the regulation of the gliding motors. In fact, although excitation by PE depends on the Dif pathway, adaptation relies on Frz, as *frz* mutants never adapt after PE excitation but are still capable of suppressing reversals (Kearns and Shimkets, 1998). Indeed, exposure to PE has been shown to increase FrzCD methylation (McBride *et al.*, 1992). The Dif and Frz systems can probably sense the presence of PE independently (Xu *et al.*, 2008). These signals from two pathways are antagonistic at the level of FrzCD modification: FrzCD shows increased methylation in *dif* mutants relative to wildtype. It is suggested that the signals from the Dif pathway suppress adaptation by modulating FrzCD methylation and this suppression functions to prolong excitation to ensure a meaningful response on the spatial scale of the slow moving *M. xanthus* (Xu *et al.*, 2008).

The Che4 chemosensory system

The Che4 system (Fig. 11.1) functions to control S-motility (Vlamakis *et al.*, 2004). For example, in A^-S^+ strains, but not in a wild type strain, a deletion of the entire *che4* operon or *cheY4* resulted in enhanced swarming. This suggests that the Che4 system negatively regulates swarm expansion under these conditions. In contrast, deleting only *mcp4* caused reduced swarming. Single cell analysis revealed a previously unknown inverse correlation between swarming rate and reversal frequency: cells that swarmed faster (*cheY4*) showed a reduced reversal frequency; cells with a reduced swarming rate (*mpc4*) showed an increased reversal frequency (Vlamakis *et al.*, 2004). Since the Frz system has been shown to be the major regulator of reversal frequency in *M. xanthus*, these results suggest that crosstalk exists between the Frz and Che4 systems. More work needs to be done to elucidate the possible crosstalk networks between the multiple *M. xanthus* chemosensory systems that may lead to regulation of motility.

Concluding remarks

In the past 30 years, many discoveries have helped to uncover the molecular mechanisms by which signals are transduced through the cores of at least three of the eight *M. xanthus* Che systems. Clearly, the complexities of motility and its regulation in *M. xanthus* are still far from being well understood. However, the novel nature of the *M. xanthus* chemotaxis promises to provide numerous insights towards our understanding of the cell biology of these bacteria and their adaptation to their soil habitat. The use of *M. xanthus* as a genetically tractable system and the emergence of high-resolution fluorescence microscopy now presents exciting opportunities for new discoveries.

Acknowledgements

Research in our laboratories is funded by a Marie Curie Career Integration Grant (CIG) to E.M.M. (BMC-CIG-293952) and National Institutes of Health award to D.R.Z. (GM020509).

References

Astling, D.P., Lee, J.Y., and Zusman, D.R. (2006). Differential effects of chemoreceptor methylation-domain mutations on swarming and development in the social bacterium *Myxococcus xanthus*. Mol. Microbiol. *59*, 45–55.

Berleman, J.E., and Kirby, J.R. (2009). Deciphering the hunting strategy of a bacterial wolfpack. FEMS Microbiol. Rev. *33*, 942–957.

Berleman, J.E., Scott, J., Chumley, T., and Kirby, J.R. (2008). Predataxis behavior in *Myxococcus xanthus*. Proc. Natl. Acad. Sci. U.S.A. *105*, 17127–17132.

Berleman, J.E., Vicente, J.J., Davis, A.E., Jiang, S.Y., Seo, Y.-E., and Zusman, D.R. (2011). FrzS regulates social motility in *Myxococcus xanthus* by controlling exopolysaccharide production. PLoS ONE *6*, e23920.

Black, W.P., and Yang, Z. (2004). *Myxococcus xanthus* chemotaxis homologs DifD and DifG negatively regulate fibril polysaccharide production. J. Bacteriol. *186*, 1001–1008.

Blackhart, B.D., and Zusman, D.R. (1985). 'Frizzy' genes of *Myxococcus xanthus* are involved in control of frequency of reversal of gliding motility. Proc. Natl. Acad. Sci. U.S.A. *82*, 8767–8770.

Bonner, P.J., and Shimkets, L.J. (2006). Phospholipid directed motility of surface-motile bacteria. Mol. Microbiol. *61*, 1101–1109.

Bonner, P.J., Xu, Q., Black, W.P., Li, Z., Yang, Z., and Shimkets, L.J. (2005). The Dif chemosensory pathway is directly involved in phosphatidylethanolamine sensory transduction in *Myxococcus xanthus*. Mol. Microbiol. *57*, 1499–1508.

Bustamante, V.H., Martinez-Flores, I., Vlamakis, H.C., and Zusman, D.R. (2004). Analysis of the Frz signal transduction system of *Myxococcus xanthus* shows the importance of the conserved C-terminal region of the cytoplasmic chemoreceptor FrzCD in sensing signals. Mol. Microbiol. *53*, 1501–1513.

Curtis, P.D., Geyer, R., White, D.C., and Shimkets, L.J. (2006). Novel lipids in *Myxococcus xanthus* and their role in chemotaxis. Environ. Microbiol. *8*, 1935–1949.

Hartzell, P.L. (1997). Complementation of sporulation and motility defects in a prokaryote by a eukaryotic GTPase. Proc. Natl. Acad. Sci. U.S.A. *94*, 9881–9886.

Hartzell, P., and Kaiser, D. (1991). Function of MglA, a 22-kilodalton protein essential for gliding in *Myxococcus xanthus*. J. Bacteriol. *173*, 7615–7624.

Hodgkin, J., and Kaiser, D. (1979). Genetics of gliding motility in *Myxococcus xanthus* (Myxobacterales): Two gene systems control movement. MGG Molecular & General Genetics *171*, 177–191.

Inclán, Y.F., Vlamakis, H.C., and Zusman, D.R. (2007). FrzZ, a dual CheY-like response regulator, functions as an output for the Frz chemosensory pathway of *Myxococcus xanthus*. Mol. Microbiol. *65*, 90–102.

Inclán, Y.F., Laurent, S., and Zusman, D.R. (2008). The receiver domain of FrzE, a CheA–CheY fusion protein, regulates the CheA histidine kinase activity and downstream signalling to the A- and S-motility systems of *Myxococcus xanthus*. Mol. Microbiol. *68*, 1328–1339.

Kearns, D.B., and Shimkets, L.J. (1998). Chemotaxis in a gliding bacterium. Proc. Natl. Acad. Sci. U.S.A. *95*, 11957–11962.

Kearns, D.B., and Shimkets, L.J. (2001). Lipid chemotaxis and signal transduction in *Myxococcus xanthus*. Trends Microbiol. *9*, 126–129.

Kearns, D.B., Venot, A., Bonner, P.J., Stevens, B., Boons, G.J., and Shimkets, L.J. (2001). Identification of a developmental chemoattractant in *Myxococcus xanthus* through metabolic engineering. Proc. Natl. Acad. Sci. U.S.A. *98*, 13990–13994.

Keilberg, D., Wuichet, K., Drescher, F., and Søgaard-Andersen, L. (2012). A Response Regulator Interfaces between the Frz Chemosensory System and the MglA/MglB GTPase/GAP Module to Regulate Polarity in *Myxococcus xanthus*. PLoS Genet. *8*, e1002951.

Kirby, J.R., and Zusman, D.R. (2003). Chemosensory regulation of developmental gene expression in *Myxococcus xanthus*. Proc. Natl. Acad. Sci. U.S.A. *100*, 2008–2013.

Leonardy, S., Miertzschke, M., Bulyha, I., Sperling, E., Wittinghofer, A., and Søgaard-Andersen, L. (2010). Regulation of dynamic polarity switching in bacteria by a Ras-like G-protein and its cognate GAP. EMBO J. *29*, 2276–2289.

Li, Y., Sun, H., Ma, X., Lu, A., Lux, R., Zusman, D., and Shi, W. (2003). Extracellular polysaccharides mediate pilus retraction during social motility of *Myxococcus xanthus*. Proc. Natl. Acad. Sci. U.S.A. *100*, 5443–5448.

Luciano, J., Agrebi, R., Le Gall, A.V., Wartel, M., Fiegna, F., Ducret, A., Brochier-Armanet, C., and Mignot, T. (2011). Emergence and modular evolution of a novel motility machinery in bacteria. PLoS Genet. *7*, e1002268.

McBride, M.J., and Zusman, D.R. (1993). FrzCD, a methyl-accepting taxis protein from *Myxococcus xanthus*, shows modulated methylation during fruiting body formation. J. Bacteriol. *175*, 4936–4940.

McBride, M.J., Weinberg, R.A., and Zusman, D.R. (1989). 'Frizzy' aggregation genes of the gliding bacterium *Myxococcus xanthus* show sequence similarities to the chemotaxis genes of enteric bacteria. Proc. Natl. Acad. Sci. U.S.A. *86*, 424–428.

McBride, M.J., Köhler, T., and Zusman, D.R. (1992). Methylation of FrzCD, a methyl-accepting taxis protein of *Myxococcus xanthus*, is correlated with factors affecting cell behavior. J. Bacteriol. *174*, 4246–4257.

McCleary, W.R., and Zusman, D.R. (1990). FrzE of *Myxococcus xanthus* is homologous to both CheA and CheY of Salmonella typhimurium. Proc. Natl. Acad. Sci. U.S.A. *87*, 5898–5902.

Maddock, J.R., and Shapiro, L. (1993). Polar location of the chemoreceptor complex in the *Escherichia coli* cell. Science *259*, 1717–1723.

Mauriello, E.M.F. (2013). Cell biology of bacterial sensory modules. Front. Biosci. (Landmark Ed) *18*, 928–943.

Mauriello, E.M.F., Nan, B., and Zusman, D.R. (2009a). AglZ regulates adventurous (A-) motility in *Myxococcus xanthus* through its interaction with the cytoplasmic receptor, FrzCD. Mol. Microbiol. 72, 964–977.

Mauriello, E.M.F., Astling, D.P., Sliusarenko, O., and Zusman, D.R. (2009b). Localization of a bacterial cytoplasmic receptor is dynamic and changes with cell–cell contacts. Proc. Natl. Acad. Sci. U.S.A *106*, 4852–4857.

Mauriello, E.M.F., Mignot, T., Yang, Z., and Zusman, D.R. (2010a). Gliding motility revisited: how do the myxobacteria move without flagella? Microbiol. Mol. Biol. Rev *74*, 229–249.

Mauriello, E.M.F., Mouhamar, F., Nan, B., Ducret, A., Dai, D., Zusman, D.R., and Mignot, T. (2010b). Bacterial motility complexes require the actin-like protein, MreB and the Ras homologue, MglA. EMBO J. *29*, 315–326.

Miertzschke, M., Koerner, C., Vetter, I.R., Keilberg, D., Hot, E., Leonardy, S., Søgaard-Andersen, L., and Wittinghofer, A. (2011). Structural analysis of the

Ras-like G protein MglA and its cognate GAP MglB and implications for bacterial polarity. EMBO J. *30*, 4185–4197.

Mignot, T., Merlie, J.P., Jr, and Zusman, D.R. (2005). Regulated pole-to-pole oscillations of a bacterial gliding motility protein. Science *310*, 855–857.

Mignot, T., Shaevitz, J.W., Hartzell, P.L., and Zusman, D.R. (2007). Evidence that focal adhesion complexes power bacterial gliding motility. Science *315*, 853–856.

Nan, B., Mauriello, E.M.F., Sun, I.-H., Wong, A., and Zusman, D.R. (2010). A multi-protein complex from *Myxococcus xanthus* required for bacterial gliding motility. Mol. Microbiol. *76*, 1539–1554.

Porter, S.L., Wadhams, G.H., and Armitage, J.P. (2011). Signal processing in complex chemotaxis pathways. Nat. Rev. Microbiol. *9*, 153–165.

Reichenbach, H. (1999). The ecology of the myxobacteria. Environ. Microbiol. *1*, 15–21.

Roberts, M.A.J., Papachristodoulou, A., and Armitage, J.P. (2010). Adaptation and control circuits in bacterial chemotaxis. Biochem. Soc. Trans. *38*, 1265–1269.

Scott, A.E., Simon, E., Park, S.K., Andrews, P., and Zusman, D.R. (2008). Site-specific receptor methylation of FrzCD in *Myxococcus xanthus* is controlled by a tetra-trico peptide repeat (TPR) containing regulatory domain of the FrzF methyltransferase. Mol. Microbiol. *69*, 724–735.

Segall, J.E., Manson, M.D., and Berg, H.C. (1982). Signal processing times in bacterial chemotaxis. Nature *296*, 855–857.

Shi, W., and Zusman, D.R. (1993). The two motility systems of *Myxococcus xanthus* show different selective advantages on various surfaces. Proc. Natl. Acad. Sci. U.S.A. *90*, 3378–3382.

Shi, W., Köhler, T., and Zusman, D.R. (1993). Chemotaxis plays a role in the social behaviour of *Myxococcus xanthus*. Mol. Microbiol. *9*, 601–611.

Shi, W., Köhler, T., and Zusman, D.R. (1994). Isolation and phenotypic characterization of *Myxococcus xanthus* mutants which are defective in sensing negative stimuli. J. Bacteriol. *176*, 696–701.

Sourjik, V., and Berg, H.C. (2000). Localization of components of the chemotaxis machinery of *Escherichia coli* using fluorescent protein fusions. Mol. Microbiol. *37*, 740–751.

Sourjik, V., and Berg, H.C. (2002a). Binding of the *Escherichia coli* response regulator CheY to its target measured *in vivo* by fluorescence resonance energy transfer. Proc. Natl. Acad. Sci. U.S.A. *99*, 12669–12674.

Sourjik, V., and Berg, H.C. (2002b). Receptor sensitivity in bacterial chemotaxis. Proc. Natl. Acad. Sci. U.S.A. *99*, 123–127.

Spormann, A.M., and Kaiser, D. (1999). Gliding mutants of *Myxococcus xanthus* with high reversal frequencies and small displacements. J. Bacteriol. *181*, 2593–2601.

Sun, M., Wartel, M., Cascales, E., Shaevitz, J.W., and Mignot, T. (2011). Motor-driven intracellular transport powers bacterial gliding motility. Proc. Natl. Acad. Sci. U.S.A. *108*, 7559–7564.

Szurmant, H., and Ordal, G.W. (2004). Diversity in chemotaxis mechanisms among the bacteria and archaea. Microbiol. Mol. Biol. Rev. *68*, 301–319.

Taylor, R.G., and Welch, R.D. (2008). Chemotaxis as an emergent property of a swarm. J. Bacteriol. *190*, 6811–6816.

Vlamakis, H.C., Kirby, J.R., and Zusman, D.R. (2004). The Che4 pathway of *Myxococcus xanthus* regulates type IV pilus-mediated motility. Mol. Microbiol. *52*, 1799–1811.

Wall, D., and Kaiser, D. (1999). Type IV Pili and cell motility. Mol. Microbiol. *32*, 1–10.

Willett, J.W., and Kirby, J.R. (2011). CrdS and CrdA comprise a two-component system that is cooperatively regulated by the Che3 chemosensory system in *Myxococcus xanthus*. MBio 2.

Wu, S.S., and Kaiser, D. (1995). Genetic and functional evidence that Type IV Pili are required for social gliding motility in *Myxococcus xanthus*. Mol. Microbiol.*18*, 547–558.

Xu, Q., Black, W.P., Mauriello, E.M.F., Zusman, D.R., and Yang, Z. (2007). Chemotaxis mediated by NarX-FrzCD chimeras and nonadapting repellent responses in *Myxococcus xanthus*. Mol. Microbiol. *66*, 1370–1381.

Xu, Q., Black, W.P., Cadieux, C.L., and Yang, Z. (2008). Independence and interdependence of Dif and Frz chemosensory pathways in *Myxococcus xanthus* chemotaxis. Mol. Microbiol. *69*, 714–723.

Yang, Z., Ma, X., Tong, L., Kaplan, H.B., Shimkets, L.J., and Shi, W. (2000). *Myxococcus xanthus dif* genes are required for biogenesis of cell surface fibrils essential for social gliding motility. J. Bacteriol. *182*, 5793–5798.

Yuan, J., Branch, R.W., Hosu, B.G., and Berg, H.C. (2012). Adaptation at the output of the chemotaxis signalling pathway. Nature *484*, 233–236.

Zhang, Y., Franco, M., Ducret, A., and Mignot, T. (2010). A bacterial Ras-like small GTP-binding protein and its cognate GAP establish a dynamic spatial polarity axis to control directed motility. PLoS Biol. *8*, e1000430.

Zhang, Y., Guzzo, M., Ducret, A., Li, Y.-Z., and Mignot, T. (2012). A Dynamic Response Regulator Protein Modulates G-Protein-Dependent Polarity in the Bacterium *Myxococcus xanthus*. PLoS Genet. *8*, e1002872.

Zusman, D.R. (1982). 'Frizzy' mutants: a new class of aggregation-defective developmental mutants of *Myxococcus xanthus*. J. Bacteriol. *150*, 1430–1437.

Zusman, D.R., Scott, A.E., Yang, Z., and Kirby, J.R. (2007). Chemosensory pathways, motility and development in *Myxococcus xanthus*. Nat. Rev. Microbiol. *5*, 862–872.

The Biophysics of *Myxococcus xanthus* Motility

12

Fabian Czerwinski and Joshua Shaevitz

Abstract

Biophysical approaches have become increasingly important to the study of *Myxococcus xanthus* from the generation of force during motility to the statistical mechanics of multi-cell pattern formation. In this chapter, we introduce a number of relevant biophysical concepts and techniques, focusing on their relevance for bacterial cells. Biophysical measurements have played a major role in the study of the *Escherichia coli* flagellar motor. Through analogy with this well-studied system, we hope to highlight current work in the field of *M. xanthus* biophysics and inspire the next generation of scientists interested in understanding the amazing behaviours of this fascinating organism.

Introduction

Watching a colony of *Myxococcus xanthus* swarming on a plate for the first time inevitably produces a feeling of awe over the capabilities of this single-celled organism. Why and how do these cells move and form large groups? Do cells have an individuality and do pioneers care about the colony as a whole? For decades, scientists have followed up on these questions and others, but only recently have we been able to ask these questions in a rigorous, quantitative way from a biophysical perspective. While significant work has been undertaken on the genetics frontier, many key questions regarding *M. xanthus* are physical in nature and require a combination of genetics and biophysical probes to unravel. How do cells convert metabolic chemical energy into productive mechanical work that can move a whole cell? How do neighbouring cells coordinate their activities and why does this cooperation lead to the correct patterns of thousands of cells in time and space?

In this chapter, we focus on these and other biophysical questions relating to *Myxococcus* motility, highlighting key open questions and similarities to other motile organisms. We begin by introducing general physical concepts at the scale of a bacterium. Then we discuss the variety of biophysical tools that have been used to improve our understanding of *M. xanthus*. We chose to focus on the two motility mechanisms of *M. xanthus* where the most biophysical research has been done. Twitching and gliding motility are both relatively well-studied, and we want to highlight in particular those studies that combine genetics and biophysical techniques. The diverse modes of movement allow *Myxococcus* to exhibit a complex grouping behaviour. The last section describes experiments and results that focus on the role the cell motility in group behaviours.

Biophysics at the bacterial scale

To achieve directed motility, cells must employ protein machines to control and harness chemical energy and produce propulsive forces. Because of their minuscule size, cell must use very different strategies for moving through fluid or pushing off surfaces compared to macroscopic organisms. As such, their very mechanisms of movement can often seem strange. We will begin by reviewing some of the relevant physical quantities that cells encounter.

It is important to begin by defining the spatial and temporal scales relevant to the field of prokaryotic biology and motility in particular. Bacterial cells themselves are typically a few microns (10^{-6} m, μm) in size. The proteins that make up the machinery inside the cells measure a few nanometres (10^{-9} m, nm) but can oligomerize into substantially larger structures. One particularly important example are polymer filaments which enable the formation of long length-scale order inside cells. With few exceptions, bacterial cell size is limited to a few microns and larger spatial scales are produced via multicellular groupings such as biofilms, swarms, and fruiting bodies. Events inside cells happen on various time scales. Molecules inside a cell typically diffuse through the whole cell in about 0.1–1 s, although this is slower for proteins embedded in lipid membranes (Li and Xie, 2011). Enzyme-driven chemical reactions can happen on a variety of time scales, but typical chemical rates vary broadly from 3 per minute for chymotrypsin to millions per second for catalase in *Pseudomonas* (Ingles and Knowles, 1966; Elkins *et al.*, 1999; Nelson, 2004a). These typical scales set limits on how fast physical processes that affect motility can work. If a molecular motor moves one protein-sized step, say 10 nm, per 1/100 s, we get a typical 'motor' speed of 1 μm/s. Of course, the cell can regulate events on much longer time scales such as the division time.

Cells have various sources of energy available to them. One of the most common are molecules of ATP, which store chemical energy in a high-energy phosphate bond. A number of different mechano-enzymes are able to harness this energy by coupling ATP hydrolysis to physical motion and force generation. The amount of energy that can typically be released through the cleavage of an ATP molecule into an ADP molecule inside a cell is of the order of 100 pN nm (10^{-19} J). This kind of mechanism is used by the conventional motor proteins kinesin, myosin, and dynein as well as bacterial motors such as the gliding motor in *Mycoplasma mobile* (Uenoyama and Miyata, 2005). Another source of energy for motility is the polymerization of proteins into polymers. In the case of the biofilament actin, polymerization at cellular concentrations produces about 40 pN nm per subunit addition. The electrical potential difference across the cellular membrane in many cells produces a third type of energy source used by molecular motors. The passage of a proton from the extracellular medium to the cytoplasm, driven by this potential, releases energy of ~30 pN nm (Pilizota *et al.*, 2009). This mechanism is used by the bacterial flagellar motor as well as the *Myxococcus xanthus* gliding motors as discussed in Chapter 9 of this book.

Another form of energy that is ever-present at small scales is thermal energy which drives the motion of all molecules with an energy of $k_B T$ that is approximately 4 pN nm at physiological temperatures. These random molecular fluctuations drive everything from biochemical reactions to protein folding to the shapes of flexible biological molecules. Furthermore, the aggregated effect of thermal energy in a fluid produces the fluctuating displacement of larger objects, a phenomenon known as Brownian motion.

In order to move, cells must produce extracellular forces that are transduced to the surroundings. In discussing motility, two different modalities must be considered: swimming in fluid, which relies on viscous forces, and motility on a solid substrate, which involves direct mechanical interactions with an elastic material. At small scales, swimming takes place in the 'low Reynolds number' regime in which the effects of inertia can be ignored (Nelson, 2004b). As a result, the trajectory of motion of a microorganism in fluid is fully described by the series of postures the body adopts and not the speed of the postural dynamics. The organism expends energy in order to produce these bodily deformations which are resisted by the fluid, and potentially the mechanics of the cell body, whereas swimming is a by-product of the fluid resistance. Examples of this include the rotation of a helical bacterial flagellum which propels the cell forward, and the chiral body undulations of *Spiroplasmas* which swim in a helical trajectory. On the other hand, movement on a solid surface involves the application of a force between an elastic element inside the cell, e.g. a cytoskeletal filament or a cell wall, and the substrate. This typically involves both the binding of cellular molecules to the substrate and the generation of force between these bound molecules to generate traction on the substrate. This type of motion can involve pulling, as in the retraction of type IV pili from the *Myxococcus xanthus* S-motility system which adhere at their ends to the substrate (see Chapter 10), or pushing, as in the A-motility gliding system which appears to involve a directed force at specifically adhered locations (see Chapter 9).

All mechanisms of motility necessarily involve the intricate coordination of force generation mechanisms. When moving on a surface, for example, it is typically not enough to simply bind part of the cell to the substrate and then apply force. To create directional movement, the cell must either generate an oriented force relative to the cell body or it must generate an asymmetric set of forces along the cell body. In pilus-based motility, the pilus retraction presumably generates more force than the frictional connections between the cell body and the substrate so that the cell moves forwards. In gliding motility, it is not clear how this asymmetry is ultimately achieved at the molecular level.

Biophysical tools for studying motility

A variety of techniques have been developed to probe biophysical aspects of cellular motility. Two important categories include imaging approaches, which allow for the precise localization and characterization of the positions of molecules and cells, and force microscopy, which enables the measurement of physical forces involved in motion production. We briefly outline here some of the most common methods that are applied to *M. xanthus*. A more in-depth discussion of how computer analyses of images and modelling can be used to study population-scale motion is provided in Chapter 8.

Imaging

Biological imaging has matured into a field that encompasses any visualization-based measurement from the localization of single fluorescent molecules to MRI-based methods for whole brain studies. Typically, a tradeoff exists between the resolution and size of the field of view. For example, studies of group behaviour in *Myxococcus xanthus* need a wider field of view than single-cell studies where the highest resolution is required to gather reasonable data on intracellular details. In addition, optimal experimental design attempts to combine sample type, the imaging modality, and the quantitative analysis techniques that can be used to extract meaningful biophysical data (van Teeffelen *et al.*, 2012).

Light microscopy

A number of light microscopy techniques are available and fairly standard for trained researchers. Good resources to comprehend the basics of light microscopy are Murphy's book and Nikon's MicroscopyU website (Murphy, 2001; MicroscopyU).

Bright-field microscopy is the most common type of microscopy. A white light source, often a mercury or xenon lamp, illuminates the sample through a condenser that is ideally placed into Koehler alignment. At the sample, light is scattered and absorbed in optically dense areas. An objective typically projects the image onto a conjugated focal plane where the spatially resolved intensity pattern is either viewed via an eyepiece or recorded on a pixelated camera. The specifications of the objective and the illumination pattern determine the resolution that can be achieved. The major limitation for bright field microscopy of bacterial samples is the low contrast, since the optical density of cells is often only a few per cent higher than the density of the surrounding buffer.

Contrast-enhancing microscopy solves this limitation and has two prominent implementations: PHASE CONTRAST (PC) and differential interference contrast (DIC). These are both extensions to bright field microscopy, but instead of relying on intensity differences due to absorption in the sample, they utilize interference and the relative retardation of light rays through the sample.

In PC microscopy, a pair of metallic (phase) rings embedded in the objective and condenser lenses housings enhances the contrast between tiny differences in optical density in areas of very close proximity, for example, in cell membranes surrounding the cytoplasm. This leads to high-contrast images that lack any directional biases. However, one drawback is the installation of the permanent phase rings. Since they are made of a highly absorbing material, it is often difficult to use PC with laser-based techniques that could damage the objective lens.

In DIC microscopy, linearly polarized light is split into two orthogonally polarized propagation directions before reaching the sample and then recombined and interfered after the sample by a pair of prisms. The analysed image substantially enhances edges between structures of different optical properties. Characteristically, the image appears to have a shadow with respect to the direction in which the polarizations are splayed. Therefore, contrast and resolution depend on the orientation of the sample locally in space. This is a major drawback for computer-assisted segmenting (Li and Kanade, 2009). Still, to resolve single bacteria, differential interference contrast microscopy can be an excellent choice as it often provides superior contrast to PC.

Wet-SEEC is a surface-based technology that has very recently resolved the thin polysaccharide-containing slime that *M. xanthus* deposits on surfaces (Ducret *et al.*, 2012). Contrast enhancement happens directly at the sample surface by a thin transparent layer with a refractive index slightly smaller than glass. For the specific coating used, only images of material as close as a few tens of nanometres experience contrast enhancement. Although only very thin surfaces on top of the coating can be used and the height calibration cannot be done directly, the superb sensitivity of Wet-SEEC and the absence of any stain represents a huge leap forward in visualizing surface bound substrates.

Fluorescence microscopy

Modern microbiology has greatly benefited from advances in fluorescence microscopy techniques over the last few decades. Fluorescence measurements enable a direct readout of, e.g. protein localization, levels of protein expression, and even protein exchange among cells (Mauriello *et al.*, 2009; Wei *et al.*, 2011). Fluorescence is induced by excitation of a fluorescent molecule at a given wavelength. It emits a limited number of photons at a longer wavelength when leaving the excited state. Fluorescent probes can be as diverse as engineered quantum dots, immunofluorescent dyes or fluorescent proteins like GFP. Due to the inherent limitations of fluorescence microscopy such as limited staining efficiency, photon count, and photostability, sample and measurement design depend more on each other than in other techniques. For bacterial samples, this is especially true because of the small size of the sample and the small number of fluorescently labelled proteins that are usually present. Furthermore, the post-acquisition image analysis is crucial, and should be considered in the experimental design process (van Teeffelen *et al.*, 2012).

Several methods have been developed to label and image the localization of proteins in *M. xanthus* cells including immunofluorescence and fluorescent fusions. Immunofluorescence uses fluorophore-conjugated antibodies but requires the use of chemically fixed, permeabilized cells. The use of fixed cells introduces potential artefacts from fixation and precludes the studies of dynamic localization. As such, immunofluorescence is of limited use for biophysical studies of motility, an inherently dynamic phenomenon.

Genetically encoded fusions of a protein of interest to derivatives of naturally occurring fluorescent proteins such as GFP from the jelly fish can be used for dynamic studies of protein localization in live cells. At the fastest scales, time resolution is limited by the number and brightness of the fluorescent probes, the tendency of these proteins to emit a small number of photons before entering into a long-lived dark state, and potential photochemical damage caused by exposure to the fluorescent excitation light. At the longest scales, dynamical measurements are often inhibited by slow drift of the sample relative to the image. This can arise from a number of sources including mechanical motions driven by environmental temperature changes, gravitational creep of microscope optics (particularly the object mount), and changes in hydration of the sample such as a swelled agar pad. These effects can often be corrected for either in data processing or through active stabilization of the microscope itself.

One common problem is the extended depth of focus of fluorescent microscopy when imaging either thick samples, or when attempting to focus on surface structures. In thicker samples, confocal microscopy can enhance the rejection of out-of-focus light in the images which improves the effective depth of field. At the surface, total internal reflection fluorescence microscopy (TIRFM) can be used to limit excitation to a very small region (50 to 100 nm) near the coverslip surface. In addition, the tendency of fluorescent probes to shift into a dark state has been used as a tool to study dynamics. In a technique called fluorescence recovery after photobleaching (FRAP), fluorophores are first bleached, and then the recovery of fluorescence can be used to investigate protein dynamics and synthesis.

Force microscopy

The understanding of propulsive force generation is a key aspect of biophysical studies of motile systems. Techniques to probe forces generated by molecules and cells have been perfected over the last several decades. These efforts have resulted in instruments capable of achieving spatial resolutions better than 1 nm and force sensitivities on the order of tens of femtonewtons (Abbondanzieri *et al.*, 2005; Carter *et al.*, 2009; Czerwinski *et al.*, 2011). Here, we highlight several of the force microscopy techniques that allow for localized, *in vivo* studies of bacterial motors (Fig. 12.1).

Optical traps use a tightly focused laser beam to generate forces. These instruments have proven to be a potent tool for force measurements and manipulation of bacterial motors. For example, they played a key role in understanding the response of the flagellar motor to external load (Pilizota *et al.*, 2009; Bai *et al.*, 2010), energy usage by the outer-membrane bound lambda-receptor (Winther *et al.*, 2009), and the contribution of the MreB protein to the cellular bending stiffness of rod-shaped cells (Wang *et al.*, 2010).

Light is comprised of photons which carry momentum in the direction of propagation. Objects that deflect light, such as a mirror or a lens, feel a reaction force equal in magnitude and opposite in direction to the change in momentum of the light. Practically, objects at the focal plane of a microscope that have a refractive index denser than the surrounding medium can be stably held by an optical trap. At the laser focus, the energy landscape of the interaction between the object and the spatial intensity pattern mimics

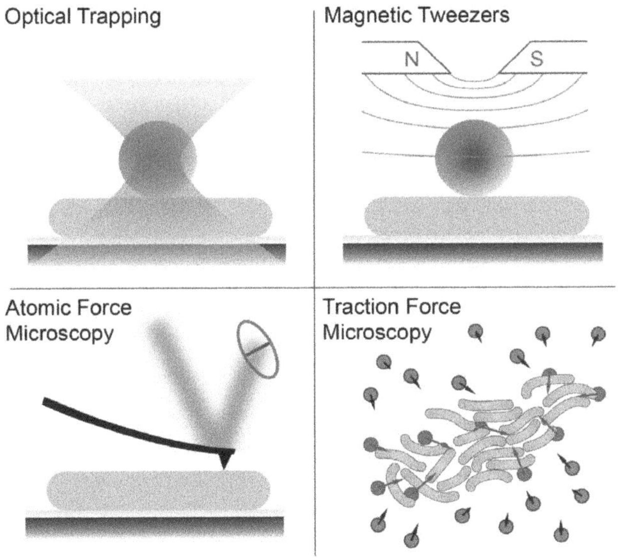

Figure 12.1 Schematic illustration (not to scale) of four biophysical force microscopy techniques as they might be applied to *M. xanthus*. An optical trap is formed by a focused laser strong enough to trap a polystyrene bead on top or at the side of an *M. xanthus* cell. Magnetic tweezers are formed by a polystyrene bead with a magnetic core that can be controlled by a pair of external magnets. These magnets are mounted outside the sample. An atomic force microscope translates the interaction between the surface and a tiny tip into an image. The deflection of the tip is amplified by the cantilever which also reflects a detection laser onto a photodiode. In traction force microscopy, the displacement of reporter beads within a solidified gel is linear to the force applied to the substrate, for example by a gliding *M. xanthus* group. A colour version of this figure is located in the plate section at the back of the book.

a harmonic spring in each of the three dimensions. For small deviations of the object from the centre of the trap, the restoring gradient force is linear and proportional to the laser intensity. This technique is usually seen as non-invasive for transparent biological samples, but a few considerations are important for *in vivo* measurements. First, it is beneficial to choose a laser wavelength at which the absorption of photons by water is minimal; 1064 nm is a typical choice (Svoboda and Block, 1994). Also, optical traps require laser powers of up to 1 W at the sample which could potentially produce free (oxidative) radicals in the vicinity of the trap that can effect enzyme kinetics and weaken molecular bonds. Optical traps work best on large, micron-sized objects such as glass or polystyrene beads. As such, molecular studies require a bio-compatible 'handle' to link the trapping force to a specific biological molecule.

Magnetic tweezers use micron-sized beads with a magnetic core that are controlled by an external magnetic field. By moving the external magnets, or changing the current in an electro-magnetic setup, the experimenter can manipulate the force at the substrate. In particular, magnetic tweezers have proven useful for single-molecule torque measurements (Lipfert *et al.*, 2010) and measurements of cell-surface interactions (Walter *et al.*, 2006). A limitation of this technique lies in the metallic cores of the beads: often these are not uniform and the beads can vary in shape and size. Furthermore, the metallic cores interfere with many laser-based imaging methods due to absorption and heating.

Atomic force microscopy (AFM) is a technique in which a nanometre-sized tip is scanned over the sample to produce topographical images and mechanical measurements. The tip is microfabricated onto the end of a tiny cantilever which can be actuated in three dimensions. By tracking a laser that reflects off the back of the cantilever, the bending force of the cantilever as well as the vertical tip position can be measured. A combination of laterally scanning the tip across the sample while measuring the tip height generates a topographical image, whereas vertical scanning of the tip can generate mechanical measurements such as sample stiffness and binding force. In addition, the tip can be biologically functionalized and used to measure the specific interaction between two binding proteins or the internal structure of bacterial filaments (Andersson *et al.*, 2006; Axner *et al.*, 2011). For *M. xanthus*, atomic force microscopy was used to visualize type IV pili as well as to determine cell-wall elasticity (Pelling *et al.*, 2005). In general, atomic force microscopy is typically used for applications that need large forces applied over short time spans.

Traction force microscopy (TFM) measures a spatially resolved map of surface stresses by recording the position of reporter particles, such as fluorescent beads, embedded in a gel substrate (Sabass *et al.*, 2008). TFM is routinely used to investigate the force generation in large, adherent eukaryotic cells (see, for example, Gardel *et al.*, 2008). The implementation of TFM is challenging, in particular the computational aspects of calibrating the surface forces. But, the potential for bacterial systems is high: TFM could quantify the force fields generated by moving cell groups with single-cell or better resolution. Adapting TFM to studies of multi-cellular bacterial systems may prove useful for studies of phenomena such as aggregation and predation.

Bacterial motility

Of the different ways that bacteria use to move, swimming via flagellar rotation is by far the best understood. We will briefly discuss the current state of our knowledge of *E. coli* swimming as a guide to the kinds of questions and understanding we wish to have for *M. xanthus*. This parallel is even more striking when one realizes that many components of the *Myxococcus* motility system appear to be related, at least partly, to those from the flagellar system.

Pioneering work from Berg, Macnab, and others demonstrated that *E. coli* and many other bacteria swim via the rotation of long helical filaments called flagella (Berg, 1988). As discussed above, the motion of small organisms, such as bacteria, in a fluid is dominated by viscous drag. To produce propulsive forces, a bacterium must employ a unique strategy that is cyclical, so as to facilitate a repetitive swimming stroke, but not reciprocal, which would result in zero net motion in the over-damped viscous regime. Rotation of a helical filament is one of several solutions to this problem that generates significant propulsion (Nelson, 2004b). By rotating the flagellar motor in the counter-clockwise direction with a stable frequency, *E. coli* cells swim in a straight path at speeds up to 30 μm/s (Darnton *et al.*, 2007). Interestingly, some bacteria, like *E. coli*, have multiple flagella that bundle together hydrodynamically into a single helical superstructure that overcomes phase differences between the motors in milliseconds.

Decades of work have led to a comprehensive understanding of the flagellar motor system including detailed insights into its mechanical design and control. The flagellum itself is a hollow extracellular organelle, 20 nm in width and up to 10 μm in length, that is connected to the rotary motor via a curved hook. This rotary engine is powered by the gradient in proton concentration across the inner membrane and translocation of individual protons drives a conformational change in the MotA and MotB stator proteins. These motions within the motor complex cause the flagellum to rotate by 13.8 degrees (Sowa *et al.*, 2005). Because the stators are arranged in a circular ring, 26 strokes of the stator proteins lead to one full revolution of the flagellum. Many consecutive strokes give rise to the processive rotation of the flagellum.

The flagellar motor has two gear settings, controlled by the phosphorylation state of the sensory output protein CheY. One gear setting leads to counterclockwise rotation of the flagella and straight swimming. The second gear produces a reverse, clockwise rotation that causes the multiple flagellar filaments to splay out. Flagellar rotation in this scenario leads to a random tumbling of the body axis in space. By tuning the timing and length of straight runs versus direction-randomizing tumbles, *E. coli* is able to move towards or away from sources of different chemicals.

The biochemical network that *E. coli* uses to search for a source of chemicals via swimming and tumbling, a process called chemotaxis, is perhaps the best understood signal-processing circuit from any biological system (Sourjik, 2004). Temporal changes in external concentration, detected by a readout that utilizes extracellular receptor occupancy, give rise to a phosphorylation and methylation cascade within the cell that affects sophisticated processes of amplification, adaptation, and filtering. The biochemical output of this processing step affects the methylation state of the receptors and thus the temporal bias between clockwise and counterclockwise rotation.

Unlike swimming organisms, bacteria that move on solid substrates must generate thrust in a very different manner. Whereas swimmers couple motion to a surrounding fluid hydrodynamically, gliders must employ elastic forces and orchestrate a delicate balance between surface attachment kinetics and force production such that forward-directed forces are used effectively. In essence, when a cell pulls or pushes on the surface, something has to break the symmetry between forwards and backwards. For surface gliding keratocytes and fibroblasts, for example, the temporal dynamics of focal adhesion maturation lead to an asymmetry in the binding strength to the surface along the cell axis. As contractile forces are generated within the eukaryotic actin cytoskeleton, attachment points at the rear are more likely to give way such that overall the centre of the cell moves forward during contraction. In order to effectively explore and inhabit surfaces, *Myxococcus xanthus* has developed a complex motility system composed of two different mechanisms, twitching and gliding, that can each be used to move on a solid substrate.

Twitching motility

The twitching motility system in *M. xanthus* utilizes the extension and retraction of filamentous pili to generate propulsion. The type IV pilus, an extracellular organelle is a highly dynamic polymer that constantly elongates and retracts. These filaments are thin, 5–7 nm in diameter, but can extend to several microns in length (Pelling *et al.*, 2005). Molecularly, the pilus is composed of pilin, a protein encoded by the *pilA* gene. Unlike the hollow structure of the bacterial flagellum, pilin subunits stacks into a solid rod that is stiffer, thinner, and a more dynamic structure. We direct the reader to Chapter 10 for a more detailed view on S-motility, the pilus organelle, and the proteins involved in regulation.

Polymerization and depolymerization of pili is controlled by the PilB and PilT enzymes. The folded pilin protein contains a helical hydrophobic domain that embeds it directionally in the inner membrane. The ATPase PilB recruits unbound pilin from within the inner membrane and synthesizes it into a stiff pilus filament whereas the PilT protein has been linked to pilus disassembly and retraction (Jakovljevic *et al.*, 2008). During assembly, the nascent pilus traverses the periplasmic space and passes through the outer membrane via a ring of the PilQ secretin. This specialized exit pore has been hypothesized to serve two purposes: (i) the stabilization of the outer membrane around the piercing pili, and (ii) the reduction of molecular fiction between the outer membrane and pilus during fast extension and retraction.

The stiffness of a polymer can be characterized by its persistence length, a characteristic length over which thermal fluctuations significantly bend the filament backbone. Experimentally, the persistence length of certain filaments has been observed to change with contour length (Pampaloni *et al.*, 2006). It provides a crucial control parameter of the dynamic rigidity between relative 'floppiness' and 'stiffness'. Although for *M. xanthus* there have been no direct measurements, the persistence length of the pilus can be estimated to be on the order of microns comparable to pili from *Pseudomonas aeruginosa* (Skerker and Berg, 2001).

Little is known about how individual pili bind to surfaces or if and how this binding might be detected by a cell. *M. xanthus* pili have been shown to attach to polystyrene, agar and agarose. The most prominent anchor for pili attachment is the exo-polysaccharid (EPS) the cells produce themselves and deposit (Kaimer *et al.*, 2012; Gibiansky *et al.*, 2013). There have been no measurements that exclusively addressed the binding strength of pilus attachment to a specific surface. Indirectly though, the strength of the retraction motor revealed that a single pilus can remain bound to a carboxylated polystyrene bead even under forces of 150 pN for tens of seconds (Clausen *et al.*, 2009a). This value might therefore serve as a lower boundary on the attachment force.

Interestingly, *M. xanthus* cells appear to require their neighbours for long-range pilus-driven motility (Shimkets *et al.*, 1982). Prior to persistent forward twitching, *M. xanthus* cells often jostle back and forth on the scale of about 400 nm. When an approaching group of cells breaks this symmetry, the cell in question begins to move productively with the others. This leads to the intriguing possibility that in the twitching system, symmetry breaking does not occur in a single cell but is instead a multi-cellular process.

Biophysical measurements of twitching motility

To date, there is only one published work concerning the forces associated with the dynamic assembly and disassembly of *Myxococcus* type IV pili (Clausen *et al.*, 2009b). Clausen *et al.* trapped a carboxylated polystyrene bead using an optical trap and positioned it near the pole of a *Myxococcus* cell immobilized on a polystyrene-coated glass slide. As the cells attempted to perform twitching motility, a pilus would inevitably bind to the bead. Upon pilus retraction, the bead was pulled away from its equilibrium position in the centre of the trap. Within a few seconds, the motor retracting the pilus built up a force larger than 100 pN. At some point, the retraction force was balanced by the optical trap, i.e. the motor could not work any harder against the optical force on the bead, called a stall. The distribution of stalling forces showed a mean of 149 ± 14 pN. When the *pilT* gene was deleted, the stalling force measured in this way decreased to 70 ± 10 pN, perhaps indicating that filament disassembly provides a significant amount of the free energy released during retraction.

In a complementary assay during the same study, Clausen *et al.* (2009b) used a force clamp to study the velocity of pilus retraction under constant loads. In these experiments, the distance between bead and trap centre was kept constant via software control to maintain a constant applied force and the authors observed a bimodal behaviour in the observed retraction velocities. A low velocity mode, with speeds of ~ 700 nm/s, corresponding to a pilus depolymerization rate of about 700 pilin subunits per second, was found to be independent of the applied load, whereas a faster set of velocities, around 1.3 μm/s, was found only for small forces less than 30 pN.

Comparison to measurements on pili in other organisms

The forces that *Myxococcus xanthus* can exert by pilus retraction are in the same order of magnitude as other bacteria. Single pilus of *Neisseria gonorrhoeae* can generate up to 160 pN, with a mean around 120 pN (Maier *et al.*, 2004). For bundles of up to 10 pili, *N. gonorrhoeae* can produce coordinated retraction in the order of nanonewtons (Biais, *et al.* 2009). Interestingly, in this organism, external loads are sufficient to elongate the pilus. When multiple pili are active, the same pilus might experience a retraction force from its PilT motor and an extension force from motors pulling in the opposite direction at the other cell pole. This tug-of-war can result in an effective elongation regulated by the force at the pilus (Holz *et al.*, 2010). It is believed that this type of behaviour is absent in *Myxococcus xanthus* as the retraction motor PilT and the elongation motor PilB localize in an asymmetrical bipolar pattern (Bulyha *et al.*, 2009). Therein, PilB primarily functions at the leading and PilT at the lagging pole. Periodically, PilT proteins also localize at the leading pole. This assembly of the motor ATPases is highly regulated by the Frz system which is not the case in *N. gonorrhoeae*. This feature emphasizes the importance of the regulation of directionality in *Myxococcus*. The directionality of twitching for *N. gonorrhoeae* is determined by the outcome of a tug-of-war between the retraction of pili on opposite sides of the cell body (Holz *et al.*, 2010). In optical tweezers studies, Holz *et al.* compared the stalling force of PilT (distributed around 120 pN, maximum stalling force 180 pN) with the unbinding force of pili from silica and BSA-coated polystyrene beads. These unbinding forces were 10 pN for silica, with rare maximum unbinding forces up to 80 pN, and 25 pN, up to 145 pN, for BSA-coated polystyrene beads. The symmetry breaking by this tug-of-war might lead to persistent unidirectional twitching motility.

Another species that exhibits twitching motility is *Pseudomonas aeruginosa*. In this organism, the retraction of 25 μm long pili generated a force of 10 pN (Skerker and Berg, 2001), substantially smaller then the forces generated by the two organisms described above. *P. aeruginosa* also displays a regulated localization of elongation and retraction motors. Surprisingly though, PilB and PilT localize bipolarly and the symmetry is broken by a second retraction motor, PilU, that localizes to only one pole (Burrows, 2012).

Unresolved problems

There are a number of interesting questions whose answers can bolster our knowledge about twitching motility in *M. xanthus*. We know the diameter and contour length of individual pili, but lack a measurement of the persistence length. Also, those experiments should cover the impact of the contour length on the persistence length. Twitching motility involves the directional extension of pili from the leading pole and there should be a functional relationship between the persistence length and the contour length of typical extended pili. Another problem is the lack of kinetic measurements about the pili attachment to substrates. We currently do not know how *M. xanthus* type IV pili bind and if/how they sense specificity to the substrate. Do these extracellular organelles signal to the cell that they are bound? Similarly, how do pili detach and is detachment controlled?

Another block of questions is related to the cell as a whole rather than to a particular subset of the twitching machinery. *M. xanthus* has been shown to respond to confinement with a more persistent forward motion (Meel *et al.*, 2012). Is this a purely passive effect due to limited degrees of freedom, or does the cell actively sense the confinement? Also, what happens to twitching cells in the confinement of a cell group? How does persistent group motion depend on the twitching motility of individual cells? Do the cells bind to each other via pili? How do cells in groups determine their direction of forward twitching?

Gliding motility

For twitching motility the hydrolysis of ATP at the retraction motor PilT is sufficient to generate enough force to pull the cell forward. For gliding *M. xanthus* cells, it was recently shown that intracellular transport, fuelled by the proton motive force, generates traction along the cell body (Sun *et al.*, 2011). The force parallel to the cell axis is generated at multiple sites and changes direction as a response to the regulatory Frz system. The gliding machinery, thus, exhibits a completely different architecture than the twitching one. It is distributed along the cell body and must provide a physical connection between the cell and the extracellular environment. One can think of it as consisting of four main subdivisions: a potential link to the cytoskeleton, the force-generating unit within the cytoplasmic membrane, a bridging complex in the periplasm, and the trans-OM binding unit to the extracellular environment. Each of these subdivisions plays a crucial role in the physical picture of the gliding machinery. For more details about the molecular picture of A-motility we advise Chapter 9.

The actin homologue MreB potentially forms a kind of cytoskeleton in *M. xanthus* as it has been seen to polymerize in a helical fashion along the cell length (Mauriello *et al.*, 2010). It is an ongoing debate as to which of the about 40 proteins involved in A-motility (Youderian *et al.*, 2003; Xanthusbase) provided the physical link between the force generating motors and the cytoskeleton. Even with this uncertainty regarding the crucial link-forming protein complex, we do know that the protein AgmU colocalizes with MreB (Nan *et al.*, 2011), and that the regulatory protein AglZ colocalizes with the protein AglQ, a crucial component of the gliding motor proton channel (Sun *et al.*, 2011). Interestingly, AgmU and AglZ colocalize in A-motile cells on agar (Nan *et al.*, 2010). The core force-generating unit of the gliding machinery is a proton channel embedded in the cytoplasmic membrane. It is a trimer formed by the proteins AglR, AglQ, and AglS, although the exact stoichiometry among these proteins remains unclear. As discussed in Chapter 9, these proteins are also homologous to the *E. coli* proteins MotA and MotB that power flagella rotation. The translocation of proteins, especially AglZ, along the length of the cell produces force that ultimately leads to a linear movement of the cell relative to the substrate (Sun *et al.*, 2011). This force is presumably generated between an intracellular, potentially cytoskeletal link, and the extracellular substrate. Thus, a periplasmic linker might play a crucial role in converting protein-based conformational changes into cellular movements. To date, very little is known biophysically about how force is transmitted to the exterior of the cell and the underlying substrate. Surface binding is fairly unspecific as gliding cells can move on agarose and agar and the motor complexes can transmit force to polystyrene beads.

The role of slime

Time-lapse videos of swarming *M. xanthus* reveal that gliding cells preferentially move along paths that other bacteria have laid down and used before (Kaiser, 2003; Gibiansky *et al.*, 2013). Those tracks contain a deposited slime that *M. xanthus* produce (Ducret *et al.*, 2012). At the individual cell level, the gliding machinery might be optimally adapted to slime in two ways: (i) the slight change in viscoelastic properties at the surface might benefit the force transduction, and (ii) enhanced binding kinetics improve the conversion of motor drive into thrust. At the cell-group level it might be an efficient way to aid cooperation between cells. Interestingly, *M. xanthus* secretes slime independent of the motility mechanism in use at that moment, but it appears to colocalize with the gliding machinery and not the twitching apparatus.

It has been difficult for researchers to probe the chemical and structural composition of the deposited slime. Because slime trails are tiny in dimension (about 5 nm in height) conventional phase contrast or differential interference microscopy fail to resolve them in any significant detail. The claim of indirectly observed slime trails for gliding *M. xanthus* on agar pads is problematic since there could also be simply indentations in the pad. Recent direct proof of surface-deposited slime required the use of Wet-SEEC imaging which features strong contrast enhancement at the surface. In combination with staining techniques, Ducret *et al.* (2012) identified LPS (lipo-polysaccharide) and EPS (exo-polysaccharide) as important slime constituents. They also showed that slime mediates adhesion and is deposited by the motility complex during motility.

Biophysical measurements of gliding motility

Mignot *et al.* gave the first quantitative estimate of forces that drive *M. xanthus* gliding (Mignot *et al.*, 2007). In this elegant experiment, they used the drug cephalexin to prevent cells from dividing.

These elongated cells glided over agarose pads whilst the number of AglZ clusters, linked to the gliding motor complexes, could be determined in the fluorescent channel. The number of gliding complexes correlated with the force production needed by cells to move at a given velocity. In order to investigate the distribution of motors along the cell body, Sliusarenko *et al.* (2007) used a similar assay of cephalexin-treated, filamentous cells, but focused on sharply bent cells. Front portions of the cells moved whilst cell rears lagged behind. At the same time, the number of bends and, therefore, the elastic energy did not change as the cells glided forward. These results strongly indicated that forces were being generated along the sides of the cell body.

Four years later, Sun *et al.* (2011) were able to identify the proteins AglR, AglQ, and AglS as principle building blocks of the force-generating proton channel of the gliding machinery. They performed a titration assay of the influence of the drug nigericin on the cell velocity. Nigericin destroys the pH gradient, but leaves the membrane potential intact. Gradually increasing the nigericin concentration slowed the cells substantially to almost complete halt at 100 μM nigericin. This was one of the necessary proofs to identify these motors as proton channels. A similar assay with drug CCCP instead was used by Nan *et al.* (2011) to inhibit the helical motion of AgmU. An even more direct assay used optical traps to position micron-sized polystyrene beads on top of immobilized cells (Sun *et al.*, 2011). AglZ transport along the cell axis colocalized with the transport of the beads. The bead transport could be inhibited by nigericin adding further evidence that bead motion and cell motion were driven by the same system.

Unresolved problems

The gliding motility system in *M. xanthus* is less understood than the twitching motility system and as such the number of open questions is quite broad. An unresolved problem is the geometrical arrangement of the gliding motor complex. How does the proton channel, whose constituents are homologues to the stators MotAB couple together to become a linear translator? Another set of problems relates to the force generation of the motor. What is the motor's response to external load? What is the functional range of the proton gradient in which the motor produces force? And beyond these questions about individual motors, how many proton channels are in a gliding motor complex? Do different complexes cooperate with each either mechanically or chemically?

How cell polarity is converted into motor direction is another interesting line of inquiry. As far as we know, cell polarity is specified by the differential localization of proteins at the cell poles. How does a motor located in the middle of the cell know which direction to go? What happens when a cell reverses direction? If the motors are moving on an intracellular filament structure, what determines the directionality? Are the filaments bipolar, or perhaps are there two sets of oppositely polarized filaments? Current data indicate that motor complexes within a single cell can move in opposite directions at the same time, but there has not been sufficient resolution to determine if these motors are travelling on the same or different paths.

Finally, how do the motors couple force to the substrate? Do the outer-membrane proteins that interact with the substrate also interact directionally or is their role purely passive binding? Are the chemical or mechanical properties of slime necessary for efficient motion, or could cells move on a different kind of sticky substrate?

Multicellular pattern formation

Myxobacteria often move together in large groups, forming a variety of beautiful dynamic patterns in the wild. Multicellular, cooperative movement can be used for a number of purposes, including aggregation, fruiting-body formation, and collective predation. More complex than conventional quorum sensing and biofilm formation, *Myxococcus* group phenomena involve spatially coordinated cellular behaviours from which a rich set of patterns can arise. In general, cells can communicate by both chemical and mechanical interactions and current understanding of pattern formation in *M. xanthus* invokes both of these modalities.

Image-based studies of fruiting body formation, predation, and other macroscopic phenomena have been performed on *Myxococcus* for decades. Recently, a number of pioneering studies have begun to quantitatively probe the physical origins of pattern formation using imaging results. As discussed in Chapter 8, sophisticated image-analysis techniques and computer modelling have been used to describe the statistical distribution and motion of large groups of cells over time (Xie *et al.*, 2011; Zhang *et al.*, 2011; Peruani *et al.*, 2012; Starruß *et al.*, 2012). Prior work from David Zusman's lab had tracked the motion of single fluorescently labelled cells in multicellular patterns of predominantly unlabeled cells during aggregation and rippling wave motion to study the motile behaviours of individuals in a group setting (Sliusarenko *et al.*, 2006). These authors report a number of interesting findings including a large variation in group sizes as expected for a system near a critical point, pneumatic ordering (parallel alignment of rod-shaped cells in a group), density-dependent formation of 3D patterns, rotational motions, and cell layering.

Aggregation and fruiting body formation

Experimental methods to study group behaviours in *M. xanthus* have become fairly mature. To induce collective behaviour, studies typically follow an established protocol such as that in Higgs and Merlie for studying fruiting body formation (Higgs and Merlie, 2008). Liquid cell cultures grown in rich media are transferred to nutrient-depleted media and plated on agar or agarose pads. On this two-dimensional substrate, motile cells start interacting with each other by colliding and aligning with each other, and potentially via chemical signalling as well.

After three to five hours, the population of cells on the surface forms a quasi steady-state that is relatively stable on the multicellular level but in which individual cell dynamics are quite active. On a velocity scale significantly slower than the motion of single cells, large clusters of hundreds of cells move and collide forming even larger clusters. These clusters become only as large as about one thousand cells, but not significant larger, whilst constantly losing and gaining individual cells. Cluster dynamics are quite fascinating, exhibiting complex features such as stable vortices (Starruß *et al.*, 2012). Cells in which both the twitching and gliding motility mechanisms are intact require a longer time to find this dynamic steady state. One reason might be that gliding direction is rather given by sensing and binding to the surface and to slime whereas pilus binding and retraction is inherently more stochastic. Another possible explanation is that the presence of pili at the rear cell pole might occasionally slow gliding cells. The effectively reduced forward velocity of single cells could account for the additional time for cluster formation.

One of the most dramatic effects is regulated by the Frz system that is discussed in Chapter 11. In the absence of reversals, e.g. in a *frzE* deletion mutant, cells form small moving groups. However, in the wildtype, where cells reverse in a regular manner once every 8 min, cells form elongated clusters that become interconnected in a mesh structure. Thus, reversal frequency is able to control the size, dynamics, and spatial distribution of multicellular patterns.

Roughly 9 h after the transfer into starvation media, wild-type *M. xanthus* cells start organizing into fruiting bodies (Zhang *et al.*, 2011). Fruiting bodies can extend up to a millimetre in diameter. Within the protective fruiting body, individual *M. xanthus* cells can undergo morphogenesis into spores or peripheral cells. Fruiting body aggregates are remarkably stable structures, although they can experience different fates such as dispersing, splitting, or merging (Xie *et al.*, 2011). Aggregate size has been shown to be an informative predictor of stability (Zhang *et al.*, 2011). In addition, aggregates are most likely to merge with nearby neighbours and the less round an aggregate is, the more likely it is to split up into two (Zhang *et al.*, 2011). While deletion mutants have been used to study the effect of motility and signalling genes on collective behaviour, the inability to track all the cells in a dense group has made concrete conclusions about how decisions made at the individual level affect pattern formation difficult to derive. Several labs are now actively working on technology to track individuals in dense cell layers while mathematical biologists are simultaneously employing computer simulations to investigate different theories of *M. xanthus* aggregation.

Numerous results demonstrating adhesion and pneumatic ordering within cell groups indicate that physical forces play a significant role in pattern formation. While imaging studies can reveal the existence of these sorts of interactions, they can not directly probe the effect of physical forces on cellular behaviour nor the relative influence of mechanical and chemical signalling on motility. The types of biophysical measurements described above for single cells yield information about the effect of forces on motility. What is currently missing are biophysical measurements of the forces present between cells in a group and within the entire group structure. As our knowledge of cellular behaviour increases, it is likely that the next frontier of biophysical questions will address multicellular phenomena. For example, traction force microscopy could be used to quantify the pattern of forces generated by moving groups during predation or aggregation. AFM or magnetic tweezers could be used to ask what stresses are present in a group, how stresses are transmitted within a group, and how applied stress affects the kinetics and morphology of a multicellular pattern. By combining these techniques with existing chemical and genetic perturbations, it is likely that a number of the mysteries of *M. xanthus* pattern formation could be solved.

Predation

In addition to aggregation, *M. xanthus* cells use collective motility to search for and hunt bacterial prey. Cells produce the antibiotic myxalamid and the production of this chemical by a large group produces a potent enough concentration to kill larges groups of bacteria (Wenzel and Müller, 2009). Furthermore, *M. xanthus* may be able to enhance the effect of the diffusive antibiotic by spatially arranging around their prey, optimizing the necessary production rates.

Beyond the synchronized production of antibiotics, when hunting, *M. xanthus* takes advantage of another strategy, the concerted movement of groups. Cells aligned in larger groups develop a synchronized reversal pattern and adjust to the same speeds, so that the movement of the cells does not interfere with the force-generating motility mechanisms. In this scenario, the reversal frequency sets the wavelength of a set of travelling 'rippling' waves (Welch and Kaiser, 2001; Berleman *et al.*, 2008; Zhang *et al.*, 2012). This wavelength can easily span multiple body lengths. While the cell density overall exhibits travelling waves, the cells within the waves do not infiltrate and instead reflect off each other so that there is no net translocation achieved for each cell (Sliusarenko *et al.*, 2006). Combined with the ability to share other secondary metabolites that assist the digestion of prey, self-organized *M. xanthus* swarms are extremely efficient predators.

References

Abbondanzieri, E., Greenleaf, W., Shaevitz, J.W., Landick, R., and Block, S.M. (2005). Direct observation of base-pair stepping by RNA polymerase. Nature *438*, 460–465.

Andersson, M., Fällman, E., Uhlin, B.E., and Axner, O. (2006). Dynamic force spectroscopy of *E. coli* P pili. Biophys. J. *91*, 2717–2725.

Axner, O., Andersson, M., Björnham, O., Castelain, M., Klinth, J., Koutris, E., and Schedin, S. (2011). Assessing bacterial adhesion on an individual adhesin and single pili level using optical tweezers. Adv. Exp. Med. Biol. *715*, 301–313.

Bai, F., Branch, R.W., Nicolau, D.V., Pilizota, T., Steel, B.C., Maini, P.K., and Berry, R.M. (2010). Conformational spread as a mechanism for cooperativity in the bacterial flagellar switch. Science *327*, 685–689.

Berg, H.C. (1988). A physicist looks at bacterial chemotaxis. Cold Spring Harb. Symp. Quant. Biol. *53 Pt 1*, 1–9.

Berleman, J.E., Scott, J., Chumley, T., and Kirby, J.R. (2008). Predataxis behavior in *Myxococcus xanthus*. Proc. Natl. Acad. Sci. U.S.A. *105*, 17127–17132.

Bulyha, I., Schmidt, C., Lenz, P., Jakovljevic, V., Höne, A., Maier, B., Hoppert, M., and Søgaard-Andersen, L. (2009). Regulation of the type IV pili molecular machine by dynamic localization of two motor proteins. Mol. Microbiol. *74*, 691–706.

Burrows, L.L. (2012). *Pseudomonas aeruginosa* Twitching Motility: Type IV Pili in Action. Annu. Rev. Microbiol. *66*, 493–520.

Carter, A.R., Seol, Y., and Perkins, T.T. (2009). Precision surface-coupled optical-trapping assay with one-basepair resolution. Biophys. J. *96*, 2926–2934.

Clausen, M., Jakovljevic, V., Søgaard-Andersen, L., and Maier, B. (2009a). High-force generation is a conserved property of type IV pilus systems. J. Bacteriol. *191*, 4633–4638.

Clausen, M., Koomey, M., and Maier, B. (2009b). Dynamics of type IV pili is controlled by switching between multiple states. Biophys. J. *96*, 1169–1177.

Czerwinski, F., Andersson, M., and Oddershede, L.B. (2011). Optimizing active and passive calibration of optical tweezers. J. Opt. *13*, 044020.

Darnton, N.C., Turner, L., Rojevsky, S., and Berg, H.C. (2007). On torque and tumbling in swimming *Escherichia coli*. J. Bacteriol. *189*, 1756–1764.

Ducret, A., Valignat, M.-P., Mouhamar, F., Mignot, T., and Theodoly, O. (2012). Wet-surface-enhanced ellipsometric contrast microscopy identifies slime as a major adhesion factor during bacterial surface motility. Proc. Natl. Acad. Sci. U.S.A. *109*, 10036–10041.

Elkins, J.G., Hassett, D.J., Stewart, P.S., Schweizer, H.P., and McDermott, T.R. (1999). Protective role of catalase in *Pseudomonas aeruginosa* biofilm resistance to hydrogen peroxide. Appl. Environ. Microbiol. *65*, 4594–4600.

Gardel, M.L., Sabass, B., Ji, L., Danuser, G., Schwarz, U.S., and Waterman, C.M. (2008). Traction stress in focal adhesions correlates biphasically with actin retrograde flow speed. J. Cell Biol. *183*, 999–1005.

Gibiansky, M.L., Hu, W., Dahmen, K.A., Shi, W., and Wong, G.C.L. (2013). Earthquake-like dynamics in *Myxococcus xanthus* social motility. *Proceedings of the National Academy of Sciences*. Proc. Natl. Acad. Sci. U.S.A. *(PNAS Early Edition)*, 1–6.

Higgs, P.I., and Merlie, J.P. (2008). *Myxococcus xanthus*: Cultivation, Motility, and Development. In Myxobacteria: Multicellularity and Differentiation, D.E. Whitworth, ed. (Washington, DC: ASM Press), pp. 465–478.

Holz, C., Opitz, D., Greune, L., Kurre, R., Koomey, M., Schmidt, M.A., and Maier, B. (2010). Multiple Pilus Motors Cooperate for Persistent Bacterial Movement in Two Dimensions. Phys. Rev. Lett. *104*, 178104.

Ingles, D.W., and Knowles, J.R. (1966). The α-chymotryptic hydrolysis of glycine esters. Biochem. J. 99, 275.

Jakovljevic, V., Leonardy, S., Hoppert, M., and Søgaard-Andersen, L. (2008). PilB and PilT Are ATPases Acting Antagonistically in Type IV Pilus Function in *Myxococcus xanthus*. J. Bacteriol. *190*, 2411–2421.

Kaimer, C., Berleman, J.E., and Zusman, D.R. (2012). Chemosensory signaling controls motility and subcellular polarity in *Myxococcus xanthus*. Curr. Opin. Microbiol. *15*, 1–7.

Kaiser, D. (2003). Coupling cell movement to multicellular development in myxobacteria. Nature Rev. Microbiol. *1*, 45–54.

Li, G.-W., and Xie, X.S. (2011). Central dogma at the single-molecule level in living cells. Nature *475*, 308–315.

Li, K., and Kanade, T. (2009). Nonnegative mixed-norm preconditioning for microscopy image segmentation. Inf. Process Med. Imaging *21*, 362–373.

Lipfert, J., Kerssemakers, J.W., Jager, T., and Dekker, N.H. (2010). Magnetic torque tweezers: measuring torsional stiffness in DNA and RecA-DNA filaments. Nature Meth. *7*, 977–980.

Maier, B., Koomey, M., and Sheetz, M.P. (2004). A force-dependent switch reverses type IV pilus retraction. Proc. Natl. Acad. Sci. U.S.A. *101*, 10961–10966.

Mauriello, E.M.F., Nan, B., and Zusman, D.R. (2009). AglZ regulates adventurous (A-) motility in *Myxococcus xanthus* through its interaction with the cytoplasmic receptor, FrzCD. Mol. Microbiol. 72, 964–977.

Mauriello, E.M.F., Mouhamar, F., Nan, B., Ducret, A., Dai, D., Zusman, D.R., and Mignot, T. (2010). Bacterial motility complexes require the actin-like protein, MreB and the Ras homologue, MglA. EMBO J. *29*, 315–326.

Meel, C., Kouzel, N., Oldewurtel, E.R., and Maier, B. (2012). Three-dimensional obstacles for bacterial surface motility. Small *8*, 530–534.

MicroscopyU. Available at: http://www.microscopyu.com.

Mignot, T., Shaevitz, J.W., Hartzell, P.L., and Zusman, D.R. (2007). Evidence that focal adhesion complexes power bacterial gliding motility. Science *315*, 853–856.

Murphy, D.B. (2001). Fundamentals of Light Microscopy and Electronic Imaging (New York City: Wiley).

Nan, B., Mauriello, E.M.F., Sun, I.-H., Wong, A., and Zusman, D.R. (2010). A multi-protein complex from *Myxococcus xanthus* required for bacterial gliding motility. Mol. Microbiol. *76*, 1539–1554.

Nan, B., Chen, J., Neu, J.C., Berry, R.M., Oster, G., and Zusman, D.R. (2011). Myxobacteria gliding motility requires cytoskeleton rotation powered by proton motive force. Proc. Natl. Acad. Sci. USA *106*, 2498–2503.

Nelson, P.C. (2004a). Appendix B Numerical Values. In Biological Physics, (New York City: W. H. Freeman and Company), pp. 158–194.

Nelson, P.C. (2004b). Life in the Slow Lane: The Low Reynolds-Number World. In Biological Physics, (New York City: W. H. Freeman and Company), pp. 569–574.

Pampaloni, F., Lattanzi, G., Jonás, A., Surrey, T., Frey, E., and Florin, E.-L. (2006). Thermal fluctuations of grafted microtubules provide evidence of a length-dependent persistence length. Proc. Natl. Acad. Sci. U.S.A. *103*, 10248–10253.

Pelling, A.E., Li, Y., Shi, W., and Gimzewski, J.K. (2005). Nanoscale visualization and characterization of *Myxococcus xanthus* cells with atomic force microscopy. Proc. Natl. Acad. Sci. U.S.A. *102*, 6484–6489.

Peruani, F., Starruss, J., Jakovljevic, V., Søgaard-Andersen, L., Deutsch, A., and Bär, M. (2012). Collective motion and nonequilibrium cluster formation in colonies of gliding bacteria. Phys. Rev. Lett. *108*, 098102.

Pilizota, T., Sowa, Y., and Berry, R.M. (2009). Single-Molecule Studies of Rotary Molecular Motors. In Handbook of Single-Molecule Biophysics, (Heidelberg: Springer), pp. 183–216.

Sabass, B., Gardel, M.L., Waterman, C.M., and Schwarz, U.S. (2008). High resolution traction force microscopy based on experimental and computational advances. Biophys. J. *94*, 207–220.

Shimkets, L.J., and Kaiser, D. (1982). Induction of coordinated movement of *Myxococcus xanthus* cells. J. Bacter. *152*, 451–461.

Skerker, J.M., and Berg, H.C. (2001). Direct observation of extension and retraction of type IV pili. Proc. Natl. Acad. Sci. U.S.A. *98*, 6901–6904.

Sliusarenko, O., Neu, J., Zusman, D.R., and Oster, G. (2006). Accordion waves in *Myxococcus xanthus*. Proc. Natl. Acad. Sci. U.S.A. *103*, 1534–1539.

Sliusarenko, O., Zusman, D.R., and Oster, G. (2007). The motors powering A-motility in *Myxococcus xanthus* are distributed along the cell body. J. Bacteriol. *189*, 7920–7921.

Sourjik, V. (2004). Receptor clustering and signal processing in *E. coli* chemotaxis. Trends Microbiol. *12*, 569–576.

Sowa, Y., Rowe, A.D., Leake, M.C., Yakushi, T., Homma, M., Ishijima, A., and Berry, R.M. (2005). Direct observation of steps in rotation of the bacterial flagellar motor. Nature *437*, 916–919.

Starruß, J., Peruani, F., Jakovljevic, V., Søgaard-Andersen, L., Deutsch, A., and Bär, M. (2012). Pattern-formation mechanisms in motility mutants of *Myxococcus xanthus*. Interface Focus *2*, 774–785.

Sun, M., Wartel, M., Cascales, E., Shaevitz, J.W., and Mignot, T. (2011). Motor-driven intracellular transport powers bacterial gliding motility. Proc. Natl. Acad. Sci. U.S.A. *108*, 7559–7564.

Svoboda, K., and Block, S.M. (1994). Biological applications of optical forces. Annu. Rev. Biophys. Biomol. Struct. *23*, 247–285.

van Teeffelen, S., Shaevitz, J.W., and Gitai, Z. (2012). Image analysis in fluorescence microscopy: bacterial dynamics as a case study. Bioessays *34*, 427–436.

Uenoyama, A., and Miyata, M. (2005). Gliding ghosts of *Mycoplasma mobile*. Proc. Natl. Acad. Sci. U.S.A. *102*, 12754–12758.

Walter, N., Selhuber, C., Kessler, H., and Spatz, J.P. (2006). Cellular unbinding forces of initial adhesion processes on nanopatterned surfaces probed with magnetic tweezers. Nano Lett. *6*, 398–402.

Wang, S., Arellano-Santoyo, H., Combs, P.A., and Shaevitz, J.W. (2010). Actin-like cytoskeleton filaments contribute to cell mechanics in bacteria. Proc. Natl. Acad. Sci. U.S.A. *107*, 9182–9185.

Wei, X., Pathak, D.T., and Wall, D. (2011). Heterologous protein transfer within structured myxobacteria biofilms. Mol. Microbiol. *81*, 315–326.

Welch, R.D., and Kaiser, D. (2001). Cell behavior in traveling wave patterns of myxobacteria. Proc. Natl. Acad. Sci. U.S.A. *98*, 14907–14912.

Wenzel, S.C., and Müller, R. (2009). Myxobacteria – 'microbial factories' for the production of bioactive secondary metabolites. Mol. Biosyst. *5*, 567–574.

Winther, T., Xu, L., Berg-Sørensen, K., Brown, S., and Oddershede, L.B. (2009). Effect of energy metabolism on protein motility in the bacterial outer membrane. Biophys. J. *97*, 1305–1312.

Xanthusbase. Available at: http://xanthus.openmods.org.

Xie, C., Zhang, H., Shimkets, L.J., and Igoshin, O.A. (2011). Statistical image analysis reveals features affecting fates of *Myxococcus xanthus* developmental aggregates. Proc. Natl. Acad. Sci. U.S.A. *108*, 5915–5920.

Youderian, P., Burke, N., White, D.J., and Hartzell, P.L. (2003). Identification of genes required for adventurous gliding motility in *Myxococcus xanthus* with the transposable element mariner. Mol. Microbiol. *49*, 555–570.

Zhang, H., Angus, S., Tran, M., Xie, C., Igoshin, O.A., and Welch, R.D. (2011). Quantifying aggregation dynamics during *Myxococcus xanthus* development. J. Bacteriol. *193*, 5164–5170.

Zhang, H., Vaksman, Z., Litwin, D.B., Shi, P., Kaplan, H.B., and Igoshin, O.A. (2012). The Mechanistic Basis of *Myxococcus xanthus* Rippling Behavior and Its Physiological Role during Predation. PLoS Comput. Biol. *8*, e1002715.

Index

Note: Page references in *italics* refer to figures and those in **bold** to tables.

A

ABM (agent-based model), computational biology 154
actA/ActA 119
actB/ActB 109, 111
  enhancer binding protein **109**
  glycerol-induced sporulation 119
actC/ActC 47
AFM (atomic force microscopy) 216
Agent-based model (ABM) 154
Aggregate fusion 164
Aggregate splitting 164
Aggregation
  biophysics 222–223
  fruiting bodies 162, *163*, 165, *165*
aglQ/AglQ 220, 221
  A-motility 174, 175
aglR/AglR 220, 221
  A-motility 174, 175
aglS/AglS 220, 221
  A-motility 174, 175
aglZ/AglZ 220
  A-motility 176
AgmU 220
Amino acid transport/metabolism **40**
Amorphous matrix, fruiting bodies 59–60
A-motility 173–181, *174*
  cell envelope 174–176
  CglF peptides 97–98
  E. coli motility vs. 202
  engine location 177, *177*
  gene identification 173–174, *175*
  helical rotary assemblies *177*
  models of 178–179, *179*
  MotAB homologues 176–177
  MreB cytoskeleton 176
  outer membrane exchange 91–92
  stratum interactions 179–180
Anaeromyxobacter dehalogenans
  fruiting bodies 45, 46, 47
  genetic content 39, 42
  genome sequencing 32, **33**
  genome size 35
  Ser/Thr protein kinases 141, 142
  signal transduction 44, **44**
Anoxia, marine biogeography 9
asgA/AsgA
  mutation 21
  spore induction 60
asgB mutants 53–54
asgC mutants 21
Atomic force microscopy (AFM) 216
ATP 212
Autocides 71

B

Bacillus subtilis spore formation 79, *80*, 85–86
Bdellovibrio 20
Behaviour
  emergent 151
  outer membrane exchange 98–100, 101–102
  social *see* Computational biology; Fruiting bodies; Rippling; Swarming
Biased random walk, *Escherichia coli* 199–200
Biogeography 8–10
  marine biogeography 8–9
  terrestrial *see* Terrestrial biogeography
Biophysics 211–224
  aggregation 222–223
  bacterial scales 211–213
  chemical energy 212
  extracellular forces 212–213
  fluorescence microscopy 214–215
  force microscopy 215–216, *215*
  fruiting bodies 222–223
  gliding 220–221
  imaging 213–215
  light microscopy 213–214
  multicellular pattern formation 221–223
  predation 223
  spatial scales 211–212
  thermal energy 212
  time scales 212
  tools 213–216
    see also specific tools
  twitching motility 218
Bright-field microscopy 213

C

Carbohydrate transport/metabolism **40**
CarD
  FruA module inputs 114
  σ^{ecf} 139
CarF 138–139
CarG 139
CarQ 138, 139
CarR 138–139
Cas (CRISPR-associated) genes 118
Caulobacter crescentus 83–84
cbgA/CbgA
  glycerol-induced sporulation 119
  spore coat production 66–67
cbgA mutants, spore coat production 67
CBM (cell-based model) 168
CDS genes 39, **40**
Cell alignment, swarming 157–159
Cell-based model (CBM) 168
Cell–cell contact
  outer membrane exchange 99–100
  protein transfer 94–95
Cell clusters 56–57
  fruiting bodies structure 59
  swarming patterns 159, *159*
Cell cycle 79–90
  control **40**
Cell densities, phase variation 54
Cell division **40**
Cell membrane, outer *see* Outer cell membrane
Cell motility **40**
Cell numbers, swarming 158–159, *158*
Cell polarity, chemosensory systems 206
Cell reversal, chemosensory systems 206
Cellular Potts model (CPM)
  aggregation models 166, *167*
  cell clusters in swarming 159
  computational biology 154
  definition 168
Cell wall, genetic content **40**
cglB/CglB **93**
  S-motility 92
cglC/CglC **93**
cglD/CglD **93**
cglE/CglE **93**
cglF/CglF **93**
  A-motility 97–98
Chaperones **41**
che3/Che3
  CrdA 133
  two-component systems 132–133
che4/Che4 208
Cheating, social diversity 4
CheB 200
Checkpoint control, DNA replication 87–88
Chemical energy, biophysics 212
Chemosensory systems 204–208
  cell polarity 206
  cell reversal 206
  Che4 208
  Dif pathway 207–208
  Frz pathway 205–206
  phosphatidylethanolamine 207–208
Chemotaxis 200–204
  Escherichia coli 199–200
  signal transduction 43
CheR 200
cheY4 208
Chromosomes
  copy numbers 88
  partitioning **40**
Cis requirements, cargo protein transfer 93–94
Classification (of Myxobacteria) 6–8
Clonal patches, terrestrial biogeography 12
Coat
  production 65–68
  spores 62, *62*
Coenzyme transport/metabolism **40**
Colonies
  dull and textured 53
  shiny and smooth 53
Computational biology 151–173
  agent-based model 154
  cellular Potts model *see* Cellular Potts model (CPM)
  continuous models 153–154
  deterministic models 154
  discrete models 153–154
  fruiting bodies 164–166
  future work 166–168
  historical perspective 152
  individual-based model 154
  off-lattice model 154
  scales 152
  stochastic models 154
  subcellular element model 154
Conserve synteny, genome 38–39
Contact-based signalling, rippling 160–161, *161*
Continuous models 153–154
  definition 168
Continuous processes 153
Contrast-enhancing microscopy 213
Cooperative behaviour, outer membrane exchange 101–102
Corallococcus, Ser/Thr protein kinases 142
Corallococcus apiculatus 48
Corallococcus coralloides
  fruiting bodies 47
  gene regulatory network evolution 120–121
  genetic conservation 42
  genome sequencing **33**, 34
Corallococcus macrosporus 48
CorE 139
CPM *see* Cellular Potts model (CPM)
CrdA 132, 133
  che3 system 133
  two-component systems 132
CrdS 132, 133
  two-component systems 132
CRISPR-associated (Cas) genes 118
csgA/CsgA
  cell clusters 57
  Mrp module outputs 113

peripheral rods 70
programmed cell death 72
spore induction 60
C-signal-dependent genes 115–117, *116*, **117**
identification of 115
C-signalling 114–115
Cuticula, spores 63
Cystobacter fulvus 48
Cystobacterineae 32
Ser/Thr protein kinases 142
Cytoplasmic membrane, spores 62

D

Daptobacter 20
Defence mechanisms **41**
Definition of myxobacteria 1
Density-dependent growth rates 54
Density-dependent social diversity 3–4
DEs (differential equations), continuous models 153
Deterministic models
computational biology 154
definition 168
Development
essential genes, *Myxococcus xanthus* 18–19
phase variation *54*, 55
Developmental genes
enhancer binding protein 107–109
regulation 105–124
see also Gene regulatory network (GRN)
Developmental lysis 70–71
dev operon regulation 117–119, *118*
CRISPR-associated genes 118
fmgA 118
FruA 117
MrpC 117
MrpC2 117
spore coat formation 118–119
devR 118
devS 118
devT 118
FruA module inputs 114
DIC *see* Differential interference contrast (DIC) microscopy
dif 192
chemosensory systems 207–208
DifA 192
DifC 192
DifD 192
DifE 192
Differential equations (DEs) 153
Differential interference contrast (DIC) microscopy 213, 214
fruiting body development 162
DifG 192
DifX 192
digR 133, 134
exopolysaccharide accumulation 192–193
Dimethyl sulphoxide (DMSO) 203
Discrete models
computational biology 153–154
definition 169
Distribution (of Myxobacteria) 6–7
DKX (DK xanthine) 53, 55
DK xanthine (DKX) 53, 55
DMSO (dimethyl sulphoxide) 203
DnaA 87
DNA repair **40**
DNA replication
checkpoint control 87–88
fruiting bodies 87
spore formation 86, 87
dofA 115
Dot-plots, genome size *37*, 38
Dull and textured colonies 53

E

Early sporulation signalling system *see* Esp (early sporulation) signalling system
EBP *see* Enhancer binding protein (EBP)
ECF *see* Extra-cytoplasmic function (ECF) σ-factors
Emergent behaviour 151
Energy production **40**
Enhancer binding protein (EBP)
developmental gene expression 107–109
gene regulatory network 105
regulatory cascade 109–110, *110*, 111
see also specific elements
structure 107, *108*
Envelope biogenesis **40**
Envelopes, spores 61
Environment, social 2
EPS (extracellular polymeric substances) 69
Escherichia coli
biased random walk 199–200
chemotaxis 199–200
chromosome 83–84
genome sequencing 31
EspA 131–132
EspC 131–132
Esp (early sporulation) signalling system *129*, 131–132
targets 132
Essential development genes, *Myxococcus xanthus* 18–19
Evolution, experimental 19, *20*
Evolutionary analysis 5, 5
Evolution of development (Myxo Evo-Devo) 17–19
exo 65
ExoA 65, 66
ExoCD 65, 66
ExoE 66
Exopolysaccharides
accumulation regulation 192–193, *192*
fruiting bodies structure 59
motility *see* S-motility
Expansion rate levels, swarming 156–157, *157*
Experimental evolution 19, *20*
Extracellular polymeric substances (EPS) 69
Extra-cytoplasmic function (ECF) σ-factors 127–128, 136–140, *137*
M. xanthus 138–140
sequence comparisons 136–137
signalling 137–138

F

FdgA
  fruiting bodies 47
  spore coat production 65, 66
FepA homologues 56
FibA metalloproteinases 56–57
  cell clusters 57
  peripheral rods 69
Fine-scale spatial structure, terrestrial biogeography 12–14
FliM 206
Fluorescence microscopy 214–215
  fruiting body development 162
Fluorescence Recovery After Photobleaching (FRAP) 215
  A-motility *177*
fmgA/FmgA
  C-signal-dependent gene regulation 117
  dev operon regulation 118
fmgBC/FmgBC 117
fmgC/FmgC 47
fmgD/FmgD 117
fmgE/FmgE, C-signal-dependent gene regulation 117
Force microscopy 215–216, *215*
FRAP *see* Fluorescence Recovery After Photobleaching (FRAP)
fruA/FruA module 19
  cell clusters 57
  C-signal-dependent gene regulation 115–116
  dev operon regulation 117
  fruiting bodies 47
  gene regulatory network 105
  inputs 114
  Mrp module outputs 113
  outputs 114–115
  peripheral rods 69–70
  spore induction 60
Fruiting bodies 3, 57–68, 162–166
  adaptive/non-adaptive variations 13–14
  aggregation 162, *163*, 165, *165*
  biophysics 222–223
  computational models 164–166
  developmental gene expression 108
  DNA replication 87
  genetic programme 45–48
  image analysis 162–164
  phenotypic heterogeneity 52, *52*
  social diversity 3–4
  structure 58–60, *58*
  traA⁻ mutation 98–99
frzA/FrzA 205
frzB/FrzB 205
frzCD/FrzCD
  cell clusters 57
  chemosensory systems 205
  FruA module outputs 114
  social behaviour 155
frzE/FrzE 205, 206
frzf/FrzF 206
frzG/FrzG 205
frz mutation 21
Frz pathway 222
  chemosensory systems 205–206
  computational models *156*
  downstream signalling 206–207
  social behaviour 155
frzS/FrzS 57
frzZ/FrzZ 205
ftsZ/FtsZ 84

G

Gene density 39, **40**
Gene regulatory network (GRN) 105–124, *106*
  enhancer binding protein 105
  evolution of 120–121
  FruA 105
  future work 121–122
  late genes 119–120
  MrpC 105
  see also specific networks
Genetic content, conservation of 42
Genetic homogeneity, terrestrial biogeography 13
Genetic variation, terrestrial biogeography 13
Genome 31–50
  conserve synteny 38–39
  future work 48
  genetic content 39–42, **40–41**
  organization 38–39
  sequencing history 31–34
  signal transduction *see* Signal transduction
  size 35–38, **36**, *37*
Gliding
  biophysics 220–221
  MreB 220
  proton motive force 219
  slime 220
gltA/GltA 174
gltB/GltB 174
gltC/GltC 174
gltD/GltD 174, 175
gltE/GltE 174, 175
gltF/GltF 174, 175
gltG/GltG 174, 175
gltH/GltH 174, 175
gltI/GltI 174
gltJ/Gltj 174
gltK/GltK 174
Glycerol-induced sporulation 60
  late regulatory gene identification 119–120
Grid 169
GRN *see* Gene regulatory network (GRN)
Group living, social diversity 3
Growth rates, density-dependent 54
GspF 190
GspL 190

H

Haemophilus influenza 31
Haliangium ochraceum
  fruiting bodies 46, 47
  gene regulatory network evolution 120–121
  genetic conservation 42

genetic content 39
genome sequencing **33**, 34
signal transduction 44, **44**
Halophiles 8–9
Helical rotary assemblies, A-motility *177*
Helix-turn-helix xenobiotic response element (HTH-Xre) 56
HGT (horizontal gene transfer) 36
Histidine kinase (HK) module 128
Histidine protein kinase (HPK)
enhancer binding protein 107–109
two-component systems 133–134
Horizontal gene transfer (HGT) 36
HPK *see* Histidine protein kinase (HPK)
HTH-Xre (helix-turn-helix xenobiotic response element) 56

I

IAA (isoamyl alcohol) 203
IBM (individual-based model) 154
Image, fruiting bodies 162–164
Imaging, biophysics 213–215
Individual-based model (IBM) 154
Ion transport/metabolism **41**
Isoamyl alcohol (IAA) 203

K

Kin discrimination 15

L

Large-scale spatial structure, terrestrial biogeography 10, *11*, 12
Lateral signalling, rippling 160–162
Lattice gas cellular automata (LGCA)
aggregation models 165
definition 169
LGCA *see* Lattice gas cellular automata (LGCA)
Light microscopy 213–214
Lipid transport/metabolism
genetic content **40**
outer membrane exchange 96
LonD 47
Lysis, developmental 70–71

M

Magnetic tweezers 216
Marine biogeography 8–9
Marine myxobacteria cluster (MMC) 7–8
habitats 9
Mass-spring model, swarming 157–158
Mathematical models 153
MazF
Mrp module outputs 113
programmed cell death 71, 72
MBHA (myxobacterial haemagglutinin) 69
MCP *see* Methyl accepting chemosensory protein (MCP)
Mean field model 169
Mechanistic discovery 19, *20*
Metabolic changes, spore differentiation 63–64
Methyl-accepting chemosensory protein (MCP) 56–57, 200
cell clusters 57
Mettangium boletus 48
mglA/MglA
chemosensory systems 206–207
fruiting bodies 47
mglB/MglB 206–207
Microarray analysis, spore differentiation 63
Microcinematography, time-lapse 162
Microscopy
bright-field 213
contrast-enhancing 213
differential interference *see* Differential interference contrast (DIC) microscopy
fluorescent *see* Fluorescence microscopy
force 215–216, *215*
light 213–214
phase contrast 213–214
surface-enhanced ellipsometric contrast 180, 214
traction force 216
MLST *see* Multi-locus sequence type (MLST) analysis
MMC *see* Marine myxobacteria cluster (MMC)
Molecular interactions, social diversity 1
Monte Carlo model 169
off-lattice 166, *167*
Morphology, evolution of development 19
MotAB homologues 176–177
Motility 1–2, 216–217
attractant responses 203
biophysics *see* Biophysics
cell polarity changes 202–203, *203*
cell reversal *202*
E. coli motility vs. *201*, 202
protein transfer 94–95
repellent responses 203
sensory regulation 199–209ki
see also A-motility; S-motility
mpc4 208
MreB
A-motility 176
gliding 220
mrpA/MrpA
glycerol-induced sporulation 119
Mrp module inputs 112
mrpB/MrpB
enhancer binding protein **109**
Mrp module inputs 112
MrpC2
C-signal-dependent gene regulation 116
dev operon regulation 117
peripheral rods 69–70
mrpC/MrpC
cell clusters 57
C-signal-dependent gene regulation 116
dev operon regulation 117
fruiting bodies 47
gene regulatory network 105
Mrp module inputs 112
Mrp module outputs 113
peripheral rods 69, 70
programmed cell death 71, 72
Mrp module 111–115

inputs 111–113
outputs 113
mspA/MspA
glycerol-induced sporulation 119
spore coat production 67
mspB/MspB
spore coat production 67
spore resistance 68
mspC/MspC
glycerol-induced sporulation 119
spore coat production 67
spore resistance 68
Multicellular pattern formation, biophysics 221–223
Multi-locus sequence type (MLST) analysis
of social fragmentation 14, *14*
terrestrial biogeography 10, 12–13
Mutational analysis, developmental gene expression 108
MXAN0881 111
MXAN0989 111
MXAN3885 67–68
MXAN4196 **109**
MXAN4299 (*dkxG*) gene 56
MXAN4899 109
enhancer binding protein **109**
Myxobacteria (definition of) 1
Myxobacterial haemagglutinin (MBHA) 69
Myxochelin 56
Myxococcus fulvus
fruiting bodies 47
gene regulatory network evolution 120–121
genetic conservation 42
genome sequencing **33**, 34
signal transduction **44**
Myxococcus stipatus 141
Myxococcus xanthus
cell division 84–86, *85*
cell heterogeneity 51–74
chemosensory systems *see* Chemosensory systems
chemotaxis *see* Chemotaxis
chromosome organisation 80–82, *81*
chromosome replication 82–84
essential development genes 18–19
fruiting bodies 45, 46
see also Fruiting bodies
gene regulatory network evolution 120–121
genetic conservation 42, **42**
genome sequencing 32, **33**
intraspecific diversity 9–10, *11*
phenotype 10
Ser/Thr protein kinases 142–143, *143*
signal transduction **44**, *46*
spore formation 79–80, *80*
spores 60–61
two-component systems 135–136
Myxo Evo-Devo (evolution of development) 17–19
Myxovirescin *see* TA (myxovirescin)

N

Nannocystineae 32
Neisseria gonorrhoeae 219
nfs
A-motility 174
glycerol-induced sporulation 119
spore coat production 65, 66
nla4/Nla4 109, 118–199
enhancer binding protein **109**
nla6/Nla6 109, 111
enhancer binding protein **109**
nla18/Nla18 109
enhancer binding protein **109**
nla24/Nla24 **109**
nla28/Nla28 109, 111
enhancer binding protein **109**
NMD *see* Non-myxobacteria Deltaproteobacteria (NMD)
Non-fruiting strains 3
Non-myxobacteria Deltaproteobacteria (NMD)
genome size 35–36
signal transduction **44**
Nutrients
as attractants 203
swarming 159–160

O

OCSs *see* One-component systems (OCSs)
ODEs (ordinary differential equations) 153
Off-lattice model 154
Off-lattice Monte Carlo model 166, *167*
One-component systems (OCSs) 127–128
signal transduction 43
ops (Protein S-1) 66
Ordinary differential equations (ODEs) 153
ori (origin of replication) 80, 81, 83
Outer cell membrane
exchange *see below*
genetic content **40**
spores 61
Outer cell membrane exchange 91–103
A-motility 91–92
behaviour regulation 98–100
cell–cell communication 99–100
cooperative behaviour 101–102
ecological perspectives 101–102
historical aspects 91–93
lipid transfer 96
mechanisms 100–101, *100*
physiological perspectives 101–102
S-motility 91–92
see also Protein transfer

P

parAB/ParB 82
Paralogous genes, genome size 37
parB/ParB 82, 83
parS/ParS 82, 83
Partial differential equations (PDEs) 153
PDEs (partial differential equations) 153
Peptidoglycans
rod-to-sphere morphogenesis 65
spore coat production 65
Peripheral rods 68–70
extracellular polymeric substances 69

fruiting bodies structure 59
myxobacterial haemagglutinin 69
Protein S 69
structure 69
Phase contrast microscopy 213–214
Phase variation 52–56
developmental role *54*, 55
interactive cell types 53–54
molecular differences 55–56
Phase velocity 169
Phenotypic heterogeneity 51–52
Phosphatidylethanolamine 207–208
Phospholipases 71
Phospholipid biosynthesis 63–64
pH stress 4
Phylogenetic analysis (of Myxobacteria) 6–8, 7
marine-inclusive *6*, 7–8
see also Marine myxobacteria cluster (MMC)
16S rRNA 7
Phylogenetic diversity (of Myxobacteria) *34*
pilA/PilA
peripheral rods 70
terrestrial biogeography 12
twitching motility 217
pilB/PilB
S-motility 193
type-IV pili 190, 191
pilC/PilC 189–190
Pili
bacteria vs. 219
depolymerization 218
polymerization 218
type-IV *see* Type-IV pili
Pilins 188
pilM/PilM 190
PilM-PilN complex, S-motility 193
pilN/PilN 189–190
pilO/PilO, type-IV pil 189–190
pilQ/PilQ 93
type-IV pili 189
pilT/PilT
S-motility 193
type-IV pili 190, 191
Pkn1 143
Pkn2 143
Pkn4 143
Pkn5 143
Pkn6 143
Pkn8 143
Pkn9 143
Pkn14 143
PktAS 143
PktB8 127
Plasmid partitioning proteins 82
Plesiocystis pacifica 32, 34
PomZ 84–85
Post-translational modification **41**
Post-translational regulation, FruA module 114
Pph1 144
Pph3 144–145
PPs (protein phosphatases) 127, 144–145
Predataxis 204, *204*
Predation 19–21
biophysics 223
prey species 22
secondary metabolites 23–24
swarming rate 22, 23
TA 23–24
Prey species 22
Programmed cell death 70–72
autocides 71
developmental lysis 70–71
phospholipases 71
Protein C, spores 63, 67
Protein degradation, spore differentiation 63
Protein phosphatases (PPs) 127, 144–145
Protein S
peripheral rods 69
spores 63, 67
Protein S-1 (*ops*) 66
Protein transfer
cell–cell contact 94–95
Cis requirements 93–94
machinery identification 96–97
motility 94–95
Protein turnover **41**
Protein U 67–68
Protein W 68
Proton motive force, gliding 219
prw/Prw 68
Pseudomonas aeruginosa 219

R

Reciprocal best BlastP hit analysis 36, **36**, 38
Recombination **40**
RedC 130
RedD 130
RedE 130
RedF 130
Red (regulation of early development) signal transduction system 130
RelA 47
Replication **40**
Replisomes 83–84
Response regulator (RR) 128
Ribosome structure **40**
Rippling 160–162
contact-based signalling 160–161, *161*
lateral signalling 160–162
RodK/RokA 134–135
Rods, peripheral *see* Peripheral rods
Rod-to-sphere morphogenesis 64–65
RomR 206–207
RpoE1 139–140
rpoN/RpoN 106
RR (response regulator) 128
rRNA operons 39, **40**

S

SapA 68
SapB 68
sasR **109**

SCEs *see* Subcellular element model (SCEs)
sdREC modules 135
Secondary metabolites
  biosynthesis **41**
  predation 23–24
SEEC (surface-enhanced ellipsometric contrast) microscopy 180, 214
Selection, socially-mediated *see* Social mediated selection
Selfish policing evolution 15–17, *16*
  evolution of development 17–19
  social mediated re-evolution of developmental proficiency 17
  social mediated selection 15–17, *16*
Serine-threonine (Ser/Thr) protein kinases (STPKs) 43, 127–128, 140–145, *141*
  distribution of 140
  M. xanthus 142–143, *143*
sgmT/SgmT 133, 134
  exopolysaccharide accumulation 192–193
  S-motility 193
Shiny and smooth colonies 53
sigB 119, 120
sigC 119, 120
σ^{54} regulatory system 105, 106–107
σ^{ecf} *see* Extra-cytoplasmic function (ECF) σ-factors
Signal transduction 127–148
  chemotaxis systems 43
  future work 145
  gene abundance 45
  genetic content **41**
  genome 42–44
  one-component systems 43
  protein profiles **44**
  rippling 160–162
  serine-threonine protein kinases 43
  two-component systems 43
  see also specific mechanisms
16S rRNA, phylogenetic analysis 7
Slime, gliding 220
S-motility 2–3, 173, *174*, 183–195, *184*
  cglB 92
  current models 185, *185*
  E. coli motility vs. 202
  exopolysaccharides 187
  future work 193–194
  historical aspects 183–185
  outer membrane exchange 91–92
  tgl 92
  type-IV pili 185–187
  see also Type-IV pili
Social behaviour
  computational models *see* Computational biology
  fruiting bodies *see* Fruiting bodies
  rippling *see* Rippling
  swarming *see* Swarming
Social diversity 1–25
  cheating 4
  density dependence 3–4
  density-dependent 3–4
  molecular interactions 1
  reasons for 2–5
    see also specific reasons
  social mediated selection *see* Social mediated selection
Social environment 2
Social fragmentation 14–15
Social incompatibility 15
Social isolation 14–15
Social mediated selection 15–24
  selfish policing evolution 15–17
Sorangiineae 32
Sorangium cellulosum
  fruiting bodies 46, 47
  gene regulatory network evolution 120–121
  genetic conservation 42
  genetic content 39
  genome sequencing 32, **33**
  predation 20
  Ser/Thr protein kinases 142
  signal transduction **44**
Spatial scales, biophysics 211–212
Spore(s)
  appearance 61
  coat formation 118–119
  coat layer 62, *62*
  cuticula 63
  cytoplasmic membrane 62
  differentiation *see below*
  dispersal by fruiting bodies 3
  envelopes 61
  features 61–63
  formation *see below*
  heat-resistance 55
  induction 60–61
  M. xanthus 60–61
  outer membrane 61
  Protein C 63
  Protein S 63
  resistance 68
  storage 68
Spore differentiation 63–68
  coat production 65–68
  metabolic changes 63–64
  rod-to-sphere morphogenesis 64–65
Spore formation 79
  DNA replication 87
  DNA synthesis 86
State 169
Stigmatella aurantiaca
  fruiting bodies 46
  gene regulatory network evolution 120–121
  genetic conservation 42
  genome sequencing **33**, 34
  predation 20
  Ser/Thr protein kinases 142
  signal transduction 44, **44**
Stochastic models
  computational biology 154
  definition 169
STPKs *see* Serine-threonine (Ser/Thr) protein kinases (STPKs)
Streptomyces coelicolor 79, *80*

Subcellular element model (SCEs)
  computational biology 154
  definition 169
Surface-enhanced ellipsometric contrast (SEEC) microscopy 180, 214
Surface layers, fruiting bodies 59
Swarming 155–160
  cell alignment 157–159
  cell cluster patterns 159, *159*
  cell numbers 158–159, *158*
  expansion rate levels 156–157, *157*
  nutrient effects 159–160
  rates *see below*
Swarming rates
  phase variation 54
  predation 22, *23*

T

TA (myxovirescin)
  phase variation 56
  predation 23–24
Tan colonies
  phase variation 53
  yellow colonies vs. 55–56
TA (toxin-antitoxin) system, programmed cell death 71
TCA cycle 63
TCSs *see* Two-component systems (TCSs)
ter (terminus) region 80, 81, 83
Terrestrial biogeography 10–15
  fine-scale spatial structure 12–14
  large-scale spatial structure 10, *11*, 12
TetR-binding sites 82
TFM (traction force microscopy) 216
tgl **93**, *94*
  protein transfer 94–95
  S-motility 92
Thermal energy 212
Time-lapse microcinematography 162
Time scales, biophysics 212
TIRFM (total internal fluorescence microscopy) 214–215
Ton B homologues 56
Total internal fluorescence microscopy (TIRFM) 214–215
Toxin-antitoxin (TA) system, programmed cell death 71
TraA 96
TraAB 98
TraB 96
Traction force microscopy (TFM) 216
Transcription **40**
Translation **40**
tRNA genes **40**
Twitching motility 217–219
  biophysics 218
Two-component systems (TCSs) 127–128, 128–136, *129*
  che3 system 132–133
  CrdA 132
  CrdS 132
  E. coli chemotaxis 200
  histidine kinase module 128
  histidine protein kinase 133–134
  M. xanthus 130–135, 135–136
  response regulator 128
  signal transduction 43
Type-IV pili
  motor dynamics 190–191
  outer membrane transversal 189
  pilins 188
  structure 187–193, *189*
  see also S-motility

U

Ultraviolet light, spore resistance 68

W

WGD (whole genome duplication) 36, 37
Whole genome duplication (WGD) 36, 37
Wolf-pack feeders 52
Wzx 66
Wzy-dependent polysaccharide export systems 66

Y

Yellow colonies
  phase variation 53
  tan colonies vs. 55–56

Z

Z-ring 84

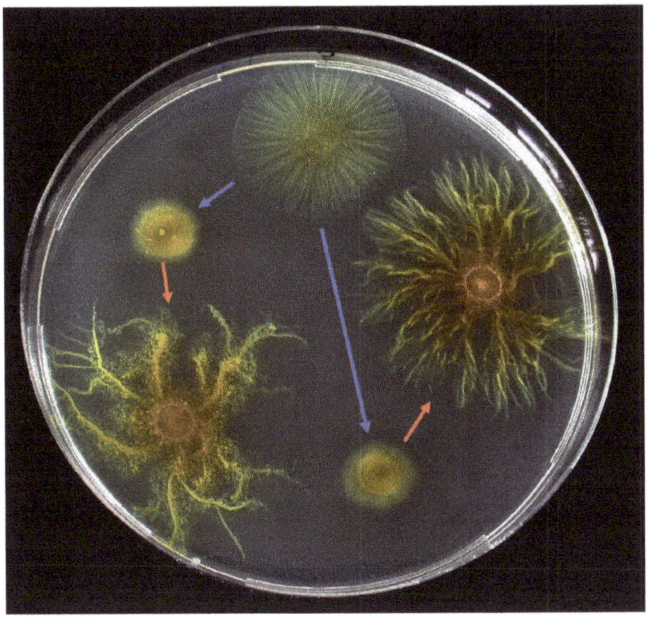

Figure 1.1 Major phenotypic change during experimental evolution. Colony phenotypes on soft agar: WT parent (top position), two ΔpilA mutants (second and fourth positions counterclockwise from top) derived from WT (blue arrows) and their lab-evolved descendants E7 and E8 (third and fifth positions counterclockwise from top, respectively, red arrows). Adapted and reprinted with permission from Yu and Velicer (2003).

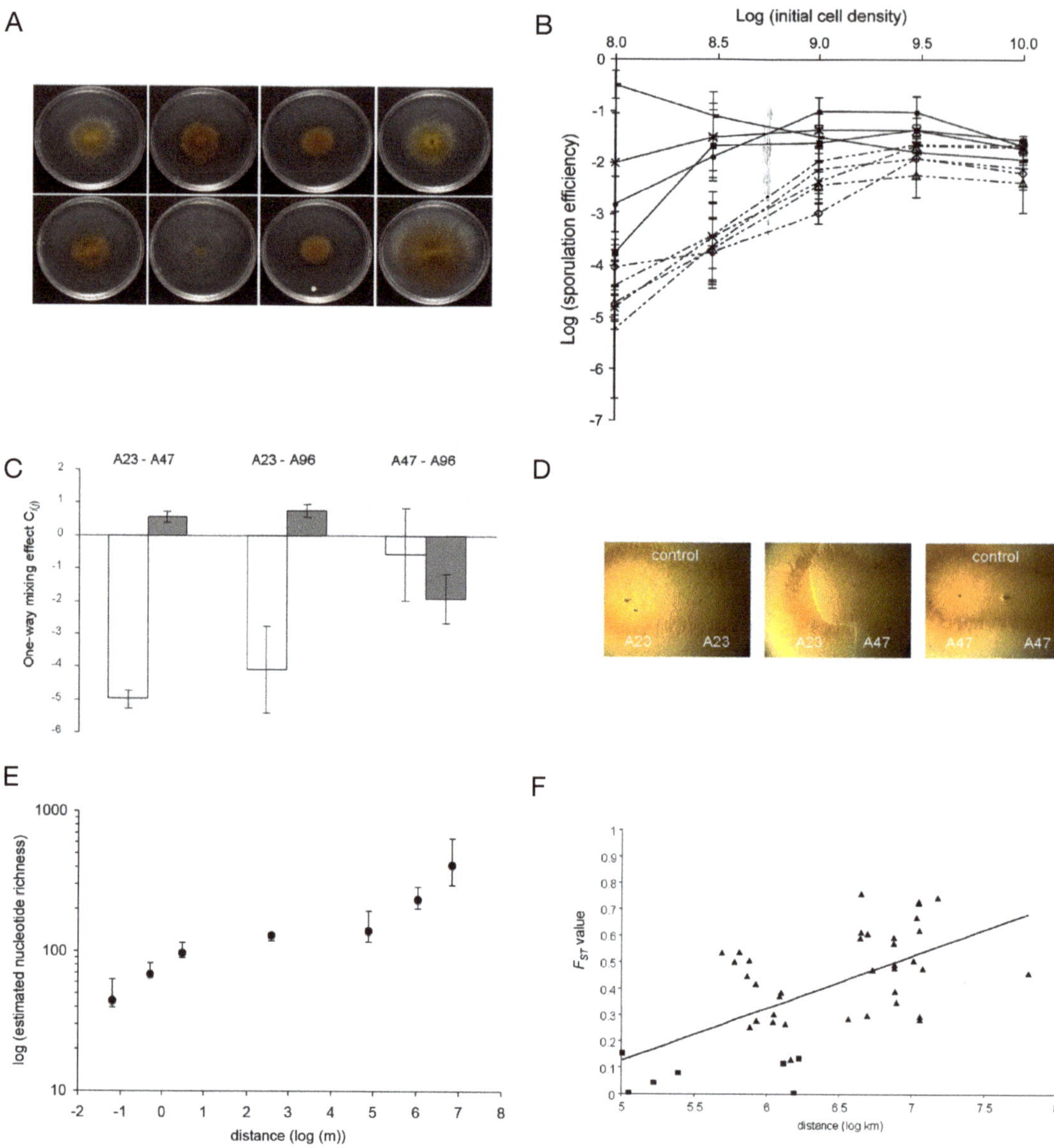

Figure 1.3 Heritable itraspecific variation at several social traits and genetic diversity in *M. xanthus*. (a) Variable swarming phenotypes of distinct clones isolated from the same fruiting body on soft agar. Reprinted with permission from Kraemer and Velicer 2011. (b) Variable relationships between density and spore production among isolates from globally distribute origins. Reprinted with permission from Kadam and Velicer 2006. (c) Variable competitiveness during development in pair-wise mixes of among three genetically similar cm-scale isolates. The y-axis shows the effect of mixing on the sporulation efficiency of individual competitors. The one-way mixing effect, Ci(j), shows the sporulation efficiency of a strain when mixed with a competitor relative to its performance in pure culture (on a log10 scale). (d) Variable social compatibility during group swarming among cm-scale isolates. The controls show swarms of the same genotype that have merged with no line of demarcation. Panels c and d reprinted with permission from Vos and Velicer 2009. (e) Nucleotide richness estimates representing the estimated number of polymorphisms in a multi-locus sequence typing (MLST) concatemer in for *M. xanthus* isolate sets across a wide range of spatial scales. (f) FST values among 10 globally distributed metre-scale populations plotted against distance between populations. FST values represent the degree of genetic differentiation between paired populations. Significantly differentiated population pairs are indicated by triangles; non-significantly differentiated population pairs are indicated by squares. (e) and (f) adapted and reprinted with permission from Vos and Velicer (2008).

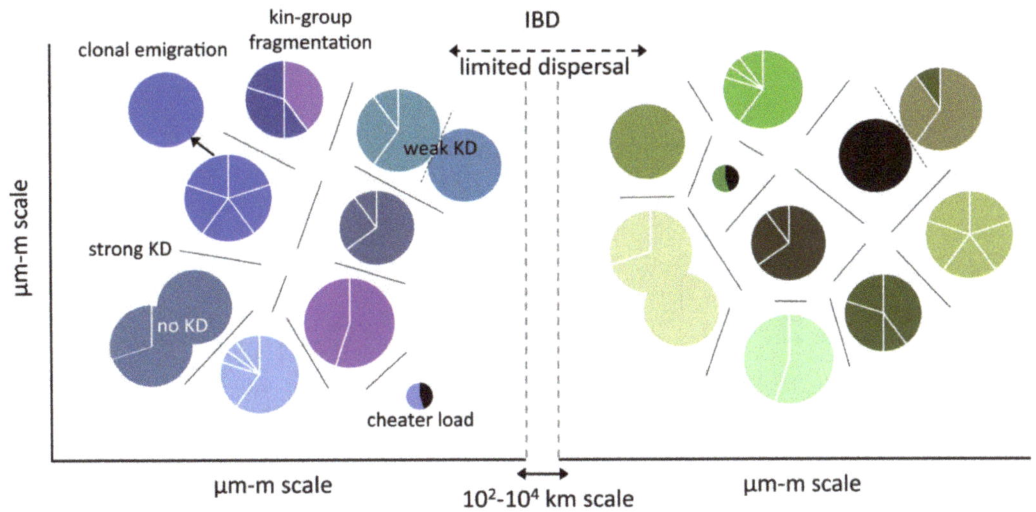

Figure 1.4 Simple hypothetical model of natural *Myxococcus* population biology. Circles represent social groups within which individuals directly interact. Sectors represent genetically distinct within-group variants. Distinctly coloured circles represent among-group genetic differentiation, with lines separating kin discrimination (KD) units. Overlapping circles represent lack of KD between highly similar, but nonetheless genetically distinct, social groups. Multicolored circles represent kin-group fragmentation (see text). Small circles with a black sector represent cheater-infected groups burdened by cheating load (Travisano and Velicer, 2004). Distinct colour ranges across left vs. right panels represent population differentiation across large spatial scales due to isolation by distance (IBD). The arrow in the left panel represents establishment of a new clonal group by clonal emigration. Reprinted with permission from Kraemer and Velicer 2011.

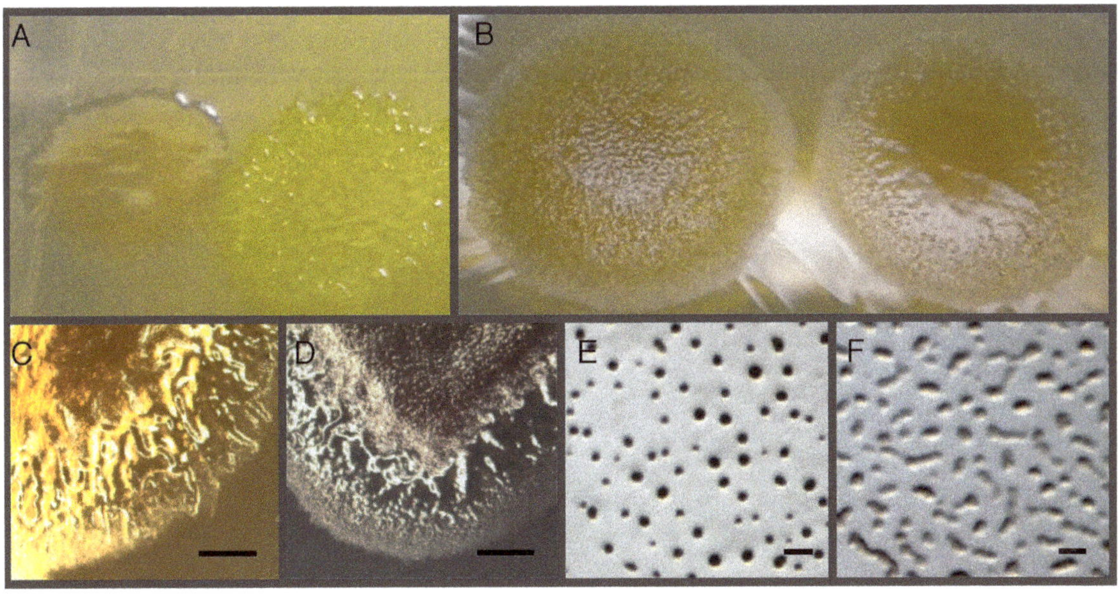

Figure 3.2 *Myxococcus* phase variation produces yellow and tan colonies. (A) Cells plated at high density show yellow (right) and tan (left) phenotypes. (B) Swarm spots (diameter 25mm) with dull matte (left) and shiny smooth (right) surface texture (Nikon coolpix image); Colony edge of a yellow (C) and tan (D) variant on CTPM agar medium (Nikon SMZ-U image toplit with a fiberoptic light; bar=2mm; Fruiting bodies after 24h at 32°C on TPM (starvation) agar for yellow variant (E) and tan variant (F); bar=0.5mm.

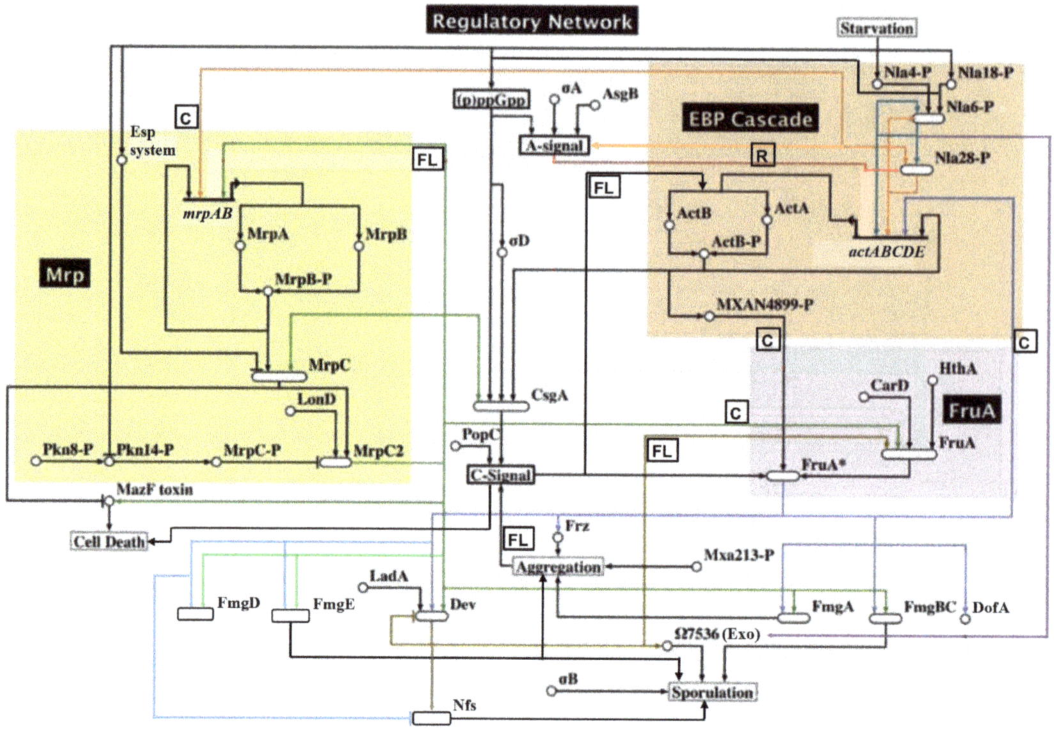

Figure 6.1 Gene regulatory network during *M. xanthus* development. Arrows indicate positive regulation and lines with barred ends indicate negative regulation, which in most cases may be at the transcriptional or post-transcriptional level. In two cases, operons are depicted (*mrpAB* and *actABCDE*) and the regulation is at the transcriptional level, although it may be direct or indirect. Some of the highly branched lines are coloured to make them easier to follow. Circles and ellipses indicate proteins or genes (no distinction is intended). Background colours distinguish the EBP cascade, Mrp, and FruA regulatory modules. Connections (C) between modules, feedback loops (FL), and reciprocal regulation (R) are indicated. The EBP cascade has three positive autoregulatory loops involving Nla6, Nla28, and ActB. The Mrp module has two positive autoregulatory loops involving MrpB and MrpC. Products of the *dev* operon (Dev) negatively autoregulate. An earlier version of this figure was published in Cold Spring Harb. Perspect. Biol. 2, a000380 (Kaiser *et al.*, 2010).

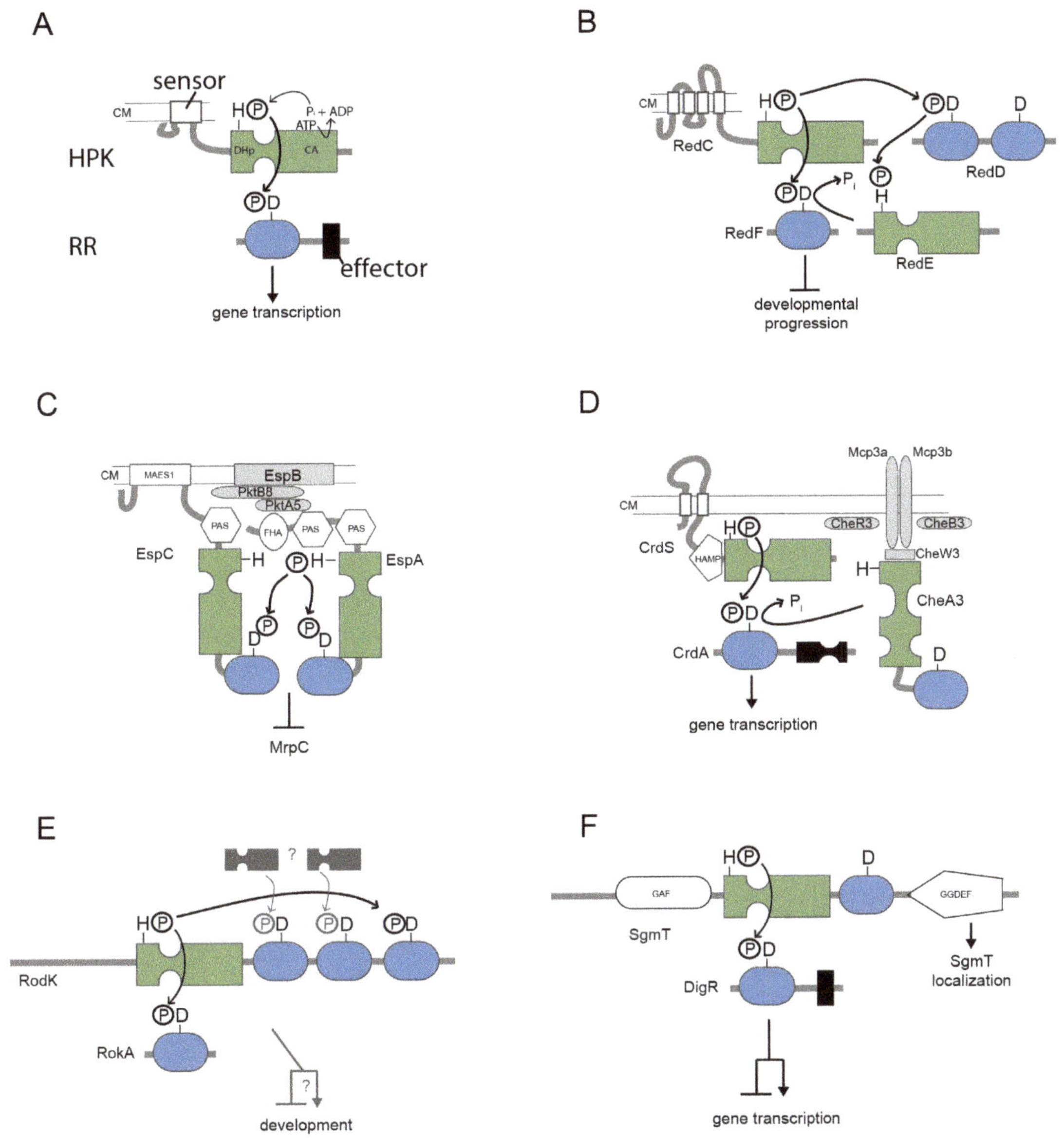

Figure 7.1 Schematic of His-Asp phosphorelay signalling mechanisms. (A) A paradigm linear two-component signalling system. Extracellular signals sensed by a HPK are transmitted to a DNA-binding RR in order to modulate gene transcription. Signal transmission occurs via His-Asp phosphorelay between conserved HK (green rectangle) and REC (blue oval) modules. The HK module consists of a DHp and a CA domain and, in response to activation by the sensor domain, autophosphorylates on an invariant His (H) residue at the expense of ATP. The phosphoryl group (P) is transferred to an invariant Asp (D) in the REC module which activates the associated DNA binding effector domain. (B) The Red signalling pathway. An HPK (RedC) and HPK-like dedicated phosphatase (RedE) and two RR lacking output domains (RedD and RedF) form a 'four-component' signalling system which controls progression through the developmental programme. (C) The Esp signalling pathway. Two HyHK (EspA and EspC), two STPKs (PktA5 and PktB8) and a transmembrane protein (EspB) form signalling complex that regulates turnover of the important transcriptional regulator of development, MrpC. (D) The Crd/Che3 multicomponent signalling pathway. The HPK CrdS and the Che3 chemosensory pathway converge on the terminal RR CrdA to regulate the developmental programme. (E) The Rod signalling pathway. The HyHK RodK, together with unidentified HPK(s) (grey rectangles) regulates the developmental programme via the single domain RR RokA and unknown downstream components. (F) The SgmT/DigR signalling pathway. The HyHK SgmT and RR DigR control gene expression of metabolic and cell surface components. SgmT localization is controlled by GGDEF dependent ci-d-GMP reception. See text for details of all signalling systems.

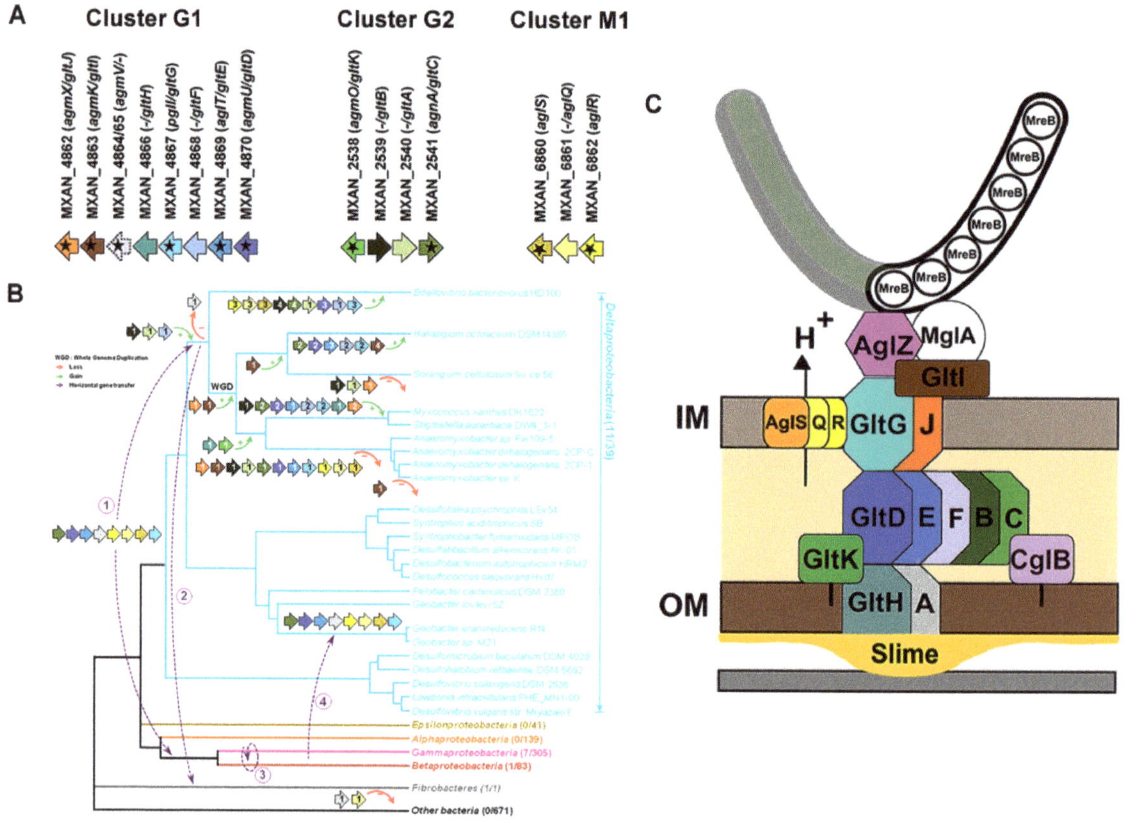

Figure 9.2 A-motility machinery is a PMF-energized envelope-spanning protein complex. (A) Genetic organization of the 15 genes composing the G1, G2 and M1 clusters encoding the core components of the A-motility machinery in *Myxococcus xanthus.* Stars indicate the genes that were hit by transposons in the Youderian *et al.* (2003) and Yu and Kaiser (2007) transposon screens. (B) Evolutionary scenario describing the emergence and evolution of the A-motility machinery in *M. xanthus*. Each gene colour is based on the colour code of the Figure 9.2A. Green and red arrows respectively indicate gene acquisition and gene loss. The number of gene copies that were acquired or lost is indicated within arrows. The purple dotted arrows represent horizontal gene transfer events of one or several components. WGD marks the putative whole genome duplication event that occurred in the ancestor of Myxococcales. (C) Architecture of the *Myxococcus* A-motility machinery. The proposed diagram is compiled from bioinformatic predictions, mutagenesis, pull down and localization studies. Note that direct physical interactions have only been shown between MglA and AglZ, AglR and GltG. The peptidoglycan layer is not shown because its layer with machinery components is not presently clear. Outer membrane proteins belonging to the system are shown integral or exposed in the periplasm based on sequence prediction, including a predicted beta-barrel fold for GltH and lipoprotein anchors for CglB and GltK. MreB may constitute the helical track but it cannot be excluded that the Glt proteins themselves form the track. It is unknown which surface protein(s) may contact the slime.

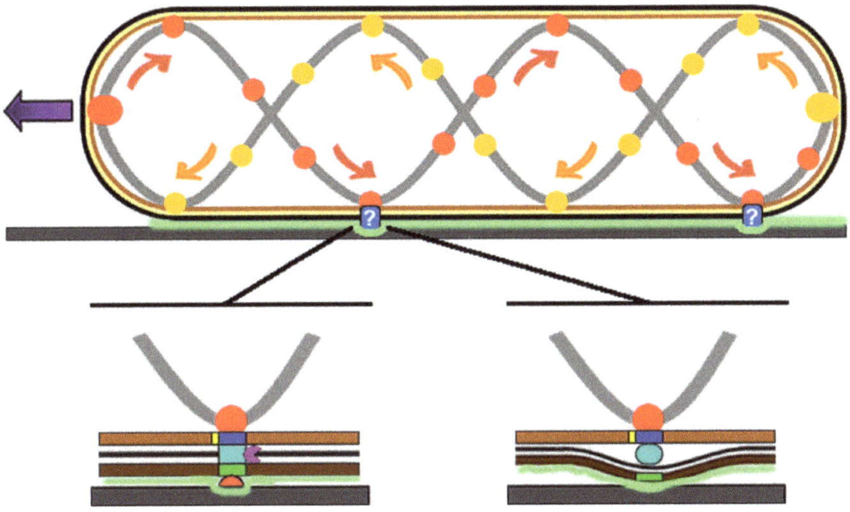

Figure 9.4 A-motility model. The purple arrow represents the direction of movement. Motility motors carrying high-drag (red circles) or low-drag (yellow circles) cargoes translocate along an endless closed loop. Only the high-drag cargoes, presumably active Glt units, are proficient for movement. The machinery could afford the rigid PG by two different ways: The motility complex may span the entire cell envelope and a PG-hydrolase may facilitate insertion of the complex through the PG (left). Alternatively, the motility complex could deform the PG, creating surface depression and drag. Outer membrane proteins may in this system reinforce local contacts at the depressions.

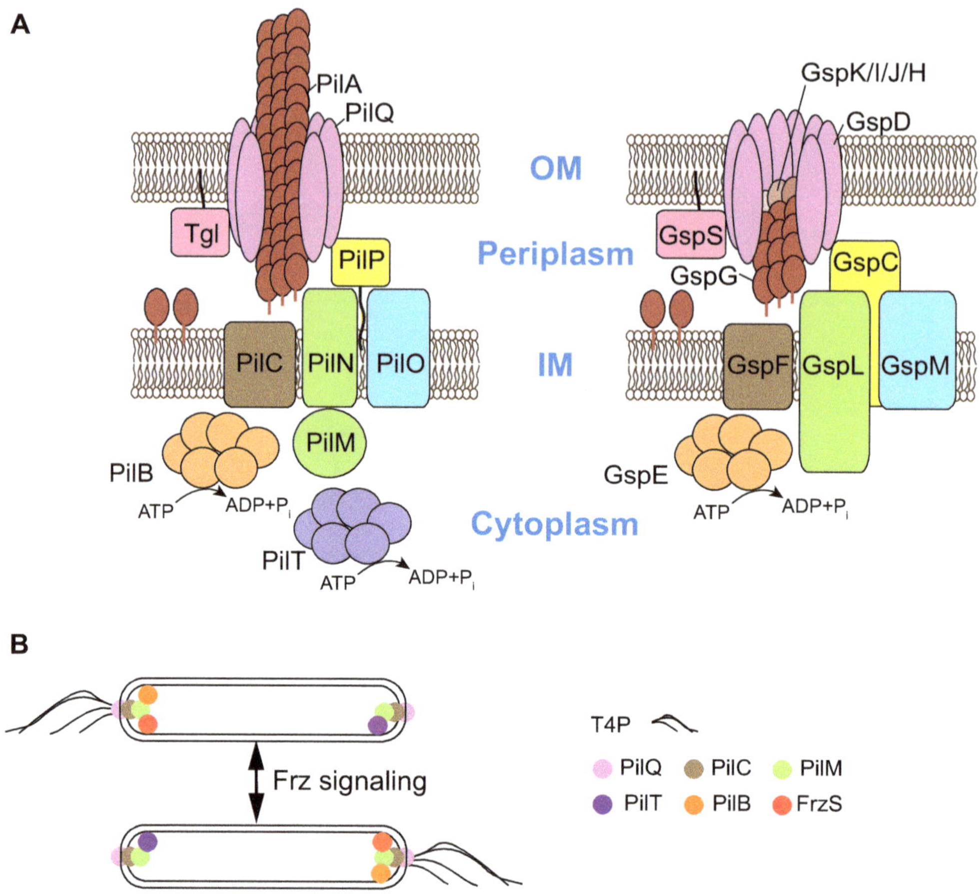

Figure 10.3 Organization of T4P and T2SS envelope structures. (A) Models for T4P (left) and T2SS (right) are shown. Homologous proteins and/or structural analogues are indicated by the same colours in the two models. Tgl and PilP are lipoproteins with attached black lines indicating their lipid moiety. For convenience, PilA here represents mature pilin instead of prepilin. See text for details. (B) Regulation of T4P localization. The two cells demonstrate the localization of T4P proteins before and after a Frz-induced reversal. The leading cell pole is the one with T4P. Shown are the cellular locations of six proteins as colour coded in the legend on the right.

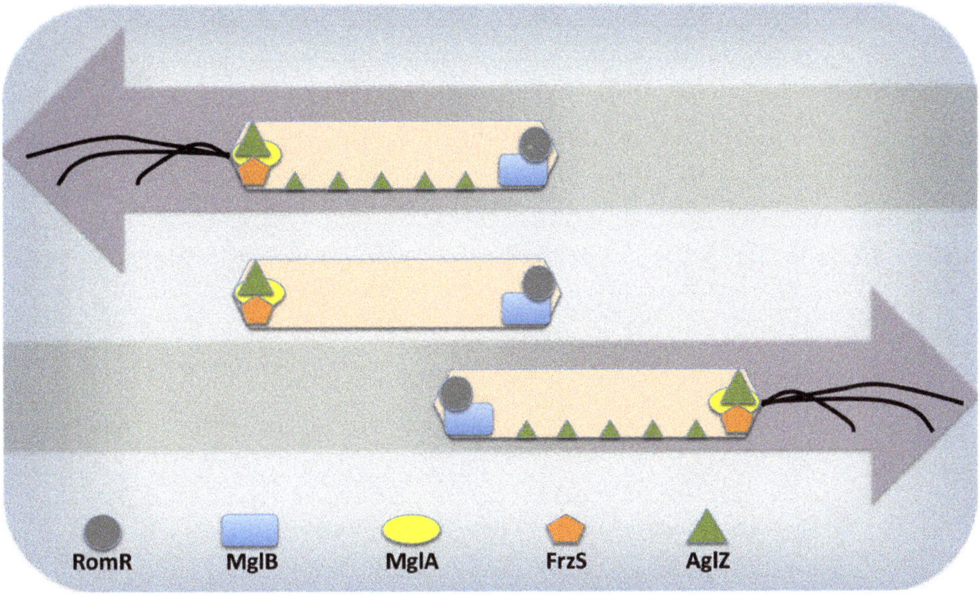

Figure 11.2 Schematic representation of a cell reversal. During a reversal, a *M. xanthus* cell make a brief pause, A and S-motility proteins move from one pole to the other and, when polarity and localization are re-established, the cell starts moving in the opposite direction.

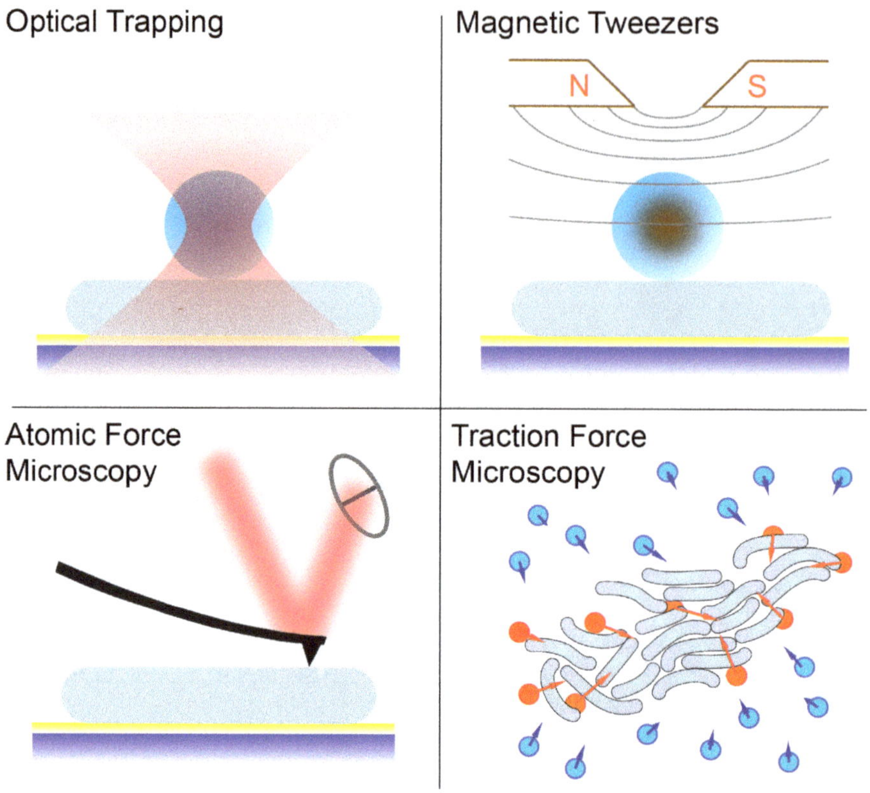

Figure 12.1 Schematic illustration (not to scale) of four biophysical force microscopy techniques as they might be applied to *M. xanthus*. An optical trap is formed by a focused laser strong enough to trap a polystyrene bead on top or at the side of an *M. xanthus* cell. Magnetic tweezers are formed by a polystyrene bead with a magnetic core that can be controlled by a pair of external magnets. These magnets are mounted outside the sample. An atomic force microscope translates the interaction between the surface and a tiny tip into an image. The deflection of the tip is amplified by the cantilever which also reflects a detection laser onto a photodiode. In traction force microscopy, the displacement of reporter beads within a solidified gel is linear to the force applied to the substrate, for example by a gliding *M. xanthus* group.